Superplastic Flow, Phenomenology and Mechanics

AF181227

Springer
Berlin
Heidelberg
New York
Barcelona
Hongkong
London
Mailand
Paris
Singapur
Tokio

K.A. Padmanabhan, R.A. Vasin, F.U. Enikeev

Superplastic Flow: Phenomenology and Mechanics

With 110 Figures and 34 Tables

Springer

Prof. Dr. K.A. Padmanabhan

Director
Indian Institute of Technology
Kanpur 208016, India

Prof. R.A. Vasin

Moscow State University
Moscow 119899, Russia

Dr. F.U. Enikeev

Institute of Metals Superplasticity Problems
Ufa 450081, Russia

ISBN 978-3-642-08740-0

Library of Congree Cataloging-in-Publication Data
Padmanabhan, K. A. (Kuppuswamy Anantha), 1945- Superplastic flow : phenomenology and mechanics/
K.A. Padmanabhan, R.A. Vasin, F.U. Enikeev.
p.cm. -- (Engineering matereials) - Includes bibliographical references and index.

1. Superplasticity. I. Vasin, R.A., 1937- II. Enikeev, F.U., 1960- III. Title. IV. Series

This work is subject to copyright. All rights are reserved, whether the whole or part of the material is concerned, specifically the rights of translation, reprinting, reuse of illustrations, recitation, broadcasting, reproduction on microfilm or in other ways, and storage in data banks. Duplication of this publication or parts thereof is permitted only under the provisions of the German Copyright Law of September 9, 1965, in its current version, and permission for use must always be obtained from Springer-Verlag. Violations are liable for prosecution act under German Copyright Law.

Springer-Verlag Berlin Heidelberg New York
a member of BertelsmannSpringer Science+Business Media GmbH

© Springer-Verlag Berlin Heidelberg 2010
Printed in Germany

The use of general descriptive names, registered names, trademarks, etc. in this publication does not imply, even in the absence of a specific statement, that such names are exempt from the relevant protective laws and regulations and therefore free for general use.

This book is dedicated to our families

Preface

Superplasticity is the ability of polycrystalline materials under certain conditions to exhibit extreme tensile elongation in a nearly homogeneous/isotropic manner. Historically, this phenomenon was discovered and systematically studied by metallurgists and physicists. They, along with practising engineers, used materials in the superplastic state for materials forming applications. Metallurgists concluded that they had the necessary information on superplasticity and so theoretical studies focussed mostly on understanding the physical and metallurgical properties of superplastic materials. Practical applications, in contrast, were led by empirical approaches, rules of thumb and creative design.

It has become clear that mathematical models of superplastic deformation as well as analyses for metal working processes that exploit the superplastic state are not adequate. A systematic approach based on the methods of mechanics of solids is likely to prove useful in improving the situation. The present book aims at the following.

1. Outline briefly the techniques of mechanics of solids, particularly as it applies to strain rate sensitive materials.
2. Assess the present level of investigations on the mechanical behaviour of superplastics.
3. Formulate the main issues and challenges in mechanics of superplasticity.
4. Analyse the mathematical models/constitutive equations for superplastic flow from the viewpoint of mechanics.
5. Review the models of superplastic metal working processes.
6. Indicate with examples new results that may be obtained using the methods of mechanics of solids.

Evidently, such a treatment has both academic and practical implications. Thus, the main purpose of this book is to lay the foundation for a *new direction of scientific research*, viz., mechanics of superplasticity, in order to rigorously study the mechanical response of superplastics.

This book is intended for a variety of readers who may be interested in the phenomenon of superplasticity for different reasons: materials scientists and physicists working in educational institutions and R&D units, those who wish to work on the applications of superplasticity, engineers in industry, students at senior undergraduate and postgraduate levels and those who wish to understand the phenomenology and mechanics of superplasticity without involvement in actual research.

A reader who has exposure to standard differential and integral calculus and elementary tensor calculus at a level taught to senior undergraduate students at a technical university should have no difficulty in following the treatments. The analytical procedures are explained in Appendixes with simple examples.

Kanpur, India
Moscow, Russia
Ufa, Russia

K. A. Padmanabhan
R. A. Vasin
F. U. Enikeev

Acknowledgements

This book is the result of a collaboration under the Integrated Long Term Programme in Science and Technology between India and Russia. The authors thank Prof. O. A. Kaibyshev, Director, Institute for Metals Superplasticity Problems (IMSP), Ufa, Russia and Dr. V.S. Ramamurthy, Mr. Y. P. Kumar and Mr. S.K.Varshney of the Department of Science and Technology, Government of India, for approving the visits of the Russian scientists to IIT Kanpur. Financial assistance was received from an Indo–US project sponsored (to KAP) by the Department of Science and Technology, Government of India and the Office of Naval Research, Washington, D.C. Drs. B.B. Rath, A. P. Kulshreshtha, A. Imam and S. Gupta are thanked in this connection.

Dr. A. G. Ermatchenko, Dr. R. V. Safiullin and Dr. A. Kruglov supplied the photographs used in the book. Mr. S. Sankaran rendered enormous help in producing a camera-ready form of the book. The meticulous proofreading of Dr. Gouthama is gratefully acknowledged.

Colleagues in the Laboratory for Mechanics, IMSP, Ufa and the Laboratory for Elasticity and Plasticity, Institute of Mechanics, Moscow State University are thanked for their interest in this work.

Finally, it is a pleasure to thank Dr. D. Merkle of Springer-Verlag, Heidelberg, for his kind invitation to write this book. Ms. P. Jantzen and Ms. G. Maas of Springer -Verlag have put in commendable effort in publishing this book.

Contents

Nomenclature

A	specimen cross-sectional area
b	absolute value of Burgers vector
$C = 1/K^n$	material constant, equation $\xi = C\sigma^n$
d	average grain size
D	diffusion coefficient
D_{gb}	grain boundary diffusion coefficient
E	Young's modulus
e	nominal or engineering strain
$\overset{o}{\vec{e}}_1, \overset{o}{\vec{e}}_2, \overset{o}{\vec{e}}_3$	basis of the convective system of coordinates at $t = 0$
$\hat{\vec{e}}_1, \hat{\vec{e}}_2, \hat{\vec{e}}_3$	basis of the convective system of coordinates at $t > 0$
G	shear modulus
g_{ij}	components of metric tensor
$\overset{o}{g}_{ij}$	components of metric tensor at $t = 0$
$\hat{g}_{ij}$	components of metric tensor at $t > 0$
h_R	length of the memory trace on the deformation trajectory
i, j, k	1, 2, 3
$\overset{0}{I}_1, \overset{0}{I}_2, \overset{0}{I}_3$	invariants of the tensors in Eulerian representation
$\hat{I}_1, \hat{I}_2, \hat{I}_3$	invariants of the tensors in Lagrangian representation
L	current specimen length
M	slope of the sigmoidal curve
k	Boltzmann's constant
K	material constant in equation $\sigma = K\xi^m$
$\overline{M}$	experimental torque
M_{max}	maximum slope corresponding to optimal strain rate
N_{exp}	number of experimental points available
m	strain rate sensitivity index, equation $\sigma = K\xi^m$
m'	strain rate sensitivity index, equation $\sigma = \sigma_0 + K'\xi^{m'}$
$n = 1/m$	stress exponent, equation $\xi = C\sigma^n$
$n' = 1/m'$	stress exponent, equation $\xi = C(\sigma - \sigma_0)^{n'}$
P	axial force (uniaxial testing)
p	gas pressure (superplastic forming processes)
p_H	hydrostatic pressure

Q	activation energy
Q_a^σ , Q_a^ξ	apparent activation energies
Q_f^σ , Q_f^ξ	fictitious activation energies
Q_{tr}^σ , Q_{tr}^ξ	true activation energies
R_i	radius of curvature of the deformation trajectory in Iljushin space
s_{dt}	length of the deformation trajectory (n = 1, 2, 3, 4)
s	Odqvist's parameter
T	absolute temperature
t	time
v	the cross head velocity
V_i	components of particle velocity in a continuum
V_{ij}	components of strain rate tensor
$\mathrm{x}^1, \mathrm{x}^2, \mathrm{x}^3$	Lagrangian or material system of co-ordinates
x^1, x^2, x^3	Eulerian system of co-ordinates

Greek symbols:

δ_{ij}	Kronecker delta ($\delta_{ij} = 1$ if i = j and $\delta_{ij} = 0$ if i ≠ j)
ξ	total strain rate (for uniaxial tensile test $\xi = v/L$)
ξ_{ij}	covariant components of the strain rate tensor
ξ_e	effective strain rate
ξ^c	creep strain rate
ξ^e	elastic strain rate
ε_{ij}	covariant components of strain tensor
ε_e	effective strain
κ_n	curvature of the deformation trajectory (n = 1, 2, 3, 4)
$\tilde{\varepsilon}_{ij}$	components of the strain tensor in Eulerian representation
$\hat{\tilde{\varepsilon}}_{ij}$	components of the strain tensor in Lagrangian representation
σ	flow stress (calculated from the primary experimental data as $\sigma = P/A$)
σ_r, ξ_r	reference point (σ_r , value of stress, corresponding to $\xi = \xi_r$)
σ_0	threshold stress (mechanical threshold)
σ_e	effective stress
σ_{opt}	flow stress corresponding to the optimal value of strain rate
ξ_{opt}	optimum strain rate
$1/\chi$	compliance of the testing machine
χ	rigidity of the testing machine
$\tau_{1/2R}$, $\tau_{1/4R}$	time intervals on the stress relaxation curve
$\tau_{1/2}$	time interval on the initial part of stress–strain curve
ε_{ij}	covariant components of strain tensor

$\overset{0}{\varepsilon}_1, \overset{0}{\varepsilon}_2, \overset{0}{\varepsilon}_3$	principal strains in Eulerian representation
$\hat{\varepsilon}_1, \hat{\varepsilon}_2, \hat{\varepsilon}_3$	principal strains in Lagrangian representation
Θ	volumetric strain
Δ	Laplace delta
ν	Poisson's ratio

Introduction

In the last three decades, many reviews on superplasticity have appeared. Some of these discuss all the main features of the phenomenon and its applications (see, e.g., monographs [1–7], reviews [8–15] and proceedings [16–23]), while others consider specific aspects, e.g., applications of superplasticity, development of new technological processes of metal working, mathematical modelling of the technological processes, properties of some superplastic materials, micromechanisms of deformation, etc. [24–31]. A recent book [6] as well as the five earlier publications [1–5] contain a comprehensive description of the metallurgical aspects of superplasticity and so there is no need to restate them in detail in the present volume. Reports on the various aspects of finite element modelling have also been published recently (see, e.g., [29–31]). Therefore, these aspects as well are not considered here in detail.

It is noted that in most of the above publications the phenomenology and mechanics of superplastic flow were investigated under uniaxial loading. Simple constitutive relationships have been used to analyse these results as also in the finite element modelling of technological processes. But less than 10% of the total number of papers presented at the major conferences of the 1990's on superplasticity [19–21] deal with analyses of constitutive equations for superplastics and/or the corresponding boundary value problems. Not withstanding this, a rudimentary form of mechanics of superplasticity does exist in the form of mechanical test data, mechanical behaviour characterisation, simple constitutive equations and solutions for mechanical working problems that use engineering methods.

Mechanics of superplastic flow as a subdivision of mechanics of solids is yet to develop fully. For example, methods of mechanics of solids are not formally used to solve problems of superplastic deformation. In the last two decades a few papers (mainly theoretical) on the mechanical response of materials for the case of finite strain behaviour, and in particular at elevated temperatures, have been published. But these reports involve special terminologies and complicated mathematics and hence are beyond the easy comprehension of many active workers in the field. This book aims to bridge this gap.

Microstructure of materials and physical and chemical processes underlying deformation are not directly taken into account in mechanics. Therefore, there is a view that this discipline cannot be useful even in principle to predict material behaviour under load and it is useful only for describing phenomenologically already known experimental facts. This view is not justified.

Metallurgists and materials scientists focus on microstructure that determines properties as being at the centre of the whole scheme. Bringing about a unification by combining the perspectives of the materials scientist and the specialist in mechanics is another aim of this book. It is hoped that this approach will lead to new findings and new perspectives that use commonly known concepts in both the disciplines and will also help to eliminate imprecise ideas and methods of investigation.

Stress, strain and strain rate are usually considered as scalar quantities while constructing physical models of plastic/superplastic deformation. Such an approach is far from reality. But the physicist captures the correct details in a model at least qualitatively. The problem of describing the macro-behaviour of materials proceeding from the behaviour of single entities is not usually considered by physicists and materials scientists, as they are mostly concerned with the micromechanisms of the physical phenomena. Instead, the values of stress, strain and strain rate in the physical equations are treated as macro-parameters while comparing the theoretical predictions with the experimental results. As a rule, a physical model allows the prediction of some aspects of macrobehaviour. Then it may be assumed that the physical model is accurate and it can be used to make predictions. But the description of non-uniaxial deformation as well as scaling from micro- (meso-) level to macrolevel are non-trivial. In fact these problems are very complicated. The development of the physically validated description of the micromechanisms of plastic/superplastic deformation into adequate constitutive equations for predicting macrobehaviour (mechanical properties) involves a statistical analysis of single events of deformation. While doing this a number of major simplifications (including for the effects of grain boundaries, segregation, etc.) are effected. All these simplifications should be validated very carefully by experiments. This is very difficult and time consuming and so validation is often done by indirect methods only.

Specialists in mechanics have considerable experience in constructing constitutive equations for different materials on the basis of a few phenomenological hypotheses. The superposition of the theoretical predictions on the corresponding experimental data indicates that these hypotheses permit in an implicit way the inclusion in the analysis of the influences of structure and structural changes resulting from plastic/superplastic deformation. But it is desirable to establish directly the relationship between the structural changes during deformation and the mechanical properties.

Moreover, there is a well-developed general theory of constitutive equations in mechanics of solids. These equations should satisfy a system of substantiated requirements that are spelt out. The methods of mathematical formulation of technological problems (as well as those of natural phenomena) are also well developed. In addition, many methods of solving the mathematical problems are available.

Exploitation of superplasticity in metalworking is done by technologists. They develop new technologies to solve practical problems. Consequently, they have valuable practical experience, which is useful in analysing the mathematical models of technological processes, e.g., the boundary conditions at contact surfaces.

Due to the following two reasons, in the last few decades significant progress has been achieved in bringing together the efforts of the physicists, metallurgists, technologists and specialists in mechanics: (i) the need to describe the technological processes of metal working with a clear understanding of the macroproperties of materials and using adequate macro-constitutive equations and boundary conditions; (ii) the desirability of taking into account the physical basis of plastic/superplastic deformation in order to construct adequate constitutive equations that have a physical basis.

In the words of Gittus et al. [32] "A central problem which has been systematically attacked in the case of metals is the relationship between the behaviour of crystal defects such as dislocations and the deformation of a large specimen or engineering component. It should be possible to produce accurate predictions of macroscopic deformation from a microscopic model and substantial progress towards this end has been made in recent years." It is clear that for achieving this goal the closest collaboration between physicists/materials scientists and specialists in mechanics is essential. This is the justification for this book on the phenomenology and mechanics of superplasticity.

In summary, the following may be mentioned. Most of the investigations on superplasticity have used standard mechanical tests. But the results have not often been analysed adequately. Therefore, investigators from different specialities may apply the methods of mechanics of solids in a study of the mechanical response of superplastics. This is of considerable practical importance because for developing technological processes of metal working it is not sufficient to understand the physical mechanisms operating in a material to be formed. It is also necessary to set up and then solve the corresponding boundary value problem so that force, power required, etc., can be estimated. The solution of the boundary value problem can also be used to optimise a technological process and develop new technologies. In this connection, it is pertinent to note that it is not necessary to solve the boundary value problems exactly, say by using a supercomputer. Sometimes it is enough to solve these problems by engineering methods. However, it is extremely important to set up the boundary value problems correctly and analyse the results thoroughly. This is the main objective of this effort – to develop the *mechanics of superplasticity*.

1 Phenomenology of Superplastic Flow

In this chapter, the phenomenology of superplastic flow is discussed. Full expositions are available in [1–6]. Other reviews consider this aspect to varying extent [7–15]. Attention is focussed here on recent results and comments are offered on the present level of understanding.

1.1 Historical

Phenomenological studies on superplasticity from Bengough [33] to Pearson [34] to Backofen et al., [35] are described in many reviews, e.g., [3, 14, 36]. Prior to 1964, the main aim was to obtain extreme elongation in uniaxial tensile tests. Technological exploitation came into focus only after the pioneering works of Backofen and co-workers [35, 37–39]. Most of the later studies on the mechanical behaviour of superplastics also have employed the uniaxial tensile test, but with emphasis on strain rate sensitivity of flow. The physical nature of superplastic deformation has been investigated thoroughly. Many physical theories have been proposed. A number of technological applications have resulted.

It is not recognised in many reviews that for a long period in history, materials akin to superplastics have been in use. Sherby and Wadsworth [14] have suggested that the history of superplasticity may date back to the early Bronze period of around 2500 BC. In that report, arsenic bronzes and the famous steels of Damascus (~ 300 BC) have been presented as examples of superplastic materials.

Gold and silver threads as well as very thin gold sheets used in many countries for centuries in many applications, as also the common glass deforming at elevated temperatures, may be regarded as exhibiting superplastics-like behaviour if superplasticity were defined as extreme elongation resulting from 'homogeneous' flow. (The mechanism of deformation is ignored in this definition.)

From a scientific/technological point of view, isolated reports on superplasticity in metals appeared in the early part of this century. The earliest report has been traced to 1912 [33], in which an $(\alpha + \beta)$ brass was extended in uniaxial tension by 163% at 700°C.

From a physical point of view, grain/interphase boundary sliding dominates superplastic flow [1–6]. If the definition is based on the operating mechanism(s), it is not possible to include the extreme elongation of gold, silver and glass mentioned above under 'superplasticity'. In 1991, the following definition of supeplasticity was evolved. "Superplasticity is the ability of a polycrystalline material to exhibit, in a generally isotropic manner, very high tensile elongation prior to failure" [19]. This definition is phenomenological as no microstructural or mechanical parameters have been included.

1.2 Mechanical Behaviour of Superplastics

1.2.1 Mechanical Tests

Mechanical response of superplastics has been studied under different types of loading.

Historically, tensile tests were first used to investigate the mechanical response of superplastics [33–35]. Torsion tests have also been employed, e.g., [40–49]. Indentation tests have been performed, see, e.g., [50–53]. Sometimes experiments that are technological in nature have been used to characterise superplasticity [54–58]. Compression tests are very common for technologists, see, e.g., [2, 3, 59–61]. In one experiment, the ring compression test was used to characterise the superplastic tendency by understanding the role of friction [3, 62]. Occasionally, the shape of the deforming specimen is also discussed as a distinct characteristic [3, 63–67].

In the above listed experiments the parameters have been changed monotonically, e.g., in a constant cross head velocity tensile test. But experiments characterised by non-monotonic loading have also been used in practice. Load relaxation tests have been used to study the mechanical response of superplastics, e.g., [3, 68–75]. Internal stresses can be estimated during load relaxation [3] or in a stress dip test [3, 76, 77]. Damping characteristics [3, 78–80] and elastic aftereffects (anelasticity) too are used to study superplasticity [81–84].

One can find a detailed description of the types and programmes of testing mentioned above in [3]. It is clear that tensile tests are the most common in a study of superplastic flow.

Mechanical tests are mostly carried out at constant temperature. Even in room temperature testing, e.g., on a tin–lead eutectic alloy sample, where the change in the dimensions of a specimen can be followed, extensometers have seldom been used to study anelasticity or the degree of strain localisation. In most other cases also no efforts have been made to calibrate specimen elongation against elongation recorded by the testing machine based on cross head movement.

Constant cross head velocity (v = constant) and constant (average) strain rate (ξ = constant) tests have often been used. Constant load (P = constant) or constant (average) stress (σ = constant) tests have been used only rarely. Strain rate jump tests have been used to determine the strain rate sensitivity of flow.

A 'standard' test specimen is yet to be defined, even though Pearson [34] used a standard tensile specimen (as per conventional wisdom) of diameter 5.08 mm and length 50.8 mm (i.e., 0.2 and 2 inch respectively) in his early classical experiments. Later tests, particularly those in which extreme elongations of 5500% and 7550% have been reported [85, 86] have employed much shorter specimens in which a uniaxial stress condition would not have been met till a late stage in deformation. Often, there is no discussion on errors in measurement. Grip flow has not been considered, even though the final dimensions clearly indicate that it had been significant. Reproducibility of results has also not been addressed.

The initial part of the stress–strain diagram is often not determined accurately as only steady state values are of consequence for plotting sigmoidal $\log\sigma - \log\xi$ curves. But it is well-known that in many problems the approach to the steady state is important.

1.2.2 Typical Experimental Results

In a uniaxial tension or compression test at constant temperature and grain size, time dependent records of axial force and elongation are converted into $\sigma - \varepsilon$ diagrams using the concepts of uniform deformation and constancy of volume (condition of incompressibility). That is,

$$\sigma = \frac{P}{A}, \quad \varepsilon = \ln\left(\frac{L}{L_0}\right), \qquad e = \frac{(L - L_0)}{L_0} \tag{1.1}$$

where σ is the flow stress, ε the true strain, e the engineering or nominal strain, P the axial force, A the instantaneous cross-sectional area ($= A_0L_0/L$ with A_0 the initial cross-sectional area) and L, L_0 are the current and initial gauge length of the specimen. Some typical results are presented in Figs. 1.1 and 1.2.

Stress–strain rate diagrams at constant strain are derived from the stress–strain plots, again assuming uniform deformation and volume constancy. Instantaneous strain rate ξ is given by

$$\xi = \frac{v}{L} \tag{1.2}$$

Strain rate sensitivity of flow is defined using a two-parameter power law

$$\sigma = K\xi^m \quad or \quad \xi = C\sigma^n \tag{1.3a,b}$$

where K is a material constant and m is the strain rate sensitivity index; $m = 1/n$ and $K = 1/C^m$.

It is to be noted that Eqs. (1.3) ignore strain hardening, which will be a reasonable assumption during superplastic flow in the absence of grain growth. Grain growth leads to what has been termed 'flow hardening' (effectively a form of strain hardening). Then, the flow stress depends also on ε and Eqs. (1.3) do not cover that situation (see also Chap. 3).

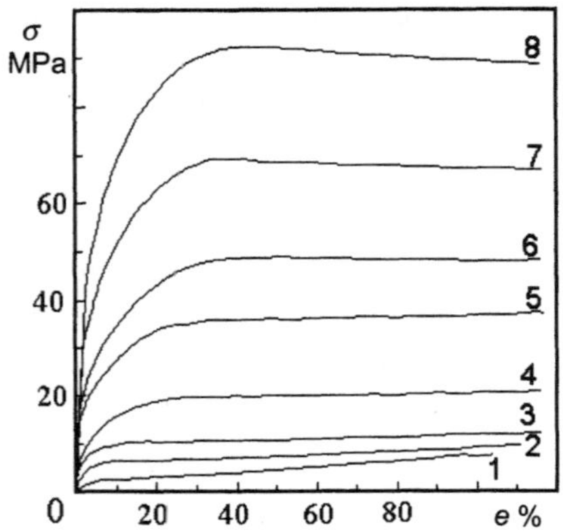

Fig. 1.1. Dependence of flow stress, σ, on engineering strain, e, in uniaxial tensile tests on Ti-6Al-4V tested at 900°C and different initial strain rates (s^{-1}) [87]:
1) $8.4 \cdot 10^{-5}$; 2) $5.5 \cdot 10^{-4}$; 3) $8.4 \cdot 10^{-4}$;
4) $1.6 \cdot 10^{-3}$; 5) $3.5 \cdot 10^{-3}$; 6) $6 \cdot 10^{-3}$;
7) $8 \cdot 10^{-3}$; 8) $2 \cdot 10^{-2}$

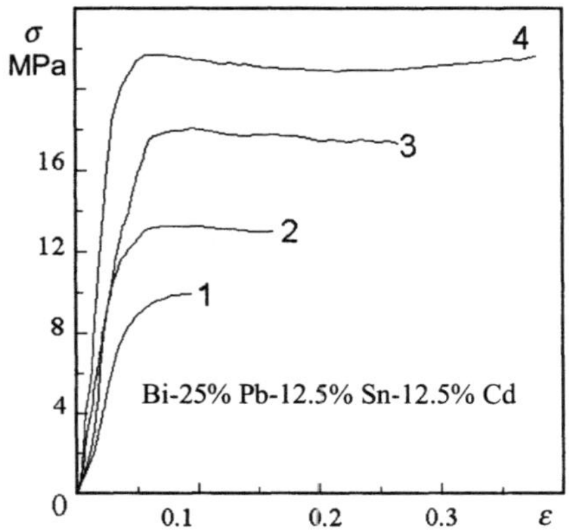

Fig. 1.2. Experimental true stress–true strain curves for Wood's alloy tested at room temperature and different initial strain rates (s^{-1}):
1) $1.28 \cdot 10^{-4}$; 2) $2.27 \cdot 10^{-4}$; 3) $3.90 \cdot 10^{-4}$;
4) $6.11 \cdot 10^{-4}$

(**Note:** In many figures and tables in this book, values reproduced from original works suggest a level of accuracy beyond experimental capability. The reader has to keep this in mind while using those results.)

If m were a (material) constant, a log – log plot of Eq. (1.3a) should be linear. But, in the case of superplastics, the $\log\sigma - \log\xi$ plot is sigmoidal (Fig.1.3a). Thus, Eqs. (1.3) are valid only over a narrow range of strain rates within which $m \cong$ constant. In general,

$$M = \frac{\partial \log \sigma}{\partial \log \xi} = \frac{\partial \ln \sigma}{\partial \ln \xi} \tag{1.4}$$

M depends on strain rate and goes through a maximum with strain rate – the curve has the so-called dome shape (see Fig. 1.3b). The maximum value of M, M_{max}, is unique for a material of given grain size and fixed temperature of deformation.

1.2.3 Conditions for Superplastic Flow

There are two types of superplasticity. The first type, environmental superplasticity, is observed in materials subjected to special environmental conditions, e.g., thermal cycling through a phase change. The second type, structural superplasticity is observed in fine grained materials. Structural superplasticity, unlike environmental superplasticity, is a universal phenomenon. It is well established now that

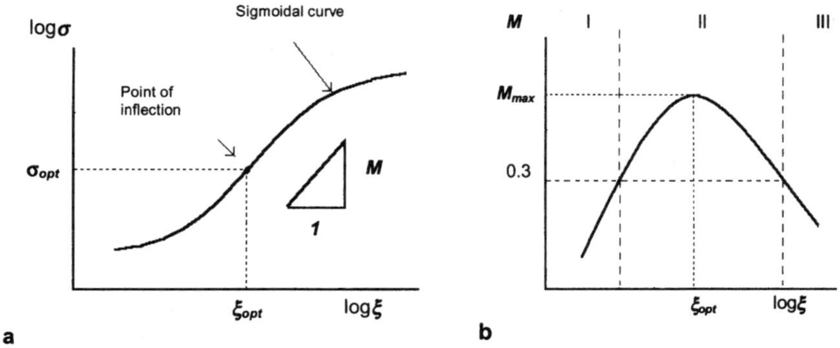

Fig. 1.3. **a** Sigmoidal $\log\sigma - \log\xi$ relationship, and **b** M (= $\partial (\log\sigma)/\partial (\log\xi)$) variation with strain rate (schematic)

most of the polycrystalline materials including metals, alloys, ceramics and glass ceramics, intermetallic compounds and metal matrix composites can be transformed into the superplastic state by appropriate structure preparation (grain refinement). In the present book, only structural superplasticity is considered in detail in view of its commercial importance.

In conventional view, the following conditions are to be satisfied for observing superplasticity:

1. The average grain size, d, should not exceed some critical value, d_{cr}, which is about 10–15 μm;
2. The temperature of deformation T should not be less than (0.4–0.5) T_m (where T_m is the melting point on the absolute scale);
3. Superplasticity is present only within a range of strain rates, usually ~10^{-5}–10^{-1} s^{-1};
4. The optimal strain rate interval in a uniaxial tensile test is conventionally defined by the empirical condition $M \geq 0.3$ (Fig. 1.3b). Significant superplasticity is present only in region II ($0.3 \leq M \leq 1$) and absent in regions I and III (in both $M < 0.3$).

The above four conditions are of different nature: the first one (on d) is microstructural, the second (on T) and third (on ξ) concern external testing conditions, while the last one defines a mechanical property. With decreasing d and/or increasing T, region II is shifted to a higher strain rate range. Significant grain growth, particularly in relation to the observed extreme elongation, is absent. Grain shape is nearly equiaxed even when the strain is large.

In one report, changing the mode of testing from tension to compression had no effect on the optimal strain rate [88], but in another [89] it was displaced to a higher strain rate level, although the test range was narrower [87]. In more recent years, attention has been focussed on high strain rate superplasticity and low temperature superplasticity achieved mainly by refining grain size into the submicrometer range. Strain rates of 1–10 s^{-1} and optimal superplastic deformation in aluminium alloys at around 200°C have been reported. The phenomenology of superplasticity is discussed comprehensively in many reviews [1–28].

1.3 Strain Rate Sensitivity of Superplastic Flow

1.3.1 Strain Rate Sensitivity Index, *m*

Strain rate sensitivity index, *m*, is considered to be the most important parameter that characterises superplastic deformation. There are a number of reports where the various experimental methods of determining the value of m are described, see, e.g. [1, 3, 4, 68].

The part of the sigmoidal curve in the vicinity of the point of inflection is rather extended. Therefore, one can use Eqs. (1.3) in the vicinity of the point of inflection to describe optimal superplasticity.

The simplest method of determining m includes experiments at a given (constant) temperature and different (but constant for each test) cross head velocities. Then, σ–ε plots and $\log\sigma - \log\xi$ diagrams at constant strain in the steady state region can be generated from the experimental *P*– *t* traces using Eqs. (1.1) and (1.2). *m* is determined as the slope of the $\log\sigma - \log\xi$ plot if it is linear or as the slope in the vicinity of the point of inflection if the diagram is sigmoidal. As all calculations are made for the steady state regime of loading the grain size is assumed to be constant when *m* is estimated.

Often, a strain rate jump test is used. In this method, the strain rate is increased in steps and the corresponding steady state (or saturated) flow stress is measured. Different variants of this method are distinguished by the way the results are handled. There are four different ways of treating the same experimental diagram shown in Fig. 1.4. They are described in detail in many reports, see, e.g., [1, 3, 4, 68]. Common assumptions for all of them are that the testing machine is absolutely rigid and the change in cross head velocity takes place instantaneously[1]. *m* is calculated from the relation

$$m = \frac{ln\dfrac{\sigma_2}{\sigma_1}}{ln\dfrac{\xi_2}{\xi_1}} \qquad (1.5)$$

where σ_1 and σ_2 are the stresses corresponding to the strain rates ξ_1 and ξ_2 respectively. By convention, the value of *m* is attributed to ξ_1 although it may also be

[1] It should be noted that these assumptions are not realistic and so will affect the results [10]. But this aspect is seldom discussed in the literature on superplasticity.

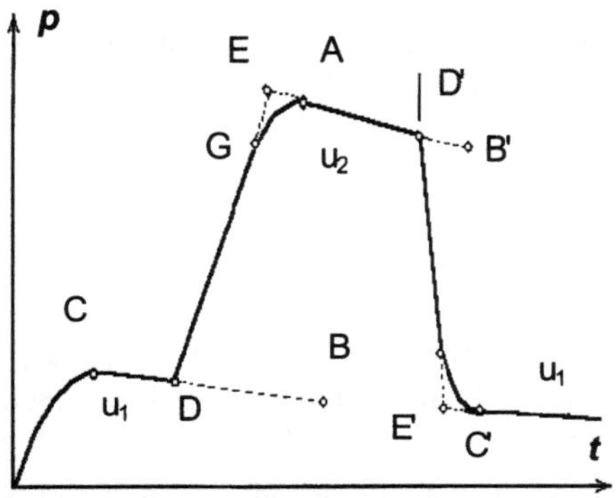

Fig. 1.4. Typical experimental time dependence of the axial load, *P*, recorded during a strain rate jump test (schematic) [68]

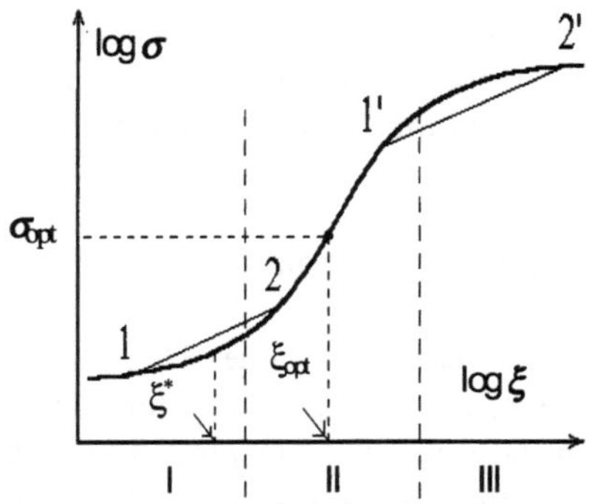

Fig. 1.5. Typical sigmoidal curve. In Region I Eq. (1.5) gives an overestimated value of the slope of the sigmoidal curve while in Region III an underestimated value [90]

attributed to $(\xi_1 + \xi_2)/2$ with justification. The difference among the 4 methods lies in the way the points in Fig. 1.4 are chosen for calculating *m*. Simple geometrical considerations show (see Fig. 1.5) that Eq. (1.5) gives an overestimate of *m* in Region I but an underestimate in Region III. The following two conclusions emerge. The value of *m* is to be assigned to some intermediate strain rate ξ^* ($\xi_1 < \xi^* < \xi_2$) where the tangent to the sigmoidal curve is parallel to the straight line connecting points 1 and 2. In practice, however, the slope of the sigmoidal curve *M* (which is strain rate dependent) is determined from Eq. (1.5) rather than from the value defined in Eq. (1.3a).

More recently, a few additional ways of evaluating *m* have been suggested [72, 91, 92]. These procedures depend on load relaxation [72], constant load application [91] and measurement of initial slopes of the $\sigma - \varepsilon$ plots [92]. The second method [91] is claimed to account for the differences in machine stiffness and inertia effects but the small strains at which the measurements are made in the third method [92] may lead to the determination of *m* in a region prior to the onset of steady state flow. Most of these methods can also be used to evaluate *M*, which in general is a function of both strain and strain rate.

Some technological tests have also been used to determine *m*, e.g., bulging of a circular [57] or a rectangular [87, 93] diaphragm under constant gas pressure. Exact solutions of some boundary value problems also may lead to an estimate for *m* [94, 95]. *M* depends on many variables like strain, strain rate, microstructure and its evolution, type of loading, etc. [96]. Hence, it is not a material constant. The difference between *m* and *M* is discussed only sometimes [3, 68, 90, 95, 97].

Equation (1.3a) involves two unknowns. If ambiguity is to be avoided, *K* may be taken as a constant and *M* equated to *m* at the point of inflection [90, 97]. A way out would be to use a (mathematically) more complete description. But, Eqs. (1.3) are very popular in the literature on superplasticity. A general form of the equation should include strain, a mechanical threshold and some structural characteristics. The problem of generalising constitutive equations to consider non-uniaxial stress–strain states is seldom addressed in the literature. Further comments are reserved for Chaps. 3, 4 and 5.

1.3.2
'Universal' Superplastic Curve

M has its maximum value, M_{max}, at the point of inflection defined by (σ_{opt}, ξ_{opt}). Values of M_{max}, σ_{opt} and ξ_{opt} for different materials are different (Fig. 1.6a). But, when the same data are plotted in normalised co-ordinates, (M/M_{max}) versus log (ξ/ξ_{opt}), the data points corresponding to different systems fall on the same curve (Fig. 1.6b). This 'universal curve' could be described by the empirical equation [98]

$$\frac{M}{M_{\max}} = \exp\left[-a^2\left\{\log\left(\frac{\xi}{\xi_{opt}}\right)\right\}\right] \tag{1.6}$$

It was assumed in [98] that $a^2 \cong 0.25$ for many materials. As Fig. 1.6b seems to describe the flow behaviour of some superplastic alloys rather well, it is meaningful to check its 'universality' as well as the relevance of the various physical models of superplastic deformation in terms of this curve.

Recently, it was noted [105] that the normalised (M/M_{max}) vs. log (ξ/ξ_{opt}) curve is to be plotted for the same normalised (homologous) temperature T/T_m, where T_m is the melting point on the absolute scale.

Careful analysis of experimental data including those shown in Fig. 1.6 has enabled the authors [105] to conclude that the (M/M_{max}) vs. log (ξ/ξ_{opt}) plots are temperature dependent–a fact not recognised in [98]. In many systems, both the values of M_{max} and ξ_{opt} are clearly temperature dependent and this would lead to the

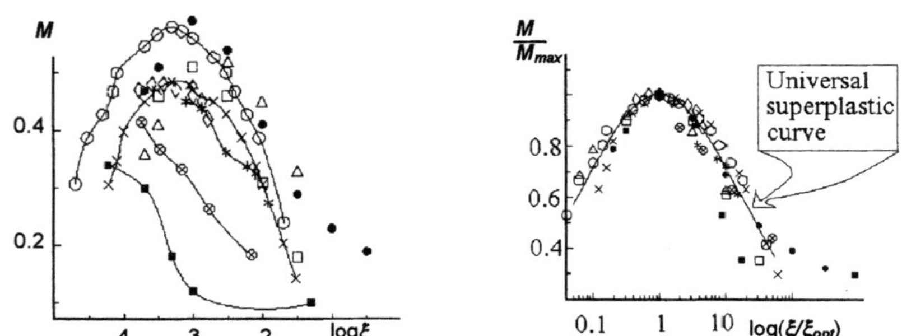

Fig. 1.6. Experimental dependence of M on strain rate in **a** normal, and **b** normalised co-ordinates [98]: ●●●● – MA21 [5]; □□□□ – VT9 [5]; ☐☐☐☐ – 0.12C18Cr10Ni 2Ti [5]; ◊◊◊◊ – TiAl [99]; ■■■■ – Bi_2O_3 [100]; ⊗⊗⊗⊗ –TiC [101]; OOOO – 5083 [102]; xxxx – Ti25Al10Nb3V1Mo [103]; **** – Ni_3Si [104]

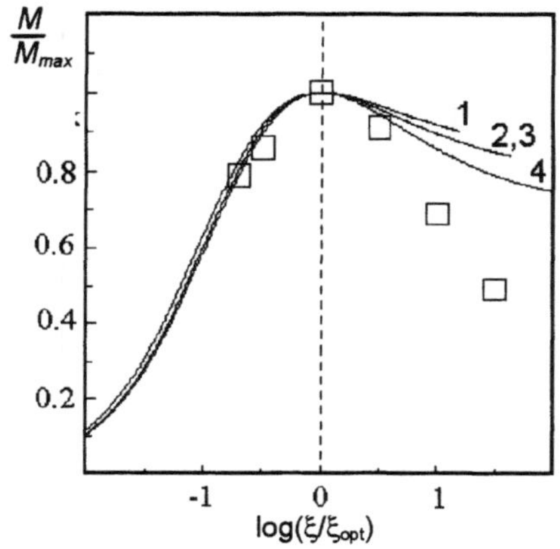

Fig. 1.7. (M/M_{max}) vs. log (ξ/ξ_{opt}) plots at comparable (T/T_m) ratio for the following alloys (solid lines):
1 – Sn-38Pb, 443 K, $(T/T_m) = 0.972$;
2 – Al-12Si, 831 K, $(T/T_m) = 0.985$;
3 – Al-33Cu, 793 K, $(T/T_m) = 0.966$;
4 – Al-33Cu-0.4Zr, 793 K, $(T/T_m) = 0.97$[105]
5 – □□□ MA21deformed, $(T/T_m) = 0.973$ [5]

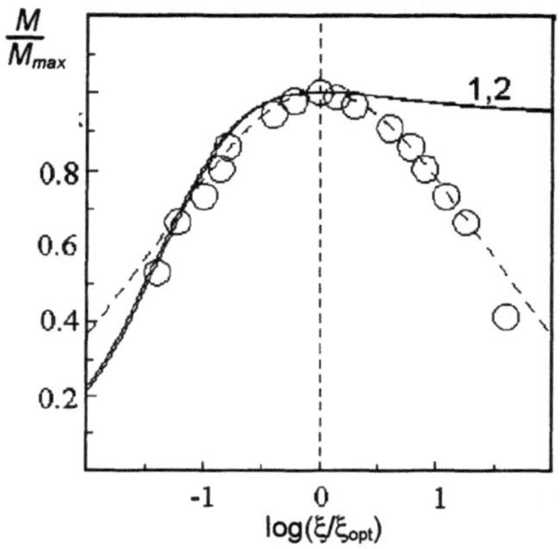

Fig. 1.8. (M/M_{max}) vs. log (ξ/ξ_{opt}) plots at comparable (T/T_m) ratio for the following alloys (solid lines):
1 – Zn-22%Al, 503 K, $(T/T_m) = 0.918$;
2–Supral alloy, 743 K, $(T/T_m) = 0.911$ [105]
OO Al5083 deformed $(T/T_m) = 0.922$ [102].
The original 'universal curve' of [98] is also shown by a dashed line

observed temperature dependence of the (M/M_{max}) vs. log (ξ/ξ_{opt}) plot. This point is driven home in Figs. 1.7 and 1.8 where the data pertaining to four aluminium alloys and those of Zn-22Al and Sn-38Pb alloys are analysed to obtain (M/M_{max}) vs. log (ξ/ξ_{opt}) plots at nearly equal (T/T_m) ratios (T = test temperature, K). The relevant curves practically coincide which is clear indication of the existence of 'universal curve'. (The original 'universal curve' of [98] is also superimposed in Fig. 1.8). It is evident that the (T/T_m) ratios for the alloys examined in [98] varied widely. When appropriate (T/T_m) values relevant to Figs. 1.7 and 1.8 were chosen the experimental points considered in [98] also fell on the universal curve. Thus, rigorous analysis [105] shows that 'a universal curve' for optimal superplastic flow exists *provided* the (M/M_{max}) vs. $\log(\xi/\xi_{opt})$ plot is made at constant (T/T_m) ratio (see also Chap. 3). Mathematically, it can be described by a temperature dependent value of a in Eq. (1.6). Typical values of a according to [105] are about 0.5–0.7 for a number of aluminium-based alloys. The existence of a 'universal' curve is interpreted in [105] as indication of a common mechanism of deformation underlying structural superplasticity.

Substituting Eq. (1.6) in Eq. (1.4) one obtains after integration

$$\log\frac{\sigma}{\sigma_{opt}} = M_{\max}\int_{0}^{\log\frac{\xi}{\xi_{opt}}}\exp\left(-a^2x^2\right)dx \equiv m\log\frac{\xi}{\xi_{opt}} \tag{1.7}$$

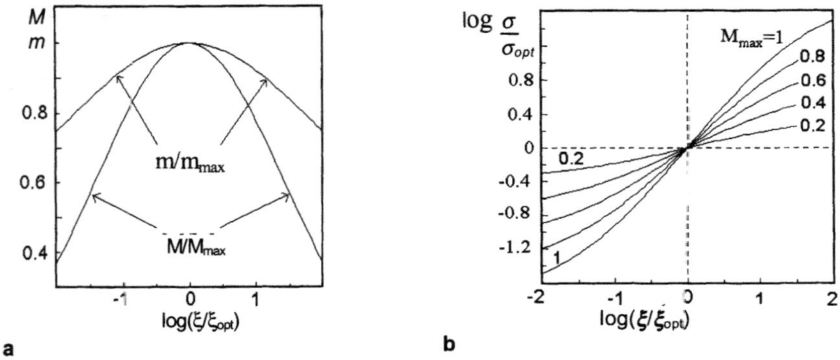

Fig. 1.9. *m, M*-strain rate, stress–strain rate normalised curves calculated according to Eqs. (1.6) and (1.7) with a = 0.5

In Fig. 1.9, the results of calculations in accordance with Eqs. (1.6) and (1.7) are presented. One can see that the difference between *m* and *M* values is significant (Fig. 1.9a) even when typical sigmoidal curves are obtained (Fig. 1.9b).

It is emphasised that the narrowing of the optimal strain rate interval (the contraction of Region II in Fig. 1.3b) with an increase in the temperature of deformation is an experimental fact. Unfortunately, only limited attention has been paid to this fact. The contraction of Region II with *T* has been reported in [106] for the intermetallic compound TiAl. This may apply to other systems also.

1.3.3 Stability of Uniaxial Superplastic Flow

Hart [107] has analysed the tensile deformation of a uniform rod. Assuming uniform flow and volume constancy and differentiating Eq. (1.1) with respect to time, *t*, one obtains

$$\frac{\dot{P}}{P} = -\left(\frac{\dot{L}}{L}\right)\left(1 + \hat{m} - \frac{\hat{n}}{\varepsilon}\right) + \hat{m}\frac{\ddot{L}}{\dot{L}} \tag{1.8}$$

where $\hat{n} \equiv (\varepsilon/\sigma)\partial\sigma/\partial\varepsilon$ and $\hat{m} \equiv (\xi/\sigma)\partial\sigma/\partial\xi$ are referred to as strain hardening and strain rate hardening indexes, respectively, the 'dot' here and throughout this book indicates a time derivative, except when otherwise stated. For the case $\dot{L} = v_0 =$ constant, the maximum strain, ε_{max}, is obtained from Eq. (1.8) using the constraint: at $\varepsilon = \varepsilon_{max}$, $P = P_{max}$ (or $dP/dt = 0$), that is,

$$\varepsilon_{max} = \frac{\hat{n}}{1 - \hat{m}} \tag{1.9}$$

which is a generalisation of the well-known condition $\varepsilon_{max} = \hat{n}$ for a strain hardening material obeying the Ludvik equation $\sigma = \widetilde{K}\varepsilon^{\hat{n}}$ [3, 108, 109].

For the case P = constant and $\hat{n} = 0$, consideration of time evolution of a local non-uniformity δA (deviation in cross-section from the average value) leads to

$$\delta A = \left(\frac{A_0}{A}\right)^{\frac{1}{\hat{m}-1}} \delta A_0 \tag{1.10}$$

where δA_0 is the local inhomogeneity at $t = t_0$. Above analysis is valid for any situation where $\hat{m} \neq 0$ and so it is useful for analysing both superplastic and non-superplastic flow. Hart [107] has pointed out the very strong dependence on $\hat{m}$ which explains why large ductility can be obtained even beyond the point of instability in materials for which m is large, that is, when m is greater than about 1/3 [107]. This analysis is included in many monographs on superplasticity and handbooks, see, e.g., [1, 3, 108].

Later, a number of other reports that enunciate the criteria for large elongation during superplastic flow were proposed, see, e.g., [3, 110–120]. Some of these are inspired by Hart's analysis, while the others are empirical/ad hoc in nature.

1.4 Superplasticity from the Point of View of Mechanics

1.4.1 On the Definition of Superplasticity

Tresca's classic experiments [121] (see also Bell [122]) on a number of materials, but notably on lead, covered a vast area – for example, forward extrusion, uniaxial compression (Fig. 1.10). He established the famous yield criterion named after him, demonstrated that solids can experience very large, rather homogeneous strains and behave like fluids under certain conditions. He also highlighted the role of hydrostatic pressure (hydrostatic component of the stress system) in enhancing ductility. If obtaining extreme deformation in *tension alone* is not defined as superplasticity, Tresca can be regarded as the discoverer of superplasticity. (Mechanisms of deformation – a point of focus of physicists and metallurgists – is not a concern of mechanics.)

In the twentieth century, Iljushin affirmed [123] that when subjected to a sufficiently high hydrostatic pressure, any material can flow infinitely and this was criticised by some scientists as 'abstraction from the real physical processes' [124, 125]. In a rebuttal [126], it was pointed out that under very high hydrostatic pressures even extremely brittle materials are seen (experimentally) to become ductile.

Based on his famous experiments, Bridgman (see, e.g., [127, 128]) concluded that "at sufficiently high pressures metals like steel become literally incapable of fracturing in elongation, any incipient fractures being pushed back into contact by the external pressure, so that indefinitely great elongations become possible". This clearly anticipates the much later use of a hydrostatic (back) pressure in enhancing the deformation of superplastic alloys (see Chap. 5)[2]. As Bridgman performed many uniaxial tensile tests also, his experiments should be regarded as pertaining to superplasticity. However, the use of the term 'isotropic' instead of 'homogeneous' or 'uniform' in the definition of superplasticity [19] will create some problem because following such large elongations in Bridgman's experiments, the properties of the material would be different in the axial and the transverse directions. In fact, in the phenomenological definition of superplasticity evolved in Osaka [19], even 'superplastic deformation in textured materials' reported in literature can not be included, as flow in those materials would be anisotropic till a late stage in deformation.

A number of similarities exist between superplasticity and creep: a) the three regions of creep: primary, secondary and tertiary creep; b) the presence of a region of flow at a constant strain rate (secondary creep); c) coincidence between the constitutive equations, e.g., the power law relation. These aspects will be discussed in Chap. 2.

As for the definition of superplasticity, the following remarks are relevant. The conditions for superplastic flow, as stated in Sect. 1.2.3, are of a mixed nature, i.e., structural and external/experimental variables have been combined with a mechanical characteristic. In terms of mechanics of solids, the following definition will hold good. *Superplastics are materials whose mechanical response during steady state (stationary) uniaxial tensile deformation can be described by the power law* $\sigma = K\xi^m$ *with* $m > 0.3$.

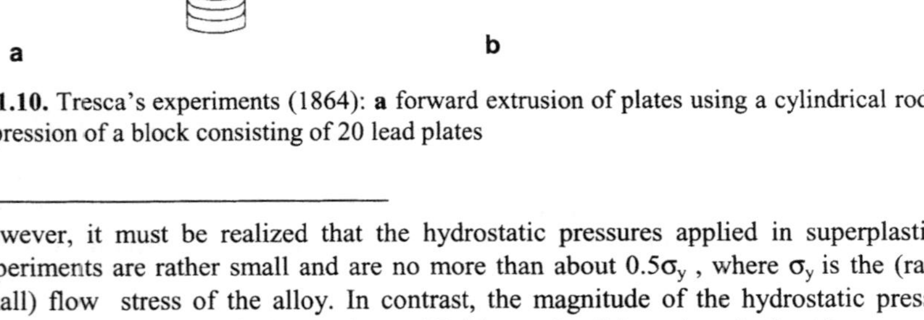

Fig. 1.10. Tresca's experiments (1864): **a** forward extrusion of plates using a cylindrical rod; **b** compression of a block consisting of 20 lead plates

[2] However, it must be realized that the hydrostatic pressures applied in superplasticity experiments are rather small and are no more than about $0.5\sigma_y$, where σ_y is the (rather small) flow stress of the alloy. In contrast, the magnitude of the hydrostatic pressure applied in experiments similar to those of Bridgman is of the order of a few kbar.

Appropriate strain rate and temperature intervals will determine the range of relevance in this definition. If the grain size is not sufficiently fine, the condition $m > 0.3$ will not be met and superplasticity will not be present. It is noteworthy that this definition will also be able to include the extreme deformation present during environmental superplasticity, in hot glass, heat softened polymers and metallic glasses. As large elongation is referred to as superplastic, it is desirable to widen the definition given in [19] and view the different classes of materials mentioned above as pertaining to sub-divisions/sub-groups of superplasticity. It is necessary to note that this definition is not fully satisfactory because, for example, it does not include the effect of hydrostatic stress (related to the first stress invariant) on superplastic flow. Unlike in plasticity, even a hydrostatic pressure less than the yield stress affects superplastic flow significantly. Further work is necessary in this regard.

Non-uniform stress–strain state present during superplastic flow has been mostly ignored. But, this feature is of immense industrial importance. The effect of temperature on optimal supeprlastic flow is not fully characterised. Also, it is well-known that the stress state as well as the presence or absence of a hydrostatic pressure shifts the optimal range of flow [4, 88, 89, 127, 128]. Thus, the simple constitutive equation $\sigma = K\xi^m$ and experimental data based only on uniaxial testing are not sufficient to understand superplastic flow of practical interest.

If superplastic flow were to be brought within the framework of mechanics of solids, constitutive equations for superplastics should be written in tensor form, as done for example in conventional plasticity for a von Mises solid where the yield criterion depends on the second tensor invariant (see Chap. 2). For superplastic flow, possibly the first stress tensor invariant (hydrostatic part) will also have to be included. This is yet to be done.

1.4.2 On Experimental Studies Concerning Superplasticity

Due to the non-availability of a standard specimen for fundamental studies in superplasticity, reproducibility and comparison of results are difficult. The problem, however, is of limited importance due to the low notch sensitivity of superplastics. (Even notches initially present disappear as a result of superplastic deformation.) Another favourable factor is that even when the initial gauge length is small in comparison with the specimen cross-section, the extreme elongation ensures that nearly uniaxial loading conditions prevail at least in the later stages of deformation. But, in the early stages flow in a short specimen of significant cross-section the use of Eqs. (1.1) and (1.2) can lead to errors. The use of these equations also ignores the multiple, diffuse necks that form during superplastic flow. Ideally, the effects of these necks on the stress – strain – strain rate curves generated assuming uniform deformation should be established by, say, numerical techniques.

In addition, the need to repeat experiments (at least three specimens per point) to establish reproducibility has been mostly ignored. The number of experimental points at a given temperature is rather small (usually not more than 5). The influence of the stiffness of the testing system is often ignored (although this was emphasised by Hart [107] in an early publication) and testing has predominantly been in the uniaxial tensile mode. Multiaxial testing and investigations in other modes like torsion, compression, load relaxation and creep have been rarely employed.

If the aim were to merely describe the results of the uniaxial tests within the framework of some physical theory, other types of tests may not be necessary. But, industrial applications of superplasticity involve different non-uniaxial stress states, non-monotonic loading, non-uniform stress–strain states, etc. So, it is essential to clearly define the limits of applicability of a constitutive equation by carrying out experiments under different loading conditions.

In mechanics as well as in materials science the analogy between a decrease in strain rate and an increase in temperature is well-known. In superplasticity, the concept of temperature-compensated strain rate is often used, viz.,

$$Z = \xi.\exp\left[\frac{Q}{RT}\right] \tag{1.11}$$

where R is the gas constant, Q is activation energy, Z is known as the Zener–Hollomon parameter [129]. In addition, a temperature–hydrostatic pressure–strain rate parameter may have to be developed for superplasticity.

1.4.3 On the Presentation of Results Obtained

The need to present and analyse results in a dimensionless form (compare Figs. 1.1, 1.5, 1.6a with 1.6b–1.8) has been mostly ignored. Errors involved in analysing the experimental results assuming uniform flow are also not discussed. Sometimes, values are reported to levels well beyond the maximum accuracy possible.

Dimensional analysis would require that Eq. (1.3a) be viewed as an approximation of the equation

$$\frac{\sigma}{\sigma_q} = \left(\frac{\xi}{\xi_q}\right)^m \tag{1.12}$$

where σ_q and ξ_q are a reference stress and a reference strain rate respectively, e.g., the values at the point of inflection in the sigmoidal curve. Unfortunately, Eq. (1.12) is seldom used in the literature, see, e.g., [95, 130, 131].

Similarly, it is worthwhile to make the grain size dimensionless by dividing it by a reference grain size, say, an arbitrarily chosen maximum grain size beyond which significant superplasticity is not seen. To use the term 'Burgers vector' for a scalar quantity is also not correct. This arises from a failure to take into account the tensorial nature of dislocation density as well as the experimental procedure used to determine the same.

The following semi-empirical equation is used in the literature on superplasticity[3]:

$$\xi = \overline{A}\left(\frac{b}{d}\right)^p \cdot \left(\frac{\sigma - \sigma_{th}}{G}\right)^n \tag{1.13}$$

where b is the absolute value of the Burgers vector, G the shear modulus (Young's modulus E can also be used), d the average grain size, σ_{th} a threshold stress, $\overline{A}$ a constant independent of d and σ and p, n are empirical constants.

When typical ranges of values for the variables are substituted, it is easy to show that Eq. (1.13) is not satisfactory for practical calculations. Formally, Eq. (1.13) is rigorous. However, for Nimonic 80A, for example, the following equation is used [132]

$$\xi = 2.7 \times 10^{21}\left(\frac{\sigma}{E}\right)^9 \tag{1.14}$$

that is, $n = 9$. Thus, to derive a typical value for ξ of about 10^{-4}–10^{-2} s^{-1} one has to deal with a number differing from ξ by greater than 20 orders of magnitude. But, a more serious problem is that an experimental inaccuracy in the value of n (and/or in p in Eq. (1.13)) leads to a major change in the value of ξ. For example, a 5% error in the value of n will lead to a change in ξ by an order of magnitude. Thus, normalisation with respect to b and G is not useful. Also, E or G enters the calculations during high temperature deformation due to quantum mechanical effects [133] while dislocation motion, diffusion and grain boundary sliding pertain to the domain of classical, albeit complicated, mechanics. This is the physical argument against normalisation with respect to E or G. (It is interesting that engineers and specialists in mechanics have always normalised the flow stress with respect to a reference stress, instead of E or G). The situation can be improved in a number of ways.

1. Significantly increase the number of experimental points used to determine the values of n and p.
2. Reject Eqs. (1.13), (1.14) and use other functional forms for the dependence $\xi = F(\sigma, d)$.
3. Normalise Eq. (1.13) in a different way. Normalisation with respect to a reference stress appears to be acceptable.

While following this course, experience accumulated in the area of creep [134, 135] is likely to be useful. An immediate consequence will be the use of numerical values for the constants with acceptable levels of accuracy; an aspect not attended to carefully so far (for publications where this aspect is ignored see, e.g., [1, 4–6, 19–22]).

---

[3] For example, in [13] this relation is used in 7 reports.

1.4.4
On Some Parameters of Superplastic Flow

1.4.4.1
Range of Optimal Flow

Identification of the optimal range of superplastic flow around the point of inflection using minimum number of experiments is of practical interest. A trial and error procedure involving detailed experimentation would allow the construction of deformation mechanisms maps. But, this is time consuming.

Recently, some alternative procedures have been suggested [105, 136, 137] which allow the identification of the strain rate for maximum superplasticity (the point of inflection in the $\log\sigma - \log\xi$ plot) with minimum number of experiments. In combination with the model of Padmanabhan and Schlipf [138, 141], this number can be reduced to three [105]. These procedures are described in Chap. 3.

1.4.4.2
Mechanical Threshold

From the point of view of mechanics, a deformable solid has a non-zero mechanical threshold ($\sigma_0 \neq 0$). For a liquid $\sigma_0 \equiv 0$. The formulation of the boundary value problems and the methods used to solve them in fluid mechanics are vastly different from those of mechanics of solids. Therefore, it is of practical significance to include correctly a mechanical threshold in a description of superplastic flow. Karim [142] considered the equation

$$\sigma = \sigma_0 + K'\xi^{m'} \tag{1.15a}$$

where σ_0, K' and m' are empirical constants ($K' \neq K$ and $m' \neq m$ when $\sigma_0 \neq 0$). K' is dependent on temperature and grain size. m' was termed the 'genuine rate sensitivity'. The equation can then be rewritten as

$$\xi = \begin{cases} 0, & \sigma < \sigma_0 \\ C'(\sigma - \sigma_0)^{n'}, & \sigma \geq \sigma_0 \end{cases} \tag{1.15b}$$

where $n' = 1/m'$, $C' = 1/(K')^{n'}$. If $\sigma_0 = 0$, then $n' = n$, $m' = m$, $C' = C$, $K' = K$. Eqs. (1.15) can be generalised as [63]

$$\sigma - \sigma_0 = \tilde{A}\xi^{p}\varepsilon^{r} \tag{1.16}$$

where σ_0, $\tilde{A}$, p, r are empirical constants. Dunlop and Taplin [143] showed that for micrograined aluminium bronze the strain hardening index $r \approx 0$. Now it is well-known that for most superplastics $r \approx 0$ so long as grain growth during flow is not significant [1–28]. It can be shown (see Chap. 3) that

$$m' = \frac{\sigma}{\sigma - \sigma_0} M \tag{1.17}$$

where M is obtained from Eq. (1.4). (The expression given by Karim [142] in this regard is erroneous.)

Burton [144] has suggested that σ_0 can be determined by extrapolating the $\log\sigma - \log\xi$ curve to $\xi = 0$. Using a miniature tensile installation, Geckinli and Barrett [145] have determined σ_0 by stress relaxation. According to them

$$v = \frac{dL}{dt} + \frac{1}{\chi} \cdot \frac{dP}{dt} \tag{1.18}$$

where P is the axial force, $1/\chi$ is the compliance of the testing machine. In a load relaxation test $v = 0$ and so

$$\xi = \frac{1}{L}\frac{dL}{dt} = -\frac{\dot{P}}{\chi L} \tag{1.19}$$

Therefore, they concluded that the strain rate was directly proportional to the unloading rate dP/dt. χ was evaluated from the initial part of loading and L corresponded to the specimen length just before commencement of relaxation. σ_0 value obtained by this procedure was vastly different from that of Burton [144]. The reasons for the significant difference are not clear.

An early procedure suggested to find σ_0 experimentally was to determine the stress in a relaxation test as $t \to \infty$. But, there will be difficulties in using this approach because the accuracy of measurements in a stress relaxation test on a standard testing machine is rather poor. Also, the temperature sensitivity of superplastic flow is very high and so even small changes in temperature can lead to large errors in measurement. As noted in [146], even in a room temperature test the ambient temperature should be carefully controlled.

Hamilton et al., [11] believe that, in general, the results of a load relaxation test will not describe the forming conditions where the strain rate is either constant or increasing. Results obtained on a Sn–Pb eutectic alloy [72, 147] have revealed significant differences in the values of the material constants (m and σ_0) corresponding to different loading conditions. Therefore, from a practical point of view, it is desirable to determine σ_0 from experiments in which the strain rate is either constant or increasing.

Mohamed [148] has suggested two different procedures for determining σ_0 in the following constitutive equation[4], which is similar to Eq. (1.13).

---

[4] Mohamed has actually described the procedures for shear mode of testing.

$$\frac{\xi kT}{D_{gb}Eb} = A'\left(\frac{b}{d}\right)^{c}\left(\frac{\sigma-\sigma_0}{E}\right)^{n'} \tag{1.20}$$

where A', D_{gb}, E, b, c and d are material constants.

In the first procedure, experimental ($\xi kT/D_{gb}Eb$) is plotted against (σ/E) (values of D_{gb} and E are taken from literature). Then, σ_0 is obtained from the difference between the linear extrapolations corresponding to regions II and I. In the second procedure, isothermal data pertaining to regions II and I are plotted on a linear scale, assuming that the value of the strain rate sensitivity index, m', is equal to the slope of the sigmoidal plot in region II. The intercept on the stress axis of a $\sigma-\xi^{m'}$ plot gives the value of σ_0. (In many studies, the second procedure has been preferred, e.g., [76, 146, 149–156].) But, the results are conflicting, see, e.g., [146]. In some cases even negative values have been reported for σ_0 [146, 151]. But, procedure 2 is erroneous, since it is based on an untenable hypothesis that $m' = M$, while the correct expression is given by Eq. (1.17). Calculations show that the value of m' for a number of aluminium alloys significantly exceeds the slope M (see Chap. 3).

Robust methods for determining the threshold stress from experimental data are described in [105, 157]. These methods do not require special investigations (e.g., low strain rate or load relaxation tests) but start with a rigorous mathematical description of superplastic flow. The methods have been verified using results on a number of Al-based superplastic alloys. These procedures are described in Chap.3.

1.4.4.3
Activation Energies

Activation energy '*Q*' is a concept useful in both physics and mechanics. For physical significance the value of *Q* determined in different modes of testing, e.g., tension, torsion, compression, should be nearly equal. If a unique physical process is dominant, *Q* should also be temperature independent. When this is not the case and there are no valid reasons to believe that the dominant operating mechanism changes with temperature, the constitutive equation will have to be improved upon.

Superplastic flow is strongly dependent on temperature; with increasing temperature the flow stress decreases and the optimal strain rate range for superplastic flow shifts to higher values. The temperature dependence of flow in a material of constant microstructure is assumed to be Maxwell–Boltzmann in character. That is, one of the following two relationships can be used. When the stress is maintained constant (a creep experiment),

$$\xi = A\cdot\exp\left(-\frac{Q_\sigma}{RT}\right) \tag{1.21a}$$

If ξ is kept constant,

$$\sigma = B \cdot \exp\left(\frac{Q_\xi}{RT}\right) \tag{1.21b}$$

Here R is gas constant, A and B are temperature independent constants. The parameters Q_σ and Q_ξ are the activation energies at constant stress and constant strain rate respectively. Evidently, Q_σ is obtained from the slope of a $\ln\xi$ vs. $(1/T)$ plot at constant stress. Likewise, Q_ξ is calculated from the slope of a $\ln\sigma$ vs. $(1/T)$ plot at constant strain rate. $Q_\sigma = Q_\xi$ *only if flow is newtonian* ($m = 1$). As superplastic flow is non-newtonian and when simple power law (Eq. 1.3) describes flow, the following relationship is obtained [3,130]

$$Q_\xi = mQ_\sigma \tag{1.22}$$

As $m < 1.0$, $Q_\sigma > Q_\xi$. Using SUPRAL alloy (Al–6Cu–0.4Zr) specimens, Bricknell and Bentley [158] have confirmed experimentally the validity of the above equation due to Padmanabhan and Davies [130]. They found the difference to be significant and have suggested that this has to be taken into account while considering the temperature dependence of superplastic flow.

Q_σ and Q_ξ depend on temperature and the stress/strain rate level at which they are determined. Therefore, they are apparent values and do not have a physical meaning. So, the concept of true activation energy is introduced. But, a true activation energy Q_{tr} depends on the constitutive equation used and one can find at least 8 different definitions of this concept, see, e.g., [76, 149, 150, 155, 159–161]. Apparent activation energies Q_a^σ and Q_a^ξ are defined unambiguously on an empirical basis [8, 161, 162] as

$$Q_a^\sigma = -R\left[\frac{\partial \ln\xi}{\partial\left(\frac{1}{T}\right)}\right]_{\sigma = const} \tag{1.23a}$$

$$Q_a^\xi = R \cdot \left[\frac{\partial \ln\sigma}{\partial\left(\frac{1}{T}\right)}\right]_{\xi = const} \tag{1.23b}$$

When two or more micro-mechanisms are dominant, Q_σ and Q_ξ are sometimes referred to as 'fictitious' activation energies (Q_f^σ and Q_f^ξ) [163] (e.g., in Region III where the contributions from dislocation creep and grain boundary sliding can be comparable). Evidently, 'fictitious' activation energies, like the apparent ones, have no physical meaning.

One can show that the equalities $Q_a^\sigma = Q_\sigma$ and $Q_a^\xi = Q_\xi$ are valid only if these activation energies are temperature independent. The situation here is similar to the situation with respect to m and M–values discussed in Sect. 1.3. (If Q_σ and Q_ξ are similar to m value, then Q_a^σ and Q_a^ξ are similar to M values. Temperature dependence of Q_a^σ and Q_a^ξ replaces the strain rate dependence of M.)

Bhattacharya and Padmanabhan [161] have considered the relationship between the apparent activation energies Q_a^σ and Q_a^ξ within the framework of an analysis due to Padmanabhan [162]. It was shown that

$$Q_a^\xi = MQ_a^\sigma \tag{1.24}$$

where M is from Eq. (1.4). It is of interest to consider the following problem. Assume that for steady state superplastic flow (when the stress ceases to vary with time/strain) the following unambiguous relationship among σ, ξ and T is available. That is,

$$\Phi(\sigma, \xi, T) = 0 \tag{1.25}$$

Evidently, Eq. (1.25) will include the power law (1.3), as well as those pertaining to a number of physical models of superplastic flow as particular cases (see Chap. 3). Can anything be said about the activation energies of a deformation process described by Eq. (1.25)? In order to answer this question Eq. (1.25) is rewritten as

$$\ln\sigma = \varphi\left(\ln\xi, \frac{1}{T}\right) \tag{1.26}$$

Then, the full differential of this function will be

$$d\ln\sigma = \varphi_{\ln\xi} d\ln\xi + \varphi_{\frac{1}{T}} d\left(\frac{1}{T}\right) = Md\ln\xi + \left(\frac{Q_a^\xi}{R}\right) d\left(\frac{1}{T}\right) \tag{1.27}$$

For a constant stress test

$$Q_a^\sigma = -R\left[\frac{\partial \ln\xi}{\partial\left(\frac{1}{T}\right)}\right]_{\sigma = const} = R\frac{\Phi'_{\frac{1}{T}}}{\varphi'_{\ln\xi}} = \frac{Q_a^\xi}{M} \tag{1.28}$$

which is the same as Eq. (1.24). Thus, the result obtained by Bhattacharya and Padmanabhan [161] is valid if the constitutive Eq. (1.25) is obeyed, i.e., it is a general rule.

The geometrical interpretation of these results will be presently examined. One can see that Eq. (1.25) defines a surface in the 3D $\sigma - \xi - T$ space. A standard sigmoidal curve (Fig. 1.3a) represents a part of this surface belonging to the plane T = constant. This sigmoidal curve can be characterised by the slope M and the index m (see Sect. 1.3). The sections of the surface Φ by planes σ = constant and ξ = constant give the ξ vs. $(1/T)$ curve and the σ vs. $(1/T)$ curve respectively. Each curve can be characterised by two parameters: Q_a^σ and Q_f^σ (for the ξ vs. $(1/T)$ curve) and Q_a^ξ and Q_f^ξ (for the σ vs. $(1/T)$ curve). Examples and calculations of activation energies for the different models known in the literature are given in Chap. 3. In the analysis of Mohamed et al. [159] the apparent activation energy is defined as

$$Q_a^\sigma = -R\left[\frac{\partial \ln \xi}{\partial\left(\frac{1}{T}\right)}\right]_{d,\sigma}, \quad Q_a^\xi = \frac{R}{m}\cdot\left[\frac{\partial \ln \sigma}{\partial\left(\frac{1}{T}\right)}\right]_{d,\xi} \tag{1.29}$$

and the true activation energy is obtained from

$$\xi = \frac{A'}{G^{n-1}T}\cdot \exp\left(-\frac{Q_{tr}^\sigma}{RT}\right) \tag{1.30}$$

where A' is a temperature independent material constant, G the shear modulus and Q_{tr}^σ is the true activation energy. The value of Q_{tr}^σ can be found as the slope of the straight-line ln $[\xi G^{n-1}T] - 1/T$. Although this method is often used it applies only when m is constant – a situation not encountered during superplastic flow.

As mentioned above, there are many other methods of introducing the concept of a true activation energy, e.g., [76, 149, 150, 155, 160, 161]. This point will be taken up further in Chap. 3.

1.4.4.4 Structure and Mechanical Response

The structural state of a material determines its mechanical response. In particular, grain refinement is accompanied by strain softening, e.g., as seen in dynamic recrystallization. Grain growth leads to strain hardening. In contrast, at low homologous temperatures (T/T_m) the Hall–Petch relationship (flow stress is inversely related to the square root of grain size) is obeyed.

The main principles of mechanics of solids do not include any structural characteristics. Drucker's criterion for material stability [164, 165], which is the theoretical basis of classical theory of plasticity (flow theory) [166, 167], precludes strain softening. But, as strain softening is a common observation during the hot deformation of polycrystalline materials, it is necessary to include this observation in the theoretical framework. This can be done in two ways: (i) extending Drucker's criterion (e.g., Iljushin's criterion [168,169] allows a material to have a descending stress–strain diagram); (ii) introducing structural characteristics in the constitutive equation, e.g., an average grain size. Structural features other than average grain size may also be considered for inclusion in a constitutive equation if they are found to be relevant, e.g., extent of cavitation, change in dislocation density.

Most constitutive equations known in literature [see Chap. 3] can be represented in a uniaxial case by

$$\sigma = f(\xi, d, T) \tag{1.31}$$

where d is the average grain size, T is the absolute temperature of deformation and f is a single-valued function. Elementary analysis based on Eq. (1.3) (see Chap. 3) shows that this equation is not useful in describing flow in the transient regime of loading. Therefore, Eq. (1.31) is to be generalised appropriately to cover that region also, see, e.g., [131]. The behaviour of materials in the transition regime of loading is of both fundamental and practical significance.

Thus, a correlation between thermomechanical history and structural evolution is essential. The effect of the loading schedule on the kinetics of structural changes in titanium alloys has been investigated [170, 171]. It was found that the loading history noticeably affected the microstructure. These results are discussed in Chap.6. Another example of the considerable influence of the loading history on the structural kinetics is concerned with the practical task of producing an ultra-fine grained structure in a polycrystalline material. A number of structure-sensitive properties of ultra-fine grained nickel have been investigated [172]. Following types of loading were used: (i) equal channel angular extrusion; (ii) torsion under pressure; (iii) second scheme after the first one. In all cases, the total strain applied was the same[5]. But, transmission electron microscopy revealed that the microstructures in the three cases were clearly different (the average grain sizes for the three cases respectively were 0.2, 0.1 and 0.02 μm). Consequently, the physical and mechanical properties were different.

1.4.5 On Stability of Superplastic Flow

The problem of stability of superplastic flow is of both theoretical and practical importance. Hart's analysis [107] (see Sect. 1.3.4) gives only a qualitative explanation for the contribution of strain rate hardening to the high ductility of superplastics subjected to uniaxial tension. Various aspects of plastic instabilities under

[5] The method used to calculate strain is not given

uniaxial tension have been considered [3, 63, 110–120, 173–175]. A stability criterion for uniaxial tensile flow when the flow stress is a unique function of strain, strain rate and temperature has been suggested. It has been shown that the point at which necking starts depends on the strain rate, temperature and (which is very important) the strain rate/temperature history. Expressions for some simple cases (e.g., isothermal tension at constant strain rate) are given in [173].

Estrin and Kubin [176] have suggested a classification of the main types of plastic instabilities considering strain hardening, strain rate hardening and temperature changes. They have given both the phenomenological and the microstructural reasons for the various types of instabilities. Investigation of stability is carried out in [176] by an extension of the linear stability analysis. The initiation of plastic instabilities of different types is investigated in [177] by taking into account finite strain behaviour. A detailed analysis of the influence of strain rate sensitivity on tensile elongation and on the growth rate of necking is given in [178].

In the last few years a number of studies on the stability of plastic flow for different constitutive equations have appeared, see, e.g., [179, 180]. Structural changes and damage accumulation are taken into account by introducing internal variables in the constitutive equations (see Chaps. 2 and 3) that are consistent with special kinetic equations. It is pertinent to note that the results of the analysis depend significantly on the choice of the constitutive equation.

Evidently, for practical purposes, plastic instabilities present during technological processing are more important than those present during tensile testing. This problem was considered for the first time in the classic works of Iljushin [181] and Ishlinski [182]. In [181], a formulation for 2D-viscoplastic flow was considered using examples from the tension/compression of a strip and a cylinder. In [182], the viscoplastic flow of a strip and a rod was investigated. Later, the problem of stability of viscoplastic flow was studied in terms of hydrodynamics, see, e.g., [183–185], where the viscoplastic flow of a cylinder, shear of a strip and necking in a sheet were considered. Investigation of stability of superplastic flow in terms of modern theories of mechanics of solids is yet to be done. This problem is very complicated.

2 Mechanics of Solids

There is no unique definition for 'mechanics'. Many regard it as an elementary branch of physics dealing with the motion and equilibrium of rigid bodies (the so-called classical dynamics) that forms the basis of many engineering disciplines. But, the term is also included in some highly specialised subjects, e.g., fluid mechanics, statistical mechanics, quantum mechanics. In this book, mechanics of solids is used to describe the macrobehaviour of superplastics. Hence, in this book unless otherwise stated 'mechanics' stands for 'mechanics of solids'.

In one view, mechanics of solids is a subdivision of mathematics devoted to the theory of approximation of experimental data. In this view, mechanics is useful in carrying out simple mechanical tests and fitting experimental data to empirical models. This view is rather naive.

In many countries, mechanics is studied in its applied aspects, e.g., mechanical engineering, applied mechanics, mechanics of metal forming. In India, this practice coexists with a study of 'theoretical and applied mechanics.' In Russia and some other countries, departments of mechanics and mathematics exist. Such an approach indicates the importance of mathematics in mechanics. A special feature of mechanics of solids is the development of mathematical methods to describe the deformation of *continua*. Many mathematical methods, including exotic ones, are successfully applied in studies in mechanics. In some cases, the requirements of mechanics have stimulated the development of new mathematical methods, e.g., the demands of viscoplasticity and viscoelasticity have led to developments in functional analysis; the need to solve extremely non-linear boundary value problems has given rise to innovative numerical methods. Even purely experimental works in mechanics, see, e.g., Bell [186], include expressions unintelligible to non-specialists. Therefore, a view has emerged that this branch of knowledge is not useful in solving *practical* problems.

In reality, mechanics of solids is the theoretical and experimental foundation of a number of engineering disciplines, e.g., strength of materials, mechanical engineering, metalworking. There are also inter-disciplinary branches that use mechanics of solids (sometimes they are considered as new subdivisions of mechanics), e.g., biomechanics, mechanics of composite materials, mechanics of natural phenomena, damage mechanics. Also, mechanics of solids has directly contributed to developments in the aerospace industry, nuclear power programme, construction of large and unique structures (dams, bridges, etc.,) and so on. The above achievements are a result of the merger of new physical approaches and development of advanced materials, on the one hand, with new methods of investigating and calculating the strength of structures, determining short- and long-time creep, fatigue, damage accumulation, stability, plasticity and reliability (all developed by specialists in mechanics of solids) on the other.

Methods for preventing accidents like the unexpected failure of ships, aeroplanes, bridges, etc. have been developed by specialists in mechanics. Difficult calculations for launching missiles have been completed. Procedures for establishing behaviour in space have been identified. Theory of explosive welding has emerged. Detailed analyses for the penetration of thick armour plates and moving earth mass by controlled explosion have been worked out. Development of crack and fracture mechanics has significantly increased the life of equipment in power stations. Shelf life of expensive items, e.g., aero-engines, has been extended based on predictions that use mechanics. Evidently, mechanics of solids can be used far more extensively than for approximating/analysing experimental data.

In this chapter, the scope of the subject is defined. Main consideration is restricted to the mechanical response of *strain rate sensitive materials*, as this is useful to specialists in superplasticity. The text is presented with minimum mathematics; some complicated mathematical procedures are summarised in Appendixes as examples. Some very important and advanced concepts in mechanics of solids are beyond the scope of this Chapter. It is assumed that the interested reader will consult specialist handbooks, whose list is given at the end of this Chapter.

2.1 The Subject

Mechanics of solids deals with models of real materials, processes and phenomena. Models are constructed on the basis of and within the framework of the concept of a *continuum* (Sect. 2.2): all materials, structures and components are regarded as continua (solids). Physical laws of conservation are stated for this medium (Sect. 2.3). Using well developed mathematical methods, the scalar, vector and tensor characteristics of the mechanical state of a continuum are examined. Different assessments of the mechanical state are made (degree of plasticity, damage accumulation, long time strength, etc.). The problems of estimating the stress–strain state, strength, compliance, stability, etc. in structures and components are solved.

The two main objectives of study are the development of (i) mathematical models of materials (constitutive equations or constitutive laws–Sect. 2.4), and (ii) mathematical models of processes and phenomena (system of equations describing a model in such a class is called the *boundary value problem* in mechanics of solids). Equations of different kinds can be included in a general formulation of the boundary value problem (algebraic, differential, integral, etc.). Some inequalities can also be introduced. The total composition of this system of equations is the same for the mechanics of continua: conservation laws, kinematic equations, constitutive equations, initial and boundary conditions. Constitutive equations (CEs) form the basis of mechanics of solids. The boundary value problems are usually classified based on the type of the constitutive equation. For example, there are boundary value problems of the theory of elasticity, theory of plasticity, viscoelasticity, viscoplasticity, mechanics of heterogeneous media, etc.

The construction of mathematical models of materials and processes (phenomena) has practical meaning only if the models are adequate. In other words, in the intervals of load and other external influences under consideration, the models should reflect correctly the mechanical behaviour of materials. They should correctly capture the principal features of the processes (phenomena). The central problem in mechanics of solids is to construct adequate constitutive equations for the materials under study. It is not possible to construct an adequate model for a process (phenomenon) if adequate constitutive equations do not exist. Thus, mechanics of solids has two principal aims: (i) development of adequate *constitutive equations* and setting up of the corresponding boundary value problems; (ii) evolving methods for solving the boundary value problems and investigating the solutions obtained. The scales of the objects of study may vary from a few nanometers (aggregates of a few hundreds of atoms) to galaxies: the only need is to apply correctly the methods of this discipline.

A spectrum of mathematical methods has been successfully used. Analytical methods as well as numerical, experimental and experimental-numerical methods are employed. The development of numerical methods of solving boundary value problems has led to a new branch of knowledge: *computational mechanics*. This is due to the complexity (significant non-linearity) of the boundary value problems that arise from the differences in the constitutive equations, types of boundary conditions, etc.

Experiments play a key role in the development of mechanics of solids (Sect. 2.5). The situation is similar to that obtainable in any branch of natural science. Mechanical tests include the experimental investigation of material properties, tests on structural elements and/or model experiments. In this book, as a rule, only the first type of mechanical tests is considered, viz., experiments used to investigate/develop constitutive equations.

Experimental observations facilitate the development of idealised mathematical models of materials, e.g., *linear elasticity* and *non-linear elasticity*, *viscoelasticity*, *plasticity*, *creep*. Experimental investigations are also needed to validate a model. Validation is done through basic experiments which allow the determination of the material constants. Special experiments are used to define the range of applicability of a constitutive equation (Sect. 2.5).

The view that mechanics of solids can only describe the results of experiments used to determine the material constants but can not describe mechanical behaviour under different loading conditions is not correct. The deformation theory of plasticity due to Hencky [166] and Iljushin [169], for example, allows the unambiguous prediction of mechanical response under different multiaxial loading conditions from results obtained in uniaxial testing, e.g., a tensile test on a cylindrical specimen. The range of applicability of the predictions is precisely stated with special theorems. Within this framework, the validity of the predictions has been confirmed in a number of experiments on different materials and structures. Such theories, however, do not involve any microscopic parameters in the constitutive equations.

Special experiments for verifying hypotheses and assumptions of a general nature (e.g., proportionality between stress and strain rate tensors, existence of a universal interdependence among the invariants of some tensors) have an important role in the development of the theory of constitutive equations. In physics, such experiments (which result in either 'yes' or 'no' answers) are referred to as critical/crucial experiments.

There is another important subdivision in experimental mechanics – the *theory of experiments*. In particular, it includes the development of methods to 'decode' the results of an experiment when a non-uniform stress–strain state is present. It is necessary to calculate the stress–strain dependence from the measured values of loads and displacements on the specimen surface. (For example, there is a well-known method to calculate the shear stress (τ) – shear strain (γ) curve from experimental torque $\overline{M}$ – angle of twist φ plot obtained in simple torsion). As it is extremely difficult to establish a uniform stress–strain state while studying the finite strain behaviour of materials, development of such methods has considerable practical importance.

Theory of experiments and extensive calculations are required while conducting experiments involving special physico-mechanical effects (change in the optical, electrical or some other physical property of a material under load), e.g., polarizable optical methods (photoelasicity, photoplasticity, photo-viscoelasticity). These methods are widely used in estimating the mechanical characteristics of different structures and components of complicated shapes, as also in modelling technological processes.

The second type of mechanical tests includes tests on some structural elements (e.g., aeroplane wings, rocket hull) or structures as a whole (e.g., aeroplanes, missiles, submarines). There are many special methods of testing structural elements. This is a special scientific direction beyond the scope of this book. Tests on full-scale structures are done to derive information on integrity, validate hypotheses and verify calculations. Such tests are also useful in identifying the hypotheses needed for a computational analysis of a given structure.

Model experiments are based on *geometrical modelling* and/or *physical modelling*. In the first case, experiments are done on the same material using typical geometrical dimensions (usually less than full scale). This method is used for modelling large constructions, e.g., dams for hydro-electric power generation, as well as natural phenomena, e.g., landslides, avalanches. In physical modelling, the mechanical behaviour of a given material is investigated using another material (which is cheaper, easier to handle, etc.). For example, the behaviour of high carbon steels during hot rolling can be successfully modelled by rolling lead ingots at room temperature. The mechanical response of superplastics can be investigated using model alloys (Zn–0.4%Al, Sn–Pb, Sn–Cd, Sn–Zn alloys, etc.). Of course, there are also combined methods of modelling, in which the geometry as well as the model material are different from those of the real construction/phenomenon.

Due to the complexity of the constitutive equations, *physical modelling* is not so widespread in mechanics of solids as in hydrodynamics. However, various physical methods of measuring mechanical parameters are in use for a long time in experimental mechanics (e.g., holographic methods of measuring surface strains). Also, well-known optical methods are widely used for investigating stress distributions. In this case, the real material is replaced by a transparent polymer (e.g., celluloid, plastic, etc.) which becomes optically anisotropic after deformation. This method allows the mapping, in three dimensions, of *stress distributions,* in a deformed body.

2.2. Basic Concepts

2.2.1 Concept of a Continuum

The concept of a continuum forms the foundation of mechanics of solids. The behaviour of real solids (as well as liquids and gases) is considered in mechanics of solids as if it pertained to that of a continuum. The concept of a continuum is introduced so as to consider infinitesimally small volume elements of a medium, infinitesimally small segments and/or areas of surfaces. The main mathematical entities are tensors of different order: zero order (scalar, e.g., density, temperature); first order (vector, e.g., velocity); second order (e.g., stress, strain) and higher orders. All these tensors are assumed to be continuously differentiable to the degree required (analytic) with respect to spatial co-ordinates and time. Therefore, one can use directly the well developed mathematical methods of field theories to describe the behaviour of continua [187, 188].

Such an emphasis on the mathematical basis of mechanics of solids may produce a 'reaction of rejection' among metallurgists and physicists who specialise in mechanisms of deformation of polycrystalline materials. Also, unlike in physics and materials science/metallurgy, structural details (crystal lattice, grains, grain boundaries, etc.) are not considered explicitly in the classical scheme of mechanics of solids. The concept of a continuum ignores the micromechanisms of plastic/superplastic deformation as also the discontinuous relative movement between grains (*grain boundary sliding*). This may raise a question with respect to the usefulness of mechanics of solids for describing superplastic flow.

According to Iljushin [188] all basic concepts in mechanics of solids have a statistical basis. In this view, the variables in mechanics of solids are similar to those in classical and statistical mechanics applied to a closed system. Historically, these concepts were introduced in mechanics of continua before the development of statistical mechanics. Some of the variables in statistical mechanics were introduced as analogues of the basic concepts in mechanics of continua, e.g., concepts of stress, strain, heat flux. One is also familiar with the rigorous statistical derivation of kinetic theory of gases.

Metallurgists and physicists use extensively the concepts of *stress*, *strain* and *strain rate* in their theories of deformation of polycrystalline materials. In such a situation, they use both the terminologies of mechanics of solids and the concept of a continuum. The main concepts and methods of mechanics of solids should be used correctly while studying the mechanical response of superplastic materials and the range of applicability of equations should be defined. Some examples in this regard were given in Sect. 2.1 and some more are given below. Basic concepts in mechanics of solids within the theory of infinitesimally small strains can be written without invoking complicated mathematics.

The concept of a *representative volume* is one of the most important abstractions. This volume is assumed to include a sufficiently large number of microstructural elements of a given material in order for them to be considered representative, but it should also be sufficiently small from a macroscopic point of view to be considered as a particle (that is, infinitesimally small volume) in a continuum. The linear dimensions of a representative volume are determined using a typical scale within which the properties can be assumed to be uniform. In other words, while introducing the concept of a representative volume a sufficiently large body is considered. Since a small part dV of this body includes a sufficient number of particles, one can use with dV macroconcepts like density, displacement, velocity, acceleration, internal energy and other parameters considered to be averaged over an ensemble of microstructural elements. The idealisation of the real physical body in mechanics of solids consists in considering the average values as true ones. In the development of mechanics of continua, the number and mathematical essence of these average variables allow a description with sufficient accuracy of the internal state and the interactions among different bodies. (For example, theory of elasticity is fundamental to many engineering disciplines, e.g., strength of materials, mechanical engineering. But, theory of elasticity is also a part of mechanics of solids, and so, a part of mechanics of continua.) In mechanics of solids, mechanical and thermal interactions are usually considered along with deformation in small volumes.

A typical feature of real metals and alloys is their microstructure, e.g., grain size. In this case, the representative volume should contain a sufficient number of grains to average out the effects of individual ones. At the same time, the representative volume should be sufficiently small compared with the dimensions of the body (the specimen) so that the idea of an infinitesimally small volume is still valid. In superplasticity it has been seen that when the specimen thickness is less than about 10 grain diameters, significant superplasticity does not develop, even when experimental conditions like temperature, strain rate and grain size are optimal [189]. So, the representative volume should be so chosen as to make the averaging of properties acceptable [190]. In this view, the deformation of a very thin specimen will have to be treated separately as the complex behaviour of an aggregate of individual grains. In such specimens, reproducibility of results will be difficult due to differences in the size and orientation of individual grains that will change from one specimen to the next. A similar situation will exist during the torsion of a specimen whose diameter is only a few grains thick. In these cases, it will be necessary to have a representative volume of a smaller size and simultaneously consider the interactions among individual grains. This approach is common in the study of composites. The composite is first considered to be a continuum (not necessarily isotropic or uniform). The stress–strain relationships are developed using standard methods of mechanics of solids. And then, the effects of reinforcing fibres etc., are introduced at another level (scale) to determine the stress–strain state in the vicinity of these elements.

Methods of continuum mechanics can be used to explain a wide range of phenomena: from the cosmic scale when the representative volume may include a number of planets down to the microscopic level, e.g., composites containing fine dispersions. The point to be kept in mind is that the representative volume should reflect the macroproperties of the medium under consideration. Thus, one can apply the methods of mechanics of solids at various levels with the representative volume being identified in an appropriate manner.

2.2.2
Stress, Strain and Strain Rate States

It is assumed that basic ideas concerning stress state, stress tensor and its invariants, principal axes, principal values, strain and strain rate states etc., are known to the reader. (See, e.g., [160, 167, 190].) Only those aspects that have a bearing on the stress–strain state in superplastics will be emphasized here. Consideration of finite deformation has great significance in superplasticity. It is known that in case of finite deformation a mathematical description of the stress–strain state becomes non-unique. Selection of measures as well as tensors for a mathematical description of the stress–strain state must satisfy certain basic requirements. Also this selection depends on the formulation of the mathematical models of materials. In other words, if some mathematical formulation exists for small strain, for finite strain this formulation can be generalised in a non-unique way. This very important fact is not discussed in the literature on superplastcity. However, considering that the mathematical tools of the theory of finite strain are quite complex, all stress–strain state problems are discussed in this book for small deformation. It is, however, to be kept in mind that for each boundary value problem being solved, its formulation as well as the methods of analysis should follow the theory of finite strain.

Another remark concerns inhomogeneous stress–strain state. As an example, consider the elongation of a specimen of varying cross-section under an axial load. In this case, generally speaking, it is not correct to determine the tensile stress (i.e. normal stress on the cross-section of the specimen) as a ratio of the axial force to the cross-sectional area of the specimen. First of all, the distribution of the normal stress along the cross-section may not be homogeneous. Secondly, there must be a non-uniaxial stress–strain state in the specimen; otherwise the equilibrium equations on the specimen surface will not be satisfied. Therefore, the stress state can not be characterized only by a scalar factor. (It is very surprising that sometimes even the stress state in the vicinity of a dislocation is described by a scalar parameter. In some physical models, the stress entering the physical equation is not clearly defined.) Sometimes non-uniaxial and inhomogeneous stress–strain state can be neglected. But, such cases will have to be individually justified.

There are many cases where an incorrect determination of stress in a reduced cross-section (e.g., a hole in a sample) has resulted in sudden and unexpected failure of ships, bridges etc. A classical example of an inhomogeneous axial stress along a cross-section of a specimen is as follows [192]. A plate uniformly stretched in the elastic region experiences an axial stress σ_0. If a circular hole were present, the axial stress in the vicinity of the hole exceeds the average value by a factor of three. It is also known that the peak stresses depend on the curvature of the hole [193, 194]. This exact solution is due to Kolosov and Inglis. Very useful information on stress concentrations in machine parts is available in [191].

Consider the equation for the experimental measurement of stress. This problem is very difficult and non-trivial. In most cases, it can not be solved at all. A solution is possible only in certain cases, for those materials (for example, polar-optical sensitive materials or piezoceramics) in which the stress tensor is related to some other physical parameters (variables) that can be measured directly. That is why, when dealing with ordinary materials stresses are never measured, but are only calculated from the measured surface loads using some hypotheses and/or by solving the boundary value problem. This fact is not always mentioned. For example, in [191] there is a section on 'measurement of the flow stress'.

Now, attention is focussed on a description of medium kinematics. It is noted that in continuum mechanics, Lagrangian and Eulerian approaches are used to describe the medium kinematics. Both the approaches are equivalent. The Lagrangian approach is popular in mechanics of solids, while the Eulerian approach is frequently used in fluid mechanics. However, for solving boundary value problems of metalworking processes the Eulerian approach has been found to be preferable in mechanics of solids also. Mixed Lagrangian–Eulerian description also is used sometimes in computational mechanics to solve such problems. Such a description combines the advantages of both the approaches. With regard to investigations in the domain of superplasticity it seems natural to describe the medium kinematics by the concept of strain rate, as in the theory of metalworking processes.

Let the fixed (immovable) Cartesian system of co-ordinates be given. According to the Eulerian approach, the motion of the medium is defined by the velocity field $V_i = V_i\,(x_i, t)$; $i = 1, 2, 3$ Where V_i are the co-ordinates of the velocity vector $\vec{V}$ of a material particle, which in time t occupies a point of space with Cartesian co-ordinates x_j. Then the strain rate tensor with components V_{ij} is introduced in the ordinary way as

$$V_{ij} = \frac{1}{2}\left(\frac{\partial V_i}{\partial x_j} + \frac{\partial V_j}{\partial x_i}\right) \equiv \frac{1}{2}\left(V_{i,j} + V_{j,i}\right) \tag{2.1}$$

(In unambiguous cases, a notation $\dot{\varepsilon}_{ij}$ for the components of the strain rate tensor is used.) If the material is incompressible,

$$div\,\vec{V} \equiv V_{i,i} = 0 \tag{2.2}$$

(Here and in what follows a rule of summation with respect to repeated indexes is assumed.) and the strain rate tensor V_{ij} is inconsistent with its deviator ξ_{ij}. The strain rate state is often characterized by a scalar value – the equivalent strain rate.

$$\xi_e = \sqrt{\frac{2}{3}\xi_{ij}\xi_{ij}} \tag{2.3}$$

With the help of ξ_e, an invariant of the strain rate deviator, another scalar characteristic of the deformation process is constructed as

$$s = \int_0^t \xi_e dt' \tag{2.4}$$

where integration is along the trajectory of motion of the individual particle.The value s (Odqvist parameter, the accumulated strain, the length of strain trajectory) is very popular in the theory of metalworking. Unfortunately, an explicit and simple method of obtaining this value from the strain rate tensor is not available. Generally speaking, the components of the strain tensor ε_{ij} can not be derived by integration with respect to t of the corresponding components V_{ij} of the strain rate tensor. (A discussion of this problem is beyond the scope of this book.) But, in some papers related to the theory of metalworking such a procedure is adopted without explanation. The problem of introduction of strain measure and strain tensor is not only complicated but is also non-unique. Many variants are possible. (An absolutely analogous situation exists with stress measures and stress tensors. In papers concerning superplasticity or in problems that consider finite (large) deformation Cauchy stress is often used without special stipulation. At the same time, some different 'Eulerian' and 'Lagrangian' stress tensors are exploited in mechanics of solids.)

Only logarithmic strain tensor is popular and clear among specialists in superplasticity/materials science/metalworking processes. One of the attractive features of the logarithmic strain measure is the simple law of strain summation (i.e., if in a tensile test a specimen has experienced a strain $\varepsilon^{(1)} = \ln l_1/l_0$ and then another strain $\varepsilon^{(2)} = \ln l_2/l_1$, then the total strain suffered $\varepsilon^{(f)} = \ln l_2/l_0 = \varepsilon^{(1)} + \varepsilon^{(2)}$). But, it should be kept in mind that an effective application of the logarithmic strain tensor is possible only when the main axes of the strain tensor do not vary in orientation with respect to those of the material fibres. In other cases, this measure loses many of its advantages and its use becomes extremely complex.

Finite strain expressions are not present in the constitutive equations (CEs) of superplastic materials (Sect. 2.4 and Chap. 3), possibly except when the principal axes of the strain tensor are immobile and are written for that case. Usually only increments of strains are included in the constitutive equations for superplastics. In mechanics of solids, in contrast, sufficiently effective variants of the description of the strain state, including cases where the principal axes of the strain tensor are rotated, are available. The interested reader can find the necessary details in specialist literature, a list of which is given at the end of this Chapter. (See also Appendix B.)

2.3 General Laws and Boundary Value Problems

The fundamental equations of mechanics of solids employ the well-known laws of conservation as applicable to a continuum (the laws of conservation of mass, impulse, momentum of impulse, energy) and also the laws of thermodynamics. Derivations are not given here as they can be found in textbooks, e.g., [195–197]. Only a general analysis is presented. (Here, as in other books, the case of infinitesimally small strains is considered for simplicity.)

The fundamental laws are divided into two groups: the first contains purely mechanics-based parameters and the second includes in addition thermodynamic parameters (temperature, heat flux, internal energy, entropy). Consider the simple situation (typical of superplasticity) when these two groups are not connected, i.e., the equations of first group may be examined separately and temperature included in them and in the solutions solely as a parameter. Then, some variables entering in the first group of equations may be given as functions of temperature and also temperature may be considered as a known function of time. Writing the equations of the first group in differential form, one obtains the equation of continuity

$$\frac{\partial \rho}{\partial t} + div(\rho \bar{V}) = 0 \quad or \quad \frac{d\rho}{dt} + \rho div\bar{V} = 0 \tag{2.5}$$

(which is reduced to Eq. (2.2) for an incompressible material) and the equations of motion

$$\rho \frac{dV_i}{dt} = \rho F_i + \frac{\partial \sigma_{ij}}{\partial x_j} \tag{2.6}$$

where ρ is the density field in a continuum, F_i the components of the given body force, σ_{ij} the components of the stress tensor.

It is obvious that the system of Eqs. (2.5) and (2.6) is not closed with respect to the unknowns ρ, V_i , σ_{ij}, i.e., these equations are not sufficient for determining these values. (Equations (2.5) and (2.6) express some general laws of a continuum but do not contain information about individual materials.)

For further development, mathematical models of the continuum are necessary. These models should include relationships among stress, deformation, deformation rate and temperature. Such relationships are referred to as 'constitutive equations' (constitutive laws). Constitutive equations along with the equations of heat transfer for non-isothermal processes complete the system of equations of mechanics of solids. Some of the simplest constitutive equations known in literature are considered below. In many investigations to describe the behaviour of a superplastic, constitutive equations of non-newtonian liquids are used.

It is known that one should set the initial and boundary conditions to solve the system of equations (for equations of elliptic type, i.e., equilibrium equations, it is not necessary to formulate the initial conditions). In general, setting the initial conditions is not a trivial problem, but it can be considerably simplified by selecting the unstressed state of a body as corresponding to the initial moment of time.

There are various types of boundary conditions. Some of them are considered in Chap. 4. Usually, the surface of the deforming body Γ is divided into different parts Γ_1, Γ_2, Γ_3: only the values of the surface stresses are given on Γ_1; only the values of surface displacements are given on Γ_2 and both of them are given on Γ_3. Examples can be found in any text book, e.g., books on the theory of elasticity. For superplastic metalworking processes the boundary conditions have specific features and these are discussed in Chap. 4. The full set of equations including equations of motion, equation of continuity, constitutive equations, heat transfer equations and some concrete form of initial and/or boundary conditions define the so-called *boundary value problem* in mechanics of solids. Examples of boundary value problems in mechanics of superplasticity are given in Chaps. 4 and 5.

Unlike the laws of conservation, the constitutive equations can have many mathematical forms. The choice of a constitutive equation plays an important role in formulating the boundary value problem. Equations characterising the state of the fields affecting a continuum are also included in a formulation of the boundary value problem, in particular, a temperature field present in a body being deformed. (Sometimes, electromagnetic fields are also considered.) For this case, two situations are possible: (i) the temperature problem is independently solved and the solution is used to calculate temperature dependent mechanical properties at any point in the body; (ii) the temperature field can not be found independently of the solution of the main mechanical problem, which depends on the temperature field (e.g., upsetting of a cylinder during which considerable adiabatic heating is present). The first situation, has been already considered. In the second case, the temperature problem can only be solved in combination with the mechanical problem. Here the full system of equations includes both mechanical variables and those involving temperature and/or electromagnetic fields.

The type of constitutive equation determines not only the nomenclature of the problem, but also the type of the mathematical problem. From a mathematical viewpoint, boundary value problems are divided into two classes: linear and non-linear. An example of a linear problem is the boundary value problem of classical theory of elasticity. Theory of plasticity deals with non-linear boundary value problems. The type of partial differential equations is well established in the theory of elasticity. When non-linear constitutive equations are used, it is difficult to determine the type of equations. In complex problems different types of equations may be useful for different parts of the same body, e.g., material constants may be different for different parts of a body being deformed. This is understandable for a materials scientist: if the deformation history is different in different parts of a body, then the microstructure and, as a consequence, the mechanical properties of the material will be different at different places. Sometimes, even the type of mathematical equations may be different for different parts of the same body. Apart from mathematical skills, an investigation of the range of application of the constitutive equations and an insight into the physical/mechanical nature of the deformation involved are essential for developing an adequate model.

According to Keunings [198], the basic difficulties associated with a numerical prediction of rheologically complex flows, in particular, the flow of viscoelastic fluids, emanates from the numerical challenges to the mathematical nature of the governing equations as well as the fundamental physical issues such as flow behaviour close to walls and singularities.

Another example is classical. In 1920, Hencky developed [166] a simple form of the deformation theory of plasticity by a generalisation of Hooke's law. This variant was seldom used in engineering calculations since the corresponding methods of solving the boundary value problem including Hencky's equations was not known. Later the same equations were written in a more convenient form by Il-jushin who defined a domain of adequacy for Hencky's equations and used typical stress–strain curves to select the parameter needed in the relationships between stresses and strains. This led him to develop an effective algorithm for solving numerical problems [169]. This theory has found many practical applications.

In conclusion, it is noted that there are many methods of analysis in mechanics of solids. Advanced branches like computational mechanics, computational plasticity and computational damage mechanics have emerged. Many numerical–experimental methods of solving problems (including the method 'complex loading – computer' [199]) have also appeared. Specialist literature exists in these areas.

2.4 Mathematical Models of Materials

2.4.1 Typical Models for Describing Mechanical Behaviour

In most materials, below the elastic limit deformation is proportional to applied load (Hooke's law). The tensor form of Hooke's law is

$$\varepsilon_{ij} = \frac{1+\nu}{E}\sigma_{ij} - \frac{\nu}{E}\sigma_{kk}\delta_{ij} \tag{2.7}$$

where E is Young's modulus, ν is Poisson's ratio. Hooke's law may be written using the other elastic constants: the shear modulus $G = E/\ [2\ (1 + \nu\ )]$ and the bulk modulus $K_v = E/\ [3(1\text{-}2\nu\ )]$. For an incompressible material, $\nu\ = 0.5$ and so G

$= E/3$ and $K_V \rightarrow \infty$. It is useful to note that in theory of plasticity sometimes the hypothesis of elastic variation in volume is made use of, i.e.,

$$\Theta = \frac{\sigma_m}{K_v} \tag{2.8}$$

(where $\sigma_m = \sigma_{ii}/3$ is the mean stress, Θ is the volume strain) instead of the condition of incompressibility (Eq.(2.2)) for which $\Theta = 0$.

The viscous behaviour of some common liquids, e.g., water, can be described by a rheological law suggested by Newton (newtonian liquids):

$$\tau = \frac{\mu \partial V}{\partial y}$$

where τ is the shear stress; $\partial V/\partial y$ is the transverse (with respect to the flow direction) derivative of the velocity V and μ is the viscosity coefficient. For Newtonian liquids, μ depends only on temperature (if the hydrostatic pressure is not too large). Then, it is a material constant. For a linear stress state, the foregoing equation can be rewritten as

$$\sigma = \lambda \xi$$

where λ is a proportionality coefficient. For an incompressible isotropic material $\lambda = 3\mu$. The constitutive equations for a non-uniaxial stress–strain rate state for this class of liquids are given by [200]:

$$S_{ij} = 2\mu\xi_{ij}\ , \xi_{ii} = 0 \text{ or } \sigma_{ij} = -p_H\delta_{ij} + 2\mu\xi_{ij},\ \ \xi_{ii} = 0 \tag{2.9}$$

where p_H is the hydrostatic pressure; ξ_{ij} are the components of the strain rate deviator. Substitution of Eq.(2.9) in Eq.(2.6) leads to

$$\rho\left(\frac{\partial V_i}{\partial t} + V_j V_{i,j}\right) + p_{H,i} - \mu\left(V_{i,jj} + V_{i,ij}\right) = \rho F_i \tag{2.10}$$

Equations (2.10) are the well-known Navier–Stokes equations. Relationships (2.9) are well-known in mechanics of solids. In the 19th century Saint-Venant proposed equations for plasticity which were similar to those used in the theory of viscous flow but had different meaning. The main difference was that the value of μ, in general, was assumed to be a function of strain rate and deformation history. These relationships were later referred to as Saint-Venant relations. They are widely used in mechanics of metalworking as well as in superplastic forming (Chaps. 4 and 5).

Mechanical behaviour of many liquids can not be described by the linear relations (the so called non-newtonian liquids). The well-known relation $\sigma = K\xi^{m}$ is widely used to describe the rheology of non-newtonian flow. For some non-linear viscous liquids (e.g., solutions of high molecular weight polymers) the values of K and $m(< 1)$ are approximately constant over wide intervals of stress and strain rate. The value of the apparent viscosity coefficient $\lambda(\equiv \sigma/\xi = K\xi_e^{m-1} = K^{1/m}\sigma^{1-1/m})$ decreases with increasing ξ. Such liquids are called 'pseudoplastic' liquids. There are also the 'dilatant' liquids, e.g., starch paste. In this case, $m > 1$ and the apparent viscosity increases with ξ. The constitutive equation for a non-linear viscous liquid can be written as

$$\sigma_{ij} = \sigma_m \delta_{ij} + \left(\frac{2K}{3} \right) \xi_e^{m-1} \xi_{ij} \tag{2.11}$$

where K and m are material constants.

In mechanics of solids, the material is considered to be deformable. It is not always possible to draw a distinction between a solid and a liquid. For example, ice is conventionally regarded as a solid; but glaciers flow like liquids. Metals and alloys behave like a liquid during hot working. From the viewpoint of mechanics of solids the distinction between a solid viscoplastic body and a liquid is established by the presence of a mechanical threshold σ_{th} in the former. If $\sigma_{th} > 0$, then a material is a solid; if $\sigma_{th} \equiv 0$, it is a liquid. As deformation may be observed even at low stresses after a very large time interval, formally the mechanical threshold is defined as the maximum stress at which no significant deformation would be seen in a specified time interval, t_s. The magnitude of t_s will depend on the nature of the problem being investigated. Many variants of the theory of plasticity are available in the literature (Sect. 2.4.3). They all feature a 'switch': only if a certain condition is satisfied a change in plastic strain is seen; otherwise the change is absent.

2.4.2 Mechanical Models/Analogues

Depending on the conditions of testing, the same material may exhibit elastic, plastic or viscous behaviour. A standard way of describing different behaviour is through the construction of mechanical models/analogues. In this approach, for the uniaxial case, the models are constructed using only three elementary 'bricks' (Table 2.1) in different combinations. In general, such models can include any number of elements that are connected in series or in parallel. It is assumed that the deformation of all elements that are connected in parallel is the same at any instant of time. Likewise, the stress in all elements that are connected in series is assumed to be the same in magnitude at any moment of time. The properties of an elastic and/or a viscous element can be chosen to be non-linear. For superplasticity, the standard model employs the two parameter power law $\sigma = K\xi^{m}$. Models that include this element are considered in Chap. 3. In this section, some classical com-

Table 2.1 Elementary 'bricks' used in rheological modelling

No.	Name	Designation	Equation
1.	Hooke's (linear elastic) element		$\sigma = E\varepsilon$ E = constant
2.	Newton's (linear viscous) element	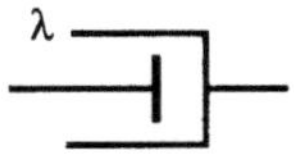	$\sigma = \lambda\xi$ λ = constant
2a.	Superplastic (non-linear viscous) element	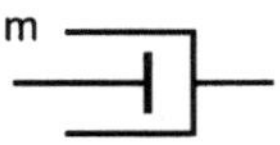	$\sigma = K\xi^{m}$ K = constant m = constant
3.	Saint-Venant (perfect plastic) element		$\xi = 0$ if $\sigma < \sigma_s$ $\xi \neq 0$ if $\sigma < \sigma_s$

binations as well as advanced models are considered. A more comprehensive treatment is available in [201].

A combination of two linear elements can be replaced by the equivalent (linear) element, e.g., two elastic elements in series, E_1 and E_2, can be replaced by the equivalent elastic element $E_{eqv} = E_1E_2/(E_1+E_2)$.

For elastoplastic behaviour (e.g., useful in cold working processes) a combination of elastic and plastic elements is used. A Prandtl body [201, 202], also known as elastic-perfectly plastic body, combines elastic and plastic elements (Fig. 2.1). The elastic and plastic elements can also be joined in parallel (Fig. 2.2) This model is not commonly used. The elasto-plastic models do not experience stress relaxation.

The combinations of viscous and plastic elements are presented in Figs. 2.3 and 2.4. The $\sigma - \xi$ diagrams for these elements are similar to those shown in Figs. 2.1–2.2.

Developments in space science and technology, e.g., use of polymer fuel in missiles, led to rapid developments in the theory of viscoelasticity. Here, two classical

a b

Fig. 2.1. The mechanical model **a** and typical σ–ε curves **b** for a Prandtl body: 1 – loading; 2 – unloading.

Fig. 2.2. Elastic and plastic elements in parallel: **a** mechanical model; **b** typical $\sigma - \varepsilon$ curves: 1–loading; 2–unloading

Fig. 2.3. Viscous and plastic elements in series: **a** mechanical model; **b** typical σ–ξ curve

Fig. 2.4. Viscous and plastic elements in parallel: **a** mechanical model; **b** typical σ–ξ curve

Fig. 2.5. Combinations of viscous and elastic elements: **a** a Maxwell body; **b** a Voigt body

models of linear viscoelasticity are considered in detail–the Maxwell body and the Voigt body (Fig. 2.5). In a Maxwell body (Fig. 2.5a), the total strain rate equals the sum of the elastic and the viscous strain rates, i.e.,

$$\xi = \frac{\sigma}{\lambda} + \frac{\dot{\sigma}}{E} \tag{2.12}$$

where $\dot{\sigma} = d\sigma/dt$ and λ is the viscosity coefficient. Under a constant stress $\sigma = \sigma_0$, a Maxwell body deforms at a constant strain rate $\xi_0 = \sigma_0/\lambda$ (like a viscous liquid). Under a constant strain rate $\xi = \xi_0$, Eq. (2.12) can be integrated using the initial condition $\sigma(0) = 0$, i.e.,

$$\sigma = \lambda\xi_0\left[1 - exp\left(-\frac{Et}{\lambda}\right)\right] \tag{2.13}$$

If the deformation is stopped at time $t = t_1$, then

$$\sigma = \sigma_1\, exp\left[-E\frac{t-t_1}{\lambda}\right];\ t > t_1 \tag{2.14}$$

with σ_1 the stress at time t_1. Thus, a Maxwell body can simulate stress relaxation. In the Voigt model (Fig. 2.5b), the total stress is equal to the sum of the elastic and the viscous stresses, i.e.,

$$\sigma = E\varepsilon + \lambda\xi \tag{2.15}$$

Integrating Eq.(2.15) with $\sigma = \sigma_0 =$ constant, one obtains:

$$\varepsilon = \left(\frac{\sigma_0}{E}\right)\left[1 - \exp\left(-\frac{Et}{\lambda}\right)\right] \tag{2.16}$$

Equation (2.16) expresses the delay time in the establishment of the elastic strain. The ratio (λ/E) is the characteristic delay time of a Voigt body. Thus, in a constant stress test, ε increases up to $\varepsilon_0 = \sigma_0/E$. If at time $t = t_2$ the load is removed ($\sigma = 0$ at $t > t_2$), further deformation will proceed according to the equation

$$\varepsilon = \varepsilon_2\, exp\left[-E\frac{t-t_2}{\lambda}\right];\quad t > t_2 \tag{2.17}$$

where ε_2 is the strain at $t = t_2$. Thus, on load removal the deformation will gradually decrease with time to 0. The phenomenon of a gradual increase in deformation on the application of a load and its gradual decrease on load removal is known as *after-effect*.

When $\sigma = \sigma_0 =$ constant, in a Voigt body $\xi \to 0$ as $t \to \infty$. In a Maxwell body, $\xi \to \sigma_0/\lambda$ as $t \to \infty$, i.e., the Maxwell body will flow infinitely at a constant strain rate $\xi = \sigma_0/\lambda$. Therefore, the Voigt body is sometimes referred to as viscoelastic hereditary solid, while the Maxwell body is designated as a viscoealstic hereditary liquid.

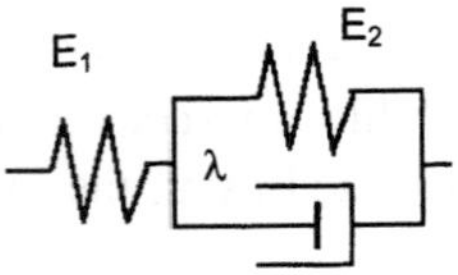

Fig. 2.6. Kelvin body

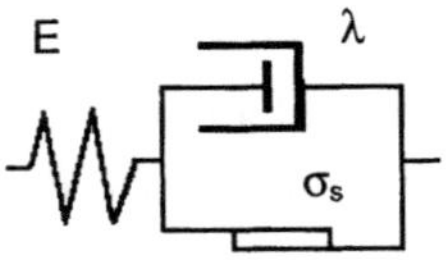

Fig. 2.7. Bingham body

A combination of spring (Hookian) and dashpot (viscous) generalises the concepts of Maxwell and Voigt bodies and is known as a Kelvin solid (Fig. 2.6). When a stress σ is applied, the strain ε_1 in the first spring will be σ/E_1. Strain in the second spring will be obtained from the relation $\sigma = E_2\varepsilon_2+\lambda\xi_2$. The total elongation is $\varepsilon = \varepsilon_1+\varepsilon_2$. Then, it follows that

$$\dot{\sigma} + \eta\sigma = E(\xi + \nu\varepsilon) \tag{2.18}$$

where E = E$_1$; $\eta = (E_1+E_2)/\lambda$, $\nu = E_2/\lambda$. In rapid loading, $\dot{\sigma}$ and ξ will dominate and so from Eq. (2.18) $\dot{\sigma} \cong E\xi$. Therefore, E_1 is called the instantaneous modulus of elasticity [12]. Under relatively slow loading, σ and ε will dominate and from Eq. (2.18) $\sigma = E\nu\varepsilon/\eta$. The quantity $E\nu/\eta = E_1E_2/(E_1+E_2)$ is referred to as the long-term modulus of elasticity [195]. When the time dependence of strain, $\varepsilon(t)$, is known, the time dependence of stress σ would be given by

$$\sigma = E\left[\varepsilon(t) - (\eta - \nu)\int_{-\infty}^{t} \varepsilon(\eta\tau)d\tau\right] \tag{2.19}$$

On the other hand, if the time dependence of stress is known, the following solution for Eq. (2.18) is obtained:

$$\varepsilon = \frac{1}{E}\left\{\sigma(t) + (\eta - \nu)\int_{-\infty}^{t} \sigma(\tau)\exp[-\nu(t-\tau)]d\tau\right\} \tag{2.20}$$

The lower limit of integration in Eqs. (2.19) and (2.20) can be made zero if at $t < 0$, $\varepsilon = 0$ and $\sigma = 0$, Evidently, the equations for Maxwell and Voigt bodies can be derived as special cases of Eq. (2.18).

The combination of elastic, plastic and viscous elements shown in Fig. 2.7 is termed the Bingham body. If the value of stress does not exceed σ_S, this body will deform elastically. Once σ equals σ_S, the viscous element λ also becomes operative. The constitutive equation for a Bingham body is written as

$$\xi = \frac{\dot{\sigma}}{E} + \begin{cases} 0, & \sigma < \sigma_s \\ \dfrac{\sigma - \sigma_s}{\lambda}, & \sigma \geq \sigma_s \end{cases} \tag{2.21}$$

where σ_s, λ and E are material constants. When $\sigma > \sigma_s$, Eq. (2.21) becomes

$$\xi = \frac{\dot{\sigma}}{E} + \frac{\sigma - \sigma_s}{\lambda} \tag{2.22}$$

For steady state plastic flow (σ = constant $\geq \sigma_s$) $\sigma = \sigma_s + \lambda\xi$. A Kelvin solid can be generalised further by including in Eq. (2.18) derivatives of higher order in the following manner:

$$a_0\sigma + a_1 \frac{d\sigma}{dt} + \ldots + a_n \frac{d^n\sigma}{dt^n} = b_0\varepsilon + b_1 \frac{d\varepsilon}{dt} + \ldots + b_m \frac{d^m\varepsilon}{dt^n} \tag{2.23}$$

It can be shown that within certain limits a medium described by Eq. (2.23) can be modelled using a number of springs and viscous elements. Likewise, any combination of a finite number of viscous and elastic elements can be reduced to simple combinations of elements of the same type. Theory of viscoelasticity is now well developed. The following general relationships are used in linear viscoelasticity [188]:

$$\begin{aligned} \sigma(t) &= E\varepsilon(t) - \frac{3}{2}\int_0^t \Gamma(t-\tau)\varepsilon(\tau)\,d\tau; \\ \varepsilon(t) &= \frac{\sigma(t)}{E} + \frac{2}{3}\int_0^t K(t-\tau)\sigma(\tau)\,d\tau \end{aligned} \tag{2.24}$$

where K and Γ are the kernels of creep and relaxation, respectively. Determination of the mathematical form of kernels K and Γ is a specific problem [188, 195], since they account for the properties of real materials. To determine the kernel of creep, the stress $\sigma = \sigma_1$ = constant is rapidly applied on a specimen and the time dependence of the strain $\varepsilon = \varepsilon_1(t)$ is recorded. The kernel of creep can then be found from the second of Eq. (2.24). If a specimen is rapidly extended to a strain ε_2 = constant and then during a stress relaxation test the response $\sigma_2(t)$ is recorded then the kernel of relaxation can be found from the first of Eq. (2.24). It is easy to show that

$$K(t) = \frac{3}{2\sigma_1} \cdot \frac{d\varepsilon_1}{dt}; \quad \Gamma(t) = -\frac{2}{3\varepsilon_2} \cdot \frac{d\sigma_2}{dt} \tag{2.25}$$

Any combination of the above mentioned elementary bricks can also be used. These models are widely used for practical calculations of strength characteristics of various constructions. The most widely used scheme to connect elements is parallel joining which was suggested by Masing [203].

Let N rods be connected in parallel. Each rod can be modelled by elastic, plastic and viscous elements, the properties of the viscous elements being assumed to be, in general, non-linear. Such a model is referred to as a generalised Masing model [203]. In a variant of this model, N elasto-perfectly-plastic rods of differing cross-sectional area S_k (k = 1, 2, ..., N) are joined in parallel. For simplicity the elastic moduli of all the rods are assumed to have the same value E. Then, the following system of equations can be stated [203]:

$$\varepsilon^k = \varepsilon \quad (k = 1, 2, \ldots, N) \tag{2.26a}$$

$$\sum_{k=1}^{N} \sigma^k S_k = \sigma S = P \tag{2.26b}$$

$$\varepsilon^k = r^k + p^k;\ r^k = \frac{\sigma^k}{E} \tag{2.26c}$$

$$\dot{p}^k \geq 0 \ \ at\ \sigma_k = \sigma_Y^k;\ \dot{p}^k \leq 0 \ \ at\ \sigma_k = -\sigma_Y^k;\ -\sigma_Y^k \leq \sigma^k \leq \sigma_Y^k \tag{2.26d}$$

where k is the number of the rod; σ^k, ε^k, r^k, p^k are its stress, total strain, elastic and plastic strains respectively; 'dot' over a symbol denotes a time derivative; σ_Y^k are the yield stresses of the rods; $S = \sum_{k=1}^{N} S_k$ is the total cross-sectional area. The average stress σ is the ratio of the total force P to S. Equations (2.26a) give the kinematic conditions; Eq.(2.26b) gives an equilibrium condition and Eqs.(2.26c, d) give the physical conditions. Equation (2.26b) can be rewritten as

$$\sigma = \frac{1}{S} \sum_{k=1}^{N} \sigma^k g_k = \left\langle \sigma^k \right\rangle$$

where $g_K = S_K/S$ are the weight coefficients; notation <...> is introduced to denote averaging. Thus, in the Masing model additional parameters S_k are considered as the cross-sectional areas of individual rods. Other physical interpretations for these parameters may also be suggested, e.g., volume fractions of phases. The Masing model can be used to describe real stress–strain curves.

While describing inverse loading, the so-called Masing's principle is used. According to this principle, a stress–strain curve in unloading is similar to that during loading, the coefficient of similarity being two [203]. For more complex modern models of materials this coefficient may depend on the strain history. The Masing model can be generalised for the case where the number of rods tends to infinity ($N \to \infty$). In a similar way, the rheological model for a viscous medium composed of non-linear viscous elements can be introduced [203]. The total strain in each element is assumed to be equal to the sum of the elastic ε_e and inelastic ε_i strains:

$$\varepsilon^k = \varepsilon = \varepsilon_e^k + \varepsilon_i^k; \quad \varepsilon_e^k = \frac{\sigma^k}{E} \tag{2.27a}$$

the value of ε_i being accumulated with a rate ξ_i:

$$\xi_i^k = f^k(\sigma, T) \tag{2.27b}$$

It is usually assumed [203] that the functions f^k for all the rods have the same functional form. The functional form $f(\sigma, T)$ can be obtained from the dependence of the secondary creep strain rate on stress and temperature. Further analysis of such a material response is given in [203].

Masing's model is useful in describing mechanical response under monotonic and non-monotonic loading conditions. It was suggested to account for non-uniformity in microstructure in real materials. Numerous experiments on different materials have clearly established the usefulness of the model/approach. In short, the Masing model has proved to be the prototype of many subsequent rheological models in mechanics of solids. In this sense, this model is as fundamental to the rheology of non-linear (viscous) behaviour as Hooke's law is in the theory of elasticity. To generalise Masing-type models to analyse any non-uniform stress–strain state is a major problem. Simulating the mechanical behaviour of superplastics with the help of mechanical models is considered in Chap. 3.

2.4.3 Theories of Plasticity

Strictly speaking, the mathematical theory of plasticity deals with elastoplastic properties of materials that do not depend on time. Such materials are referred to as scleronomous vis-a-vis the materials whose properties depend on time (rheonomous materials). The origins of the theory of plasticity are connected with the names of Tresca [121], Saint-Venant [204, 205], Levy [206, 207], Huber [208] and Mises [209]. Starting from 1920, there has been continuous development in the theory of plasticity [166, 169, 210–215]. At present, the theory of plasticity is the most complicated branch of mechanics of solids. It comprises many directions and approaches for formulating constitutive equations. Some are briefly outlined below.

The best known and widely used part of theory of plasticity is the flow theory. In addition to the above mentioned, Prandtl [216], Reuss [217], Prager [196],

Koiter, Hill and others have made very significant contributions to the development of this theory. Modern formulation of the classical flow theory is based on the following suppositions. (General form of the constitutive equations of the flow theory for strain hardening behaviour was proposed by Melan[1]).

1. There is a yield surface (see items 3–8 in [166] for details). Inside this surface the plastic strain does not change. For a perfectly plastic body the yield condition (equation of the yield surface) is given by

$$f(\sigma_{ij}) = 0 \tag{2.28a}$$

   For a material exhibiting strain hardening the yield condition becomes

$$f(\sigma_{ij}, \varepsilon_{re}^{p}, \chi_n) = 0 \tag{2.28b}$$

   where χ_n are the scalar parameters characterising the history of loading.
2. The increment of the total strain $d\varepsilon_{ij}$ consists of the increments of the elastic and plastic strains, i.e.,

$$d\varepsilon_{ij} = d\varepsilon_{ij}^{e} + d\varepsilon_{ij}^{p} \tag{2.29}$$

3. The increment of the elastic strain $d\varepsilon_{ij}^{e}$ is related to the increment in stress $d\sigma_{ij}$ through Hooke's law (see Eq.(2.7)).
4. Material is plastically incompressible, i.e.,

$$d\varepsilon_{ii}^{p} = 0 \tag{2.30}$$

   In other words, the volumetric strain is assumed to be proportional to the hydrostatic pressure (see Eq. (2.8)).
5. Associated flow rule is assumed to be valid for a smooth yield surface, i.e., the increment in plastic strain is proportional to the gradient of the yield surface, $f = 0$. In other words, the yield condition is a plastic potential. The condition for Eqs. (2.28a) and (2.28b) are written separately. In the first case, one has for a smooth yield surface

$$d\varepsilon_{ij}^{p} = d\lambda \frac{\partial f}{\partial \sigma_{ij}} \tag{2.31}$$

[1] Melan, E. (1938): Zur Plastizitaet des raumlichen Kontinuums. Ing. Arch., 9, 116.

where $d\lambda = 0$, if $f < 0$, or $f = 0, df \equiv \frac{\partial f_\alpha}{\partial \sigma_{ij}} d\sigma_{ij} < 0$; $d\lambda \geq 0$, if $f = 0$, $df = 0$.

For a singular yield surface

$$d\varepsilon_{ij}^{p} = \sum_{\alpha} d\lambda_\alpha \frac{\partial f_\alpha}{\partial \sigma_{ij}} \tag{2.32}$$

where, $d\lambda_\alpha = 0$, if $f_\alpha < 0$ or $f_\alpha = 0$, $df_\alpha \equiv \frac{\partial f_\alpha}{\partial \sigma_{ij}} d\sigma_{ij} < 0$, $d\lambda_\alpha \geq 0$, if $f_\alpha = 0$,

$df_\alpha = 0$.

Singular yield condition is formed by intersections of smooth surfaces $f_\alpha = 0$ (interior of this yield surface is given by $f_\alpha < 0$). When strain hardening is present, for a smooth yield surface one has

$$d\varepsilon_{ij}^{p} = \lambda h \frac{\partial f}{\partial \sigma_{ij}} df \tag{2.33}$$

where, $\lambda = 0$ if $f < 0$, or $f = 0$, $df \equiv \frac{\partial f\alpha}{\partial \sigma_{ij}} d\sigma_{ij} < 0$; $\lambda = 1$, if $f = 0$, $df \geq 0$. Here $h >$ 0 is the hardening function which depends on the history of loading and (in classical flow theory) not on the stress increment. This means that when $df > 0$ the increments $d\varepsilon_{ij}^{p}$ are linear functions of $d\sigma_{ij}$

The theoretical basis of the flow theory at present is Drucker's postulate [164, 218]. Consequences of this postulate are the convexity of the yield surface and the condition of normality (increment in stress is proportional to the gradient of the yield surface). As was mentioned earlier, Drucker's postulate is not valid for materials having a descending $\sigma - \varepsilon$ diagram (strain softening). Iljushin has suggested a postulate which is more general than Drucker's postulate; in particular, it is applicable to materials having a descending $\sigma - \varepsilon$ diagram [219, 220]. The simplest and the most often used yield conditions for a perfectly plastic material are the Huber–Mises criterion (regular surfaces) which can be written as

$$\sigma_e = \sigma_s = constant \tag{2.34}$$

and the Tresca yield condition (singular surface):

$$\tau_{\max} = \tau_s = constant \tag{2.35}$$

It is noted that in the principal stress space the Huber–Mises condition defines a cylinder whose axis of symmetry is equally inclined to the three main axes. The Tresca condition defines a hexagonal cylinder inscribed into the above Huber-

Mises cylinder. For uniaxial tension ($\sigma_1 = \sigma$, $\sigma_2 = \sigma_3 = 0$), the body starts to flow at $\sigma_1 = \sigma_s$ in accordance with the Huber-Mises yield condition. According to the Tresca condition this occurs when $\sigma_1/2 = \tau_s$, and so $\sigma_1 = 2\tau_s$.

The full set of Prandtl–Reuss equations, a known variant of the flow theory for a perfectly plastic body, is written as

$$\begin{aligned} d\varepsilon_{ij} &= d\varepsilon_{ij}^{e} + d\varepsilon_{ij}^{p} \\ d\varepsilon_{ij}^{e} &= \frac{1+\nu}{E} dS_{ij} + \frac{1-2\nu}{E}\frac{d\sigma_{kk}}{3}\delta_{ij} \\ d\varepsilon_{ij}^{p} &= \frac{3}{2}\frac{d\varepsilon_{e}}{\sigma_s} S_{ij};\; d\varepsilon_{e} \equiv \sqrt{\tfrac{2}{3} d\varepsilon_{ij}^{p} d\varepsilon_{ij}^{p}} \end{aligned} \tag{2.36}$$

For hardening materials, the yield condition includes (as was mentioned above) some parameters reflecting the history of loading, e.g., invariants of the plastic strain tensor, plastic work, the Odqvist parameter. Typical laws describing the evolution of the yield surface in such materials are based on isotropic hardening, kinematic hardening, isotropic-kinematic hardening. In the first case, the yield surface changes its dimensions without changing its shape and position. In the second case, the yield surface may be shifted as a rigid entity in stress space. The third case is a combination of the first and the second. The most often used law of yield surface movement is:

$$dr_{ij} = a d\varepsilon_{ij}^{p} \tag{2.37}$$

where r_{ij} characterises the position of the center of the yield surface.

At this stage, the main directions in the development of the flow theory are noted.

1. Development of new yield conditions, in particular, the development of a multi-surface flow theory. (In this case one uses a number of embedded yield surfaces.) Multi-surface flow theory is more effective in particular in describing cyclic loading.
2. The development of combined models of thermoelastoviscoplastic materials where the flow theory may be effectively used.
3. Flow theory is the basis for constructing constitutive equations that account for finite strain behaviour. This is of special importance to superplastic materials.

In 1924, Hencky [166] proposed the deformation theory of plasticity and Nadai [211] applied it to the problem of torsion. This is an extension to a plastic body the form of stress–strain relationship known in the theory of elasticity. IIjushin [212, 213, 215] developed this theory further and presently it is known as Hencky-Iljushin theory of small elastic-plastic deformations [166, 195, 214]. This theory includes the following relations. It is assumed that the stress and strain deviators are proportional, i.e.,

$$e_{ij} = \varphi S_{ij} \tag{2.38}$$

where $\varepsilon_{ij} = \varepsilon_{ij} - (\varepsilon_{kk}/3)\delta_{ij}$ are strain deviator components and $\varphi = 3\varepsilon_e/2\,\sigma_e$, $\varepsilon_e \equiv \sqrt{\frac{2}{3} e_{ij} e_{ij}}$ is the effective strain.

$$\varepsilon_{ii} = \varepsilon_{ii}^{e} = \frac{1-2\nu}{E}\sigma_{kk} \quad or \quad \varepsilon_{ii}^{p} = 0 \tag{2.39a, b}$$

Here, $\Phi(\varepsilon_e)$ is a universal function of material. Iljushin has presented this function as

$$\sigma_e = \Phi(\varepsilon_e) \tag{2.40}$$

$$\sigma_e = 3G\varepsilon_e[1 - \tilde{\omega}(\varepsilon_e)] \tag{2.41}$$

where $\tilde{\omega}$ can be considered as a minor parameter. This circumstance allowed Iljushin to develop an effective method to solve various boundary value problems including Eqs. (2.38)– (2.41). This method has been used widely in Russia to solve problems of considerable practical importance.

The constitutive equations of the theories of plasticity so far considered are phenomenological. They are useful for engineering applications but have a weak physical basis. Plastic deformation in metals results from slip in individual grains/crystallites. This has led to the concept of slip.

The first proposals of slip in single crystals are due to Taylor and Elam [221, 222] and Schmidt [223]. Schmidt stated that slip takes place when the resolved shear stress reaches a critical value (critical resolved shear stress) and that this value is unaffected by the magnitude of the normal stress on the slip plane. Taylor [224] related the properties of a polycrystalline aggregate to the properties of its constituent grains. (The fundamental idea was that arbitrary deformation can be represented as a result of shears in five independent slip systems.)

In 1949, Batdorf and Budiansky [225, 226] proposed a new theory. This theory took into account the number of slip systems present in a material. Plastic deformation starts when the yield condition is reached in certain of the slip systems. The aggregate is considered to be made up of randomly oriented crystals and after an averaging process the assumptions of homogeneity and isotropy on a macroscale are justified. Further, it is assumed that each crystal in the aggregate has only one slip system and there are no interactions within individual crystals. (The last assumption is not valid for many materials.)

Further development of these theories is due to Ciealla [227, 228] and Bishop and Hill [229, 230]. Lin [231–233] proposed a theory of slip invoking three slip directions instead of one. Finite strain behaviour was considered by Hill and Rice [234] and Asaro [235]. In Russia, the corresponding investigations were completed by Malmeister, Leonov, Rusinko, Hristianovich, Shemjakin et al: see these and other references in [167]. A feature of all theories of slip is the complexity of the mathematical expressions. That is why they are not considered here in detail. It is, however, noted that the availability of fast computers has facilitated elegant numerical calculations based on slip theories. Aspects of the microstructural basis of models of plasticity and creep as well as the relationships among the micro- and macro-approaches in mechanics of solids are discussed extensively in the literature.

Important part in the theory of plasticity is the so called theory of elastoplastic processes (TEPP). TEPP was formulated by Iljushin in 1954 for the case of small deformation of initiality isotopic material. Geometrical images of the stress and strain states in a particle of material are convenient to use for the formulation of the postulate of this theory. Normally, in the theory of plasticity the strain state is defined by the deviator e_{ij} and the volumetric deformation Θ. But the stress state is defined by both the deviator S_{ij} and the mean stress $\sigma_m = -p_H$. In this case the main idea of formulating the constitutive equations (CEs) is to determine the relations between the deviators of stress and strain – it gives five CEs and the sixth equation is either the condition of incompressibility or the law of elastic change of volume.

Strain deviator having five independent components can be brought down to a 5D vector strain deviator space E_5. Components $\ni_k$ of the 5D strain vector $\vec{\ni}$ in E_5 are expressed linearly and uniquely (one-to-one) by strain deviator components e_{ij}. Moreover, the components $\ni_k$ are selected in such a way that

$$|\vec{\ni}| = \sqrt{\ni_K \ni_K} = \sqrt{\frac{2}{3} e_{ij} e_{ij}} = \varepsilon_e .$$

Law of evolution of strain deviator $e_{ij} = e_{ij}(t)$ is expressed in the space E_5 by a curve joining the end of the vector $\vec{\ni}(t)$. This curve is called the deformation trajectory. It is known from differential geometry that the internal geometry of the deformation trajectory is characterised by its curvature $\kappa_1, \ldots \kappa_4$, which are functions of the arc length, s, of the deformation trajectory. The value of s by definition is expressed as

$$s = \int \sqrt{d \ni_i \, d \ni_i} = \int \sqrt{\frac{2}{3} de_{ij} de_{ij}} = \int_0^t \sqrt{\frac{2}{3} \xi_{ij} \xi_{ij}} \, dt' \tag{2.42}$$

which is the same as the Odqvist parameter [236]. Naturally, accompanying frame $\{\vec{p}_i\}$, i = 1, …., 5 may be constructed at each point of the deformation trajectory;

vectors $\vec{p}_i$ are defined through the vectors $d\vec{\ni}/ds, \ldots, d^5\vec{\ni}/ds^5$.

5D vector stress space Σ_5 containing 5D vectors $\vec{\sigma}(t)$ can be formulated similarly. If this space is superimposed on the E_5 space and on each point of the deformation trajectory, $s = s(t^*)$, a corresponding vector $\vec{\sigma}(t^*)$ is constructed and added to corresponding values of scalar parameters $T(t^*)$ (temperature), $(\sigma_m(t^*)\ or\ \Theta(t^*))$, $\dot{s}(t^*) \equiv \xi_e(t^*)$, then the so called image of loading process for the given particle of material is obtained. Basic postulates of Iljushin's theory consist of the following:

1. 'Image of loading process' is an invariant with respect to transformation of rotation and/or reflection of deformation trajectory in E_5 space (special postulate of isotropy).
2. *Retardation principle (delay effect)*: The orientation of the stress vector $\vec{\sigma}$ with respect to the deformation trajectory (i.e., with respect to the accompanying natural pentahedron) at any point of deformation trajectory depends on some restricted part of this trajectory of length s_R preceding this point rather than on the entire deformation history. In other words, it is assumed that there is a retardation trace (memory trace), the internal geometry of which is important (the material 'remembers' it). The value of s_R at room temperature is usually about $(3\text{–}10)\varepsilon_S$, where ε_S is the ultimate yield strain in a tensile test [237].

The value of the arc length of delay trace s_R (generally speaking, s_R depends on T and $\dot{s}$) is considered to be a material characteristic. The value of s_R is used in TEPP for the classification of deformation trajectories: if typical values of curvatures are $\kappa_k \sim 1/s_R$, then this type of trajectories is referred to as deformation trajectories of mean curvature; if $\kappa_k << 1/s_R$, then they are known as the deformation trajectories of small curvature and if $\kappa_k >> 1/s_R$, then there are known as deformation trajectory of large curvature. The direct mathematical consequence of these postulates is the correlation between stress and strain in a very general linear tensor form.

$$\vec{\sigma}_k = A_k\,\vec{p}_k \tag{2.43}$$

where $\vec{p}_k$ are vectors of accompanying frame and A_k are the functionals of inner geometry of deformation trajectory, i.e., curvatures κ_k. (In a more general case A_k also depends on T and $\dot{s}$.) They do not depend on location of deformation trajectory in E_5, i.e., on the third invariant of strain deviator. According to postulate (2) functionals A_k depend on the values of κ_k only in the range of delay trace.

Vast literature exists (see, e.g., [166, 167, 169, 220, 237–248]) which is devoted to experimental/theoretical investigations of functional of theory of plasticity and formulation of particular variants of the CEs within the frame work of TEPP.

General postulates of TEPP and the initial manner of its formulation may make it appear as irrelevant to superplasticity. This is not true.

For mechanics of superplasticity the framework of TEPP is important.

1. Deformation trajectories are classified into several groups according to the complexity. The material characteristic (material function), s_R, exists which helps in this classification.
2. Mechanical behaviour of a material and the evolution of its microstructure (and hence the CEs) may depend not only on temperature and strain rate, but also on the type of deformation trajectory. (For example, Iljushin has demonstrated that in case of simple (proportional) deformation when the deformation trajectory in E_5 is a line, the results of all classical theories of plasticity coincide. But for non-proportional (complex) loading, these theories predict different behaviour for a material. Thus, the theory of Henky-Iljushin is adequate for processes of simple deformation and those close to it. But, this theory may provide very unreliable results for very complex loading, e.g., for ε_e= constant in E_5 space). Therefore, in the formulation of a field of validity for any CE, apart from temperature and strain rate, types of deformation trajectories also will have to be mentioned.
3. The previous point (2) has to be kept in mind in case certain technological problems (e.g. metal forming processes) are to be solved. If the possible types of deformation trajectories can be guessed a priori this could be used to select the suitable CEs for solving the boundary value problem.
4. In the framework of TEPP, the concepts of scalar and vector properties of constitutive equations can be introduced. The properties of a material that are described by the relationships among the directions of the corresponding vectors in the 5D spaces $(\vec{\sigma}, \vec{\text{э}}, d\vec{\sigma}/ds, d\,\vec{\text{э}}/ds)$ are referred to as vector properties. Those properties described by the relationships among the invariants of stress, strain and strain rate tensors are known as scalar properties.

An example of a vector property is the well-known hypothesis concerning the proportionality between the stress and the strain rate tensors, popularly known as Saint-Venant's principle. An example of a scalar property is the hypothesis concerning the existence of a 'unique curve', suggested by Ludvik [238] and Malinin [239]. In this hypothesis it is assumed that there is a relationship between the equivalent stress σ_e and the equivalent strain ε_e which is independent of the type of stress state.

A very important conclusion of TEPP that directly concerns superplasticity should be emphasized. The second postulate has established that for deformation trajectories of small curvature, Eq. (2.43) may be simplified as

$$\vec{\sigma} = A_1\vec{p}_1 \quad or \quad S_{ij} = \frac{2A_1}{3\xi_e}\xi_{ij} \tag{2.44}$$

which are relations of Saint-Venant type, normally used in superplasticity. In this way, if for a superplastic material the value of s_R is known even approximately, then for this material the deformation trajectories for which the CEs (2.44) are valid can be determined. Subsequently, the coefficient, A_1, only needs to be determined. Practically all the experimental work concerning CEs and the phenomenology of superplasticity are concerned with this problem. On the other hand, if the typical curvature of the deformation trajectories in a problem being investigated is substantially greater than s_R^{-1}, then it is recommended to use CEs of more general nature than the one given as Eq. (2.44) (see Sect. 3.5).

Finally, the case of large deformation is considered. Generalization of TEPP to include finite deformation has led to the formulation of several tensor measures of stress and strain. For a representative class of tensor measures of stress–strain state, a wide variety of postulates for isotropy and the retardation principle have been suggested by Massakovsky and will be discussed later. These new formulations may set the stage for the construction of CEs for superplastic materials subjected to highly complex loading.

2.4.4 Theories of Creep

2.4.4.1 Phenomenology of Creep

Most constitutive equations proposed for superplastics may be considered as special cases of those used in the theories of creep [131, 249]. So, superplastic flow may be regarded as a particular case of high temperature creep. For this reason, theories of creep known in the literature should be examined.

The early theories of creep were developed by practising engineers. Later, the theory of creep developed as an independent branch of mechanics of solids[2]. In a standard creep test a constant tensile load is applied on a cylindrical specimen. For small strains, the change in cross-section is rather small and so the stress may be regarded as constant. Although measurement of creep resistance is easy in principle, in practice elaborate apparatus/equipment is required [250, 251]. Long time creep tests may take from several months to more than 10 years [210]. The general procedures for creep testing are covered in ASTM Specification E139–70. For analysing finite strain behaviour (as seen during superplastic flow) it is necessary to use special accessories/controls to record the curves, establish the reliability of recording over long time intervals at constant stress, and reduce axial load in proportion to the decrease in cross-section [34, 252].

[2]It is pertinent to note that mechanics of superplastic flow is not yet developed as an independent branch of mechanics of solids. One of the main objectives of writing this book is to briefly outline the contours of this discipline.

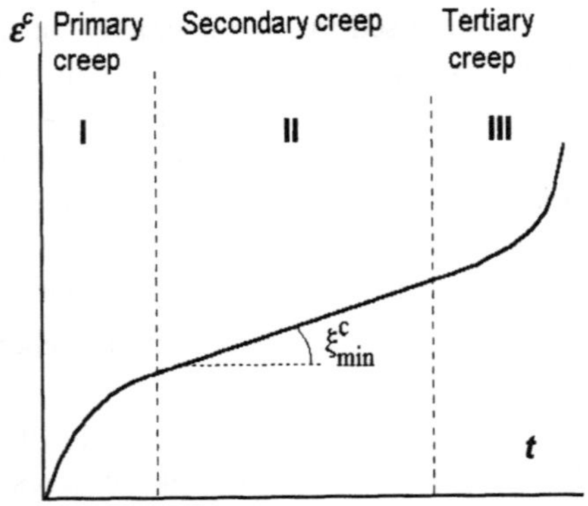

Fig. 2.8. Time dependence of strain in the three stages of creep (schematic)

The phenomenology of creep is well investigated. In the early part of the 20th century, Andrade initiated a systematic study of creep in metals [252]. He published his classic work in 1910, in which the notions of 'primary', 'secondary' and 'tertiary' creep (which are still in use) were introduced (Fig. 2.8).

In stage I (primary creep), the strain rate decreases to its minimum value ξ^c_{min}. Stage II corresponds to stationary (secondary) creep, a region in which the slope of the curve is constant: $\xi^c = \xi^c_{min}$= constant. In the third stage (tertiary creep), strain rate increases till the rupture of specimen. Depending on the experimental range some stages may be absent.

The shape of the creep curves depends on stress and temperature. It has been established that creep depends on the homologous temperature (T/T_m), where T is the absolute temperature of deformation and T_m is the absolute melting point, rather than on the absolute temperature *per se*. (This has helped physical modelling by testing lead specimens at room temperature to simulate the high temperature behaviour of metals and alloys.) Many empirical expressions have been suggested for the minimum strain rate $\xi^c_{\min}$. Bayley [253, 254] proposed the following:

$$\xi^c_{\min} = B\sigma^n \tag{2.45a}$$

This equation has been used extensively in later works. Presently, it is formally expressed as [195, 236, 239, 255]

$$\xi^c_{min} = \xi^*\left(\frac{\sigma}{\sigma^*}\right)^n \tag{2.45b}$$

where σ^* and ξ^* are a reference stress and reference strain rate respectively, n is a temperature dependent material constant. A variant of Eq. (2.45a), $\sigma = K\xi^m$, has been suggested to describe superplastic flow ($m = 1/n$). For primary creep (stage I in Fig. 2.8) also a number of descriptions are available, e.g.,

$$\frac{d\varepsilon^c}{dt} = ct^{-a} \tag{2.46}$$

where a and c are material constants ($0 < a < 1$). The value of 'a' decreases with decreasing temperature. At relatively low temperatures (less than 0.3–0.5 T_m) for a wide range of metals and alloys $a \approx 1$ and logarithmic creep is present.

$$\varepsilon^c = \alpha \ln t + c_1 \tag{2.47}$$

where α and c_1 are material constants. At these temperatures, practically the entire creep curve can be approximated by Eq. (2.47) with reasonable accuracy (logarithmic creep). At higher temperatures, the following time dependence of strain in stage I is seen.

$$\varepsilon^c = \beta t^{1/3} + kt \tag{2.48}$$

where β and k are material constants. This is the well-known Andrade law [256, 257]. (Andrade described the change in specimen length, L, by the relationship

$$L = L_0\left(1 + \beta t^{\frac{1}{3}}\right)e^{kt}$$

and assumed that β and k were functions of stress and temperature.) An equation which gives a better fit than that of Andrade and has also been verified for some materials is that proposed by Garofalo [258].

$$\varepsilon = \varepsilon_0 + \varepsilon_t\left(1 - e^{-rt}\right) + \xi^c_{\min} t \tag{2.49}$$

where ε_0 is the instantaneous strain on loading, ε_t is the limit of transient creep, r is the ratio of transient creep rate to transient creep strain. (It is to be noted that in British terminology the instantaneous strain ε_0 is considered to be an additional stage of creep.) It is believed [210] that even when the applied stress is below the yield stress, not all the instantaneous strain is elastic. Most of this strain is instantly recovered upon release of load (elastic), but a part is recovered with time (anelastic) and the rest is irrecoverable (plastic). Superplastic materials exhibit pronounced anelasticity [81–84] and the ratio of anelastic to elastic strains can be much larger than one. For secondary creep also many approximate descriptions are available. These may be regarded as special forms of Eq. (2.45a). At relatively large stresses, the relationship

$$\xi_{\min}^{c} = C_1 \exp\left(\frac{B\sigma}{T}\right) \tag{2.50}$$

with C_1 and B as constants, is used. It should be noted that some of the above examined equations contain values with dimensions (without normalization) e.g., t^{-a}, ln t etc. This is not correct.

At this stage some elementary (classical) theories of creep are considered. These theories include relationships among three independent variables out of strain, strain rate, stress and time. Here, only secondary creep is considered. Therefore the equality $\xi^c = \xi^c_{\min}$ is assumed to be valid. The presence of a steady state creep curve in which the creep rate is constant at a given stress suggests a comparison between high temperature creep in metals and quasi-viscous flow in liquids. Unlike ordinary viscous flow, quasi-viscous flow occurs with the flow rate being dependent on stress non-linearly. Neglecting elastic and instantaneous strains, the following relationship is assumed.

$$\xi^c = f(\sigma, T) \tag{2.51}$$

The following simplifications are made. The creep curves (time dependence of creep strain) are replaced by straight lines that pass through the origin. As can be seen from Fig. 2.8, such an assumption is acceptable, if stage I creep is neglected. Some experimental support for this representation is also available [195]. Creep theory, based on Eq. (2.51), is termed steady-state creep theory. To describe material behaviour during load relaxation, elastic strains have to be considered. Therefore, Soderberg [260] proposed the equation

$$\frac{d\varepsilon}{dt} = \frac{\dot{\sigma}}{E} + f(\sigma, T) \tag{2.52}$$

When $\dot{\varepsilon} = 0$, the time dependence of σ during relaxation is obtained as

$$t = -\frac{1}{E}\int_{\sigma_0}^{\sigma} \frac{d\sigma}{f(\sigma, T)} \tag{2.53}$$

A similar approach has been used with the standard power law $\sigma = K\xi^m$ in [69, 72, 95, 131]. In general, Eq. (2.53) is not in good agreement with experimental results for conventional (non-superplastic) materials [195]. Odqvist [261] suggested an important improvement in steady-state creep theory by including instantaneous plastic deformation in total strain. That is,

$$\varepsilon = \varepsilon^e + \varepsilon^p + \varepsilon^c \tag{2.54}$$

Table 2.2 The values of k_1 and k_2 in Eq. (2.56) [262]

σ	$\dot{\sigma}$	k_1		k_2	
		even m	odd m	even n	odd n
+	+	1	1	1	1
+	–	0	0	1	1
–	+	0	0	–1	1
–	–	1	–1	–1	1

where ε^e is the elastic strain, $\varepsilon^p = g(\sigma)$ is the instantaneous plastic strain and ε^c is the creep strain. Then, instead of (2.52), one obtains

$$\xi = \frac{\dot{\sigma}}{E} + kg'(\sigma)\dot{\sigma} + f(\sigma, T) \tag{2.55}$$

where k is unity when $\dot{\sigma} > 0$; otherwise $k = 0$. Hoff [262] has used a variant of this equation for analysing bending where

$$\xi = \frac{\dot{\sigma}}{E} + k_1\left(\frac{\sigma}{\mu}\right)^m \frac{\dot{\sigma}}{\mu} + k_2\left(\frac{\sigma}{\lambda}\right)^n \tag{2.56}$$

with μ, λ, E, m and n material constants. k_1 and k_2 depend on the signs of σ and $\dot{\sigma}$ as also on the values of m and n. Different situations are listed in Table 2.2. In the Ageing theory of creep , stress, strain and time are related as

$$\varepsilon = h(\sigma, t) \tag{2.57}$$

This theory is very convenient since experimental diagrams can be directly used to calculate the strain. Also, in this theory the relaxation curves agree well with Eq. (2.57) when $\varepsilon =$ constant. In another ageing theory, sometimes termed 'the flow theory' [263, 264], it is assumed that there is a definite relationship among creep strain rate ξ^c, stress and time.

$$F(\xi^c, \sigma, t) = 0 \tag{2.58}$$

The following variant is widely used.

$$\xi^c = B\left(\frac{\sigma}{\sigma^*}\right)^p \tag{2.59}$$

where B is a function of time and temperature. Integrating Eq.(2.59) with σ = constant, one obtains

$$\varepsilon^c = \left(\frac{\sigma}{\sigma^*}\right)^p \tau \tag{2.60}$$

where $\tau(t)$ is a normalised parameter, which can be treated as modified time. The relaxation problem is solved then similar to Eq. (2.13) since $\xi = \dot{\sigma}/E + \xi^c$.

Another variant of Eq. (2.58) [265, 266] is

$$\xi^c = \xi - \frac{\dot{\sigma}}{E} = S(\sigma)\tau(t) \tag{2.61}$$

where $\tau(t)$ is modified time. Then, the following relationship between strain and stress can be considered.

$$\varepsilon = \sigma\left[\frac{1}{E} + \int_0^\tau \frac{S(\sigma)}{\sigma} d\tau'\right] \tag{2.62}$$

In the strain hardening theory, it is assumed that at a given temperature there is a relationship between ξ^c, σ and ε^c. That is,

$$\Phi(\xi^c, \sigma, \varepsilon^c) = 0 \tag{2.63}$$

This theory was suggested in [238, 263, 266] and later developed in [195]. The dependence (2.63) is usually represented as

$$\xi^c (\varepsilon^c)^\alpha = f(\sigma) \tag{2.64}$$

where it is assumed that $f(0) = 0$. Primary creep is satisfactorily described by the power time dependence so that the creep strain ε^c is proportional to t^β. As in this case ξ^c is proportional to $\beta t^{\beta-1}$, from Eq. (2.64) $\alpha = (1-\beta)/\beta$. Andrade [252] found that $\beta = 1/3$; then $\alpha = 2$. It is interesting that Andrade considered his constant 1/3 to be valid for a wide range of materials. A number of physical theories have favoured such an opinion. Later works, however, have revealed that this coefficient is, in general, different for different materials and that it also depends on stress and temperature. Some new types of equations that describe creep flow are found in [259], e.g.,

$$\xi_{ij}^c = \frac{\partial \Phi}{\partial S_{ij}}, \Phi = -A\sqrt{\sigma_B^2 - \sigma_e^2}, \quad \xi_e^c = \frac{4}{9} A \frac{\sigma_e}{\sqrt{\sigma_B^2 - \sigma_\epsilon^2}} \tag{2.65}$$

where ξ_e^c is the equivalent creep strain rate. For uniaxial loading, the following relations have been suggested on the basis of experimental data [259]:

$$\xi^c = A\left(\frac{\sigma}{\sigma_B - \sigma}\right)^n \tag{2.66}$$

and

$$\xi^c = A\frac{\sigma - \sigma_0}{\sigma_B(1-\omega) - \sigma} \tag{2.67}$$

where A, n, σ_B and σ_0 are material constants; ω is damage parameter for which a kinetic equation is formulated. A typical feature of these relationships is that their construction includes fractional equations which allow adequate description of experimental curves having a point of inflection. Smirnov's formula (Chap. 3) uses a fractional expression. There is scope for using the above theories to understand superplastic flow. However, there is a need to take into account microstructural changes.

2.4.4.2 Internal Variable Approach

Various deformation mechanisms have been considered and deformation mechanisms maps plotted for creep deformation e.g., [210]. A detailed consideration of these aspects is beyond the scope of the present book. Here, these aspects will be used to develop constitutive equations of creep. Consider the following two problems: (a) The type of experimental data that can be included in CEs; (b) The methods available for including these data in CEs.

One structural parameter used in physical theories, e.g., theory of strain hardening, is dislocation density [267]. In physical theories, this is often reported as a scalar value (total length of dislocation line per unit volume or number of dislocation lines that cut through unit cross-sectional area [2]). But, a rigorous definition leads to the dislocation density tensor (see, e.g., [268]). Likewise, the Burgers vector is also used as a scalar characteristic of dislocation density. While Burgers vector can be determined experimentally, only the scalar characteristic of dislocation density is experimentally evaluated. No experimental methods are available to estimate the dislocation density tensor.

Another widely used structural characteristic is the average grain size, d. Two different methods are used to determine d. In the first method, the given microstructure is compared against a standard of known value, see, for example, [269]. (This method is useful as a first approximation.) For accurate estimates stereological techniques have to be used, see, e.g., [270, 271]. These methods allow reliable estimates of d in single phase materials of approximately equiaxed grains. For industrial alloys these procedures have serious limitations, e.g., two phase alloys of bimodal grain size distribution. Thus, the structural characteristics known in the literature (dislocation density, average grain size, Burgers vector, etc.,) can only be included in constitutive equations as phenomenological parameters. Rabotnov [272] has suggested a way of including structure changes accompanying creep. For example, if a series of micrographs $A_1, A_2, ..., A_n$ is available ($A_{n+1} \neq A_n$) from one test programme A and another set $\hat{A}_1, \hat{A}_2, ..., \hat{A}_n$ from another test programme B, the material may be regarded as a single parameter material if the following two conditions are met: (i) every micrograph $\hat{A}_s$ has an identical micrograph A_k ($\hat{A}_s = A_k$); (ii) If $\hat{A}_{s+1} > \hat{A}_s$, A_{k+1} should be greater than A_k. In this case, a single valued function $q(k)$ ($k = 1, 2,..., n$) can be introduced into the constitutive equation as a structural term/factor. This procedure was later termed as the internal variable approach as opposed to the external variables of temperature, strain rate and stress which can be manipulated independently.

Presently, the internal variables approach is widely used in mechanics of solids and in studies of superplastic behaviour [273-275, 287, 306]. In this approach, it is assumed that the current value of stress is determined by the current values of a set of internal variables. These internal variables are macroscopic manifestations of the microstructure of the deforming material which, in turn, depend on the entire history of loading. Constitutive equations with internal variables should include evolution equations for each internal variable. Evolution equations are usually chosen based on physical considerations. The following system of equations is formulated in [272] .

$$\xi_{ij}^c = \lambda \sigma_{ij}, \ \xi_e^c = f\left(\sigma_e, \varepsilon_e^c, \omega, T\right) \tag{2.68}$$

where λ is the coefficient of proportionality, ω is a damage parameter. The following evolution equation is to be added to Eq. (2.68)

$$\dot{\omega} = \varphi\left(\sigma_e, \varepsilon_e^c, \omega\right) \tag{2.69}$$

Rabotnov has used the damage parameter ω as an internal variable. In addition, a criterion of failure is also to be selected (see Chap. 4). The general scheme of the internal variables approach may be formulated as follows for the uniaxial case(see also, e.g., [273]). The current $\sigma - \varepsilon$ relationship is assumed to be dependent on the history of loading through the instantaneous values of a set of internal variables as

$$\sigma = \Phi(\xi, T, \alpha_i) \quad or \quad \xi = \Psi(\sigma, T, \alpha_i), \ i = 1, 2, ..., k \tag{2.70}$$

where α_i ($i = 1, 2, ..., k$) are internal variables. The evolution equations are

$$\dot{\alpha}_i = \Psi_i(\xi, T, \alpha_j), \ \alpha_i|_{t=0} = \alpha_i^0, \ i, j = 1, 2, ..., k \tag{2.71}$$

The usefulness of the internal variable approach in describing the mechanical response of superplastics has already been demonstrated, see, e.g., [276–279].

2.5 Experiments in Mechanics

As was mentioned in Sect. 2.1, experimental studies of material properties alone are considered in this book. Details of such experiments are briefly discussed below. Unfortunately, the very important problem related to the experimental determination of stresses and strains in a specimen with a non-homogeneous stress – strain state is beyond the scope of this book. (For some remarks in this regard see Sect. 3.6.1.)

2.5.1 Mechanical Tests on Materials

While determining the mechanical properties of materials one has to remember that the results of experiments can be affected by the dimensions of the test specimens. Also, reproducibility of results is essential for comparison. These facts are often ignored in the literature on superplasticity. Attempts to use standard specimens (see, e.g., [280]) are rare. So, some papers dealing with the effects of specimen geometry on the accuracy of the constitutive equations used have appeared (see, e.g., [63–67]). It should, however, be noted that the extreme elongations obtainable in superplastic alloys and their very low notch sensitivity do not make the above limitations very serious in the case of superplastic flow as they are for other materials. Still, it is desirable to design at the earliest standard specimens as well as procedures of testing for superplastics.

Calculations of stress, strain and strain rate from the experimental outputs of load and deformation should be carefully done taking into account errors in measurements of different kinds. In addition, one has to take into consideration the inhomogeneity of stress and strain states present in deformation. Extra care should also be taken to minimise the errors. Special techniques for interpreting the experimental results may also have to be deviced. Experiments should be repeated under identical conditions to check reproducibility (Chap.1). It is essential to evaluate mechanical response under different loading conditions, e.g., uniaxial/multiaxial loading and various regimes of loading (testing at constant stress as well as constant cross head speed/strain rate, testing under active($\sigma_{ij}\ d\varepsilon_{ij} > 0$) and passive ($\sigma_{ij} d\varepsilon_{ij} < 0$) loading conditions etc.) Only then reliable constitutive equations can be developed. Multiaxial loading can be of two types: simple (proportional) and complex (non-proportional) loading.

The fundamental method used in experimental mechanics to verify constitutive equations is to perform some *crucial* experiments. These experiments should differ *principally* from those used to determine the material constants by *basic* experiments. For example, if the material constants are determined using constant cross head velocity tests, the crucial experiment may be, e.g., a torsion test, or a load relaxation test. Unfortunately, crucial experiments have been performed only rarely on superplastics [57, 59, 147, 281].

Care has also not been taken while validating physical models at atomistic level. Often, only simple comparisons in terms of macroparameters like the value of the strain rate sensitivity index and the activation energy are made.

Finally, the elegance of a constitutive equation is assessed by the number of basic experiments necessary to evaluate all the constants in the equation. (This number should be minimal for maximum elegance.) Then, some crucial experiments should be performed to determine the range of applicability of the constitutive equation.

2.5.2 Influence of Testing Machine

A testing machine is characterised by a fixed stiffness, inertia of cross head movement, clearance etc. The influence of specimen – testing machine system is important particularly at high temperatures (due to the thermal softening of the test assembly). Thus, calibration of actual specimen elongation against elongation recorded by the testing machine (from cross head movement) is essential. Hart [107] emphasised this very early but it was largely ignored.

Often, the specimens used in tensile tests on superplastics are not filleted. The sudden change in cross-section will give rise to a stress concentration and invalidate the primary formulae for stress ($\sigma = P/A$) and strain rate $\xi = v/L$ .This has led to the use of filleted specimens. (But, this will increase grip flow.) Details of such specimen dimensions, geometry and conditions of testing can be found in many reference books.

It is desirable to make measurements directly in the gauge portion using an extensometer or by placing special markers. The extreme elongations and high temperatures of testing often make the first technique impossible. Oxidation, on the other hand, may smear out the markers. So, additional assumptions will be necessary to analyse the results. For example, if necking is not significant in a v = constant test, the final gauge length can be calculated from the final width and thickness of the specimen in the gauge portion and assuming volume constancy. Then, strain and strain rate can be calculated in the usual manner. These strains and strain rates can be significantly different from those determined from the machine output based on cross head movement.

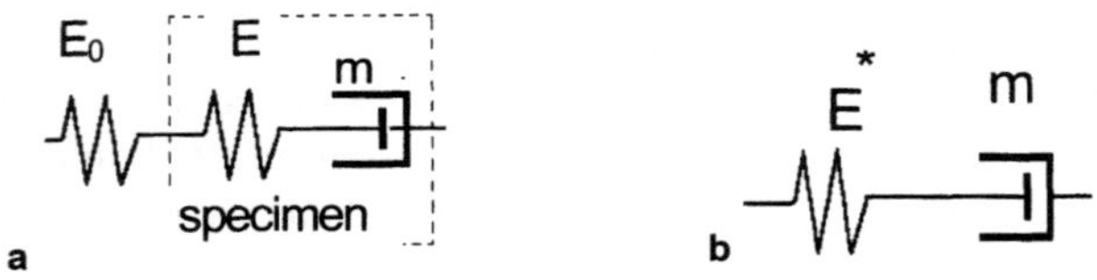

Fig. 2.9. Mechanical model of the system 'testing machine–specimen' **a** and equivalent scheme **b** E_0 is the 'elastic modulus' of the testing machine; the specimen is modelled using a generalised Maxwell body

Ghosh and Hamilton [282] have pointed out based on elementary analysis that when significant grain growth (particularly at low strain rates) is present in a superplastic alloy, a maximum in load will never be reached with strain. Hart [107] and others [69, 72, 95] have emphasised the importance of the effects of elastic deflection of the machine during loading. For this case, Eq. (1.18) will be valid. From incompressibility condition

$$\xi = \left[\frac{v}{L} - \frac{\dot{\sigma}A}{\chi L}\right]\left[1 - \frac{\sigma A}{\chi L}\right]^{-1} \tag{2.72}$$

For an absolutely rigid testing machine ($\chi \to \infty$), $\xi = v/L$. In a constant load test, it follows that $v = dL/dt$ regardless of the value of χ. This can be understood in terms of Fig. 2.9. Under a constant load, the elastic element E_0 experiences an elongation and will not 'disturb' the system further. Its influence is notable only in the unstable region where P is not constant. (Further details are available in [91].) The behaviour of the generalised Maxwell body (Fig. 2.9) is analysed in Chap. 3. The cross head of the testing machine can not be arrested instantly: some delay time t^* will be present. This time dependence may be represented by a smooth function, e.g., exponential or linear. Both these cases are considered in detail in [147, 283]. From Fig. 2.10 it is evident that the delay time t^* for cross head arrest significantly affects the measured values in a relaxation test.

In Fig. 2.11, a schematic of a compression test is presented. (Microstructural stability during deformation is assumed.) The slope of the linear part in one view will give the Young's modulus of the specimen at least to a first order approximation. In Region IV, the increase in load is due to an increase in specimen cross-section with deformation.

The following observations are in order. The Young's modulus in Region II would correspond to that of the 'specimen–testing machine' system, E^*. A standard procedure for eliminating the elasticity of the testing machine is available. First, the machine is loaded 'absolutely rigidly' (in comparison with the specimen) and the resulting curve is subtracted from that obtained with the specimen in posi -

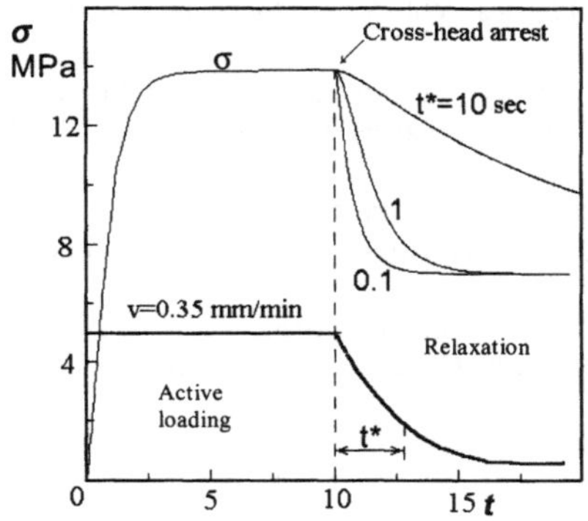

Fig. 2.10. Theoretical time dependences of stress in a relaxation test on an absolutely rigid testing machine, calculated using different delay times t^* for cross head arrest (indicated near the curves). The lower curve represents schematically the time dependence for cross head velocity v

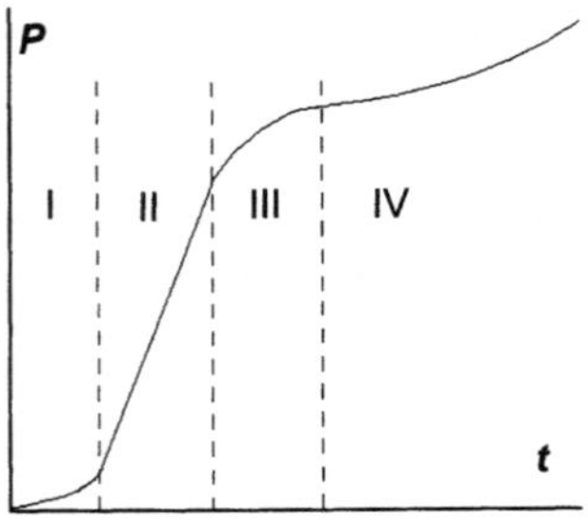

Fig. 2.11. Typical experimental diagram in a compression test on a cylindrical specimen (schematic): I – influence of gaps, clearance, etc.; II – linear part; III – transition to stable plastic flow; IV – stable flow

-tion before analysing the results further. As elastic deformation is of the order of 0,1% and typical gauge length of superplastic alloy specimens is about 10 mm, given the accuracy of modern testing machines the elastic limit of the specimen can not be measured accurately. So, the initial linear slope will not give theYoung's modulus of the specimen. Recently it has been suggested that from the initial slope it will be possible to evaluate the strain rate sensitivity index of superplastics [92]. (This is not likely to be reliable as in this region steady state flow is not present.) The strain range over which direct proportionality between stress and strain has been observed in tests on superplastic alloys is about 5–10%. This can be understood only if the extensive anelasticity present during superplastic flow is taken into account [81–84].

Recommended Reading For Chapter 2

1. Reiner M (1958) Rheology. In:Encyclopaedia of Physics. Ed. Flugge, S., Vol. VI. Elasticity and Plasticity. Springer-Verlag, Berlin Heidelberg pp. 434–550
2. Prager W (1961) Einfuhrung in die Kontinuumsmechanik. Birkhauser, Basel-Stuttgart. English translation: Introduction to Mechanics of Continua, Ginn & Co New York
3. Thomas TY (1961) Plastic Flow and Fracture in Solids. Academic Press, NY London
4. Freudenthal AM (1966) Introduction to the Mechanics of Solids. Wiley, New York
5. Drucker DC (1967) Introduction to Mechanics of Deformable Solids. McGraw-Hill, NY
6. Rabotnov YuN (1969) Creep Problems in Structural Members. North–Holland, Amsterdam
7. Mase GE (1970) Theory and Problems of Continuum Mechanics. McGraw-Hill, New York
8. Kachanov LM (1971) Foundations of the Theory of Plasticity. North-Holland, Amsterdam
9. Truesdell C (Ed.) (1972) Encyclopaedia of Physics, Vol. VIa/2. Mechanics of Solids II, Springer–Verlag, Berlin New York
10. Johnson W, Mellor PB (1972) Engineering Plasticity. Van Nostrand Reinhold, London
11. Bell JF (1973) Experimental Foundations of Solid Mechanics. Springer, Berlin Heidelberg New York
12. Peterson RE (1974) Stress Concentration Factors. John Wiley, New York
13. Slater RAC (1977) Engineering Plasticity - Theory and Application to Metal Forming Processes. John Wiley, NY
14. Dieter GE (1988) Mechanical Metallurgy. McGraw-Hill, Singapore
15. Hill R (1983) The Mathematical Theory of Plasticity. Clarendon Press, Oxford

3 Constitutive Equations for Superplastics

Most of the constitutive equations (CE) for superplastics found in the literature [284–294] are written in scalar form in terms of infinitesimally small strains. A review of such scalar CEs is made first. Some problems associated with generalising such scalar CEs for non-uniform stress–strain state [295–300] are discussed next. Finally, attention is paid to the interpretation of experimental results obtained under conditions of a non-uniform stress–strain state (e.g., torsion, superplastic forming of thin sheet materials [301–303]). An examination of the physical validity/relevance of the models is beyond the scope of the present book.

3.1 Basic Requirements of Constitutive Equations

The construction of an adequate constitutive equation (CE) is a fundamental problem. CEs are equations that describe the mechanical state of a superplastic. Development of mathematical models for technological processes of metalworking as well as evolving algorithms and methods for solving the corresponding boundary value problems are the final goals.

From the viewpoint of the general theory of constitutive laws, any CE relating stress, strain and temperature should satisfy the requirements listed below.

1. A system of CEs should be written in a closed form, i.e., the number of unknowns should be equal to the number of equations. For CEs to be useful in practical calculations, they should allow the determination of the deformation state of the material from the known stress state and vice-versa.
2. The CEs should be written in a tensor form if complex/industrially important boundary value problems are to be formulated and solved.
3. They should indicate the full set of basic experiments for evaluating all the material constants and such experiments should be realisable in practice.
4. The methods used to determine the material constants should be stable with respect to small variations in the experimental data.
5. The constitutive laws should be stable against small variations in the magnitude of the material constants.
6. The range of applicability of any CE should be experimentally determined and indicated theoretically.

Early CEs for superplasticity were written for uniaxial loading. Generalising them into a tensor form is a non-trivial problem. In particular, the problem of choosing the measures of stress and deformation states of superplastic materials needs attention [304–305]. Following Iljushin [169] one can identify the vector as well as the scalar properties of a CE. The vector properties characterise the degree of non-coaxiality of the stress and the strain tensors and their increments. The scalar properties include the relationships among the invariants of these tensors. The physical CEs known in the literature may be used to describe the scalar properties. Often, they do not include information on the vector properties. This aspect has remained uninvestigated. It is, however, recognised that some qualitative considerations of the vector properties are available. Further, it is often assumed that the vector properties of superplastics coincide with those of viscous liquids. The stress tensor is assumed to be proportional to the strain rate tensor. These assumptions are common in all calculations concerning superplastic deformation.

Serious problems arise in using some of the physical CEs proposed to describe the mechanical response of superplastics under uniaxial loading. Usually, such CEs include many material constants, whose values are determined in physical experiments at the microlevel [307–312]. The methods of determining such constants, associated errors, etc., are often not discussed. Thus, unambiguous determination of all constants in a CE is not always possible. The problem of stability of the values of the constants obtained is also not discussed. (This will include the mathematical procedure adopted to obtain the constants.) Numerical experiments that introduce small variations in the data/values of the constants to check the reliability of the solutions obtained are also necessary. The range of applicability of the CE is another issue which is not often considered (see Chap. 2 also).

To the best of our knowledge, the tensor CEs used at present to describe the mechanical response of superplastics are based on the assumption that the stress tensor is proportional to the strain rate tensor (see Sect. 3.2). (As mentioned earlier, this assumption is justified only for some simple loading conditions and infinitesimally small strains.) Then, only the scalar properties would be needed and they are determined under uniaxial loading. Ideally, the validity of the proportionality hypothesis should be verified for every practical problem before use. When this is not valid, the constitutive laws should be written in a more general form, e.g., using three terms formula as a linear relationship among stress, strain and strain rate tensors (see Sect. 3.5). Even when the constants have unambiguous physical meaning, they have to be introduced in a CE in a formal way as fitting coefficients (see Chap. 6).

3.2 Phenomenological Constitutive Equations

Phenomenological constitutive equations do not consider the micro-mechanisms of plastic/superplastic deformation. All material constants are determined from macro-experiments *only*. In principle, they should describe the stress–strain curve within the area of their applicability. In superplasticity, the main aim is to account for the sigmoidal $log\sigma - log\xi$ relationship.

3.2.1 Standard Power Law

The simple power law is often used to describe superplastic flow under uniaxial testing.

$$\sigma = K\xi^m \varepsilon^n \tag{3.1}$$

where σ is the flow stress, ξ is the strain rate, K, m, n are material constants and ε is the true strain. The true strain is estimated from

$$\varepsilon = \int_0^t \xi(\tau)\, d\tau \tag{3.2}$$

Equation (3.1) is well-known in the theory of metalworking, see, e.g., [313, 314]. Hart [107] has also used Eq. (3.1) in his classical analysis. For non-superplastic (conventional) flow $m \approx 0$, while $n \neq 0$. So Eq. (3.1) becomes

$$\sigma = K\varepsilon^n \tag{3.3}$$

In contrast, for stable superplastic flow $n \approx 0$ and so

$$\sigma = K\xi^m \tag{3.4}$$

which is identical to Eq. (1.3). Equation (3.1) is sometimes suggested to be applicable when both strain hardening ($n \neq 0$) and strain rate sensitivity of flow ($m \neq 0$) are present. Recently, the following equation (which is similar to Eq. (3.1)) has been used to describe superplastic flow in a number of Zn–Mn alloys [315].

$$\sigma = k\xi^m \varepsilon^n \left(\frac{T_m - T}{TT_m} \right)^q \tag{3.1a}$$

where T_m is the melting point. Unfortunately, the methods used to determine the material constants are not described. It is also not clear if tension or compression was used [315]. The values of K, m, n on the right hand side of Eq. (3.1) can be easily estimated by standard regression analysis. With that in view, one can consider the following goal function

$$\Phi(K,m,n)=\sum_{i=1}^{N}\left[ln\,\sigma_i - ln\,K - n\,ln\,\varepsilon_i\right]^2 \rightarrow min \tag{3.5}$$

For $\partial\Phi/\partial m = 0$, $\partial\Phi/\partial n = 0$ and $\partial\Phi/\partial K = 0$, one obtains three linear equations with three unknowns K, m, n. This system can be solved using any known procedure, e.g., Kramer's formulae. However, in this analysis [315] considerable scatter in the values of K, m and n was found. Such an analysis was used in [95] to describe the mechanical response of a superplastic Pb–Sn eutectic alloy and Wood's alloy (Bi–25%Pb–12.5%Sn–12.5%Cd). Considerable scatter in the values of K, m, n was found in [95] also.

A similar approach has been adopted in the analysis of data concerning a superplastic Al–Ca–Zn alloy [316]. Here the mathematical procedure used to evaluate the constants was sufficiently stable against small oscillations in the experimental data. Therefore, Eq. (3.1) can be used as a first approximation in estimates for technological processes when the flow is in the stable/steady state region. The procedure described can be modified easily to cover the equation $\sigma = K\xi^{\,m}(1+\varepsilon)^{\,n}$, which is sometimes used.

Taking into account the experimental fact that the values K and m in Eq. (3.4) are not constants, Padmanabhan and Davies [317] have suggested the following approximations for these values.

$$\begin{aligned} m &= f_1(\xi,d) \approx \sum_{r,s=0}^{r+s=N_1} a_{rs}x^r y^s, \\ K &= f_2(\xi,d) \approx \sum_{r,s=0}^{r+s=N_2} b_{rs}x^r y^s \end{aligned} \tag{3.6}$$

where d is the average grain size, $x = -\log\xi$, $y = d$; a_{rs}, b_{rs} are numerical coefficients. These equations have been used to account for the experimental data for the aluminium–33wt.% copper alloy [325]. It was found that the best fits were obtained with $N_1 = 4$ and $N_2 = 5$, the corresponding values for a_{rs} and b_{rs} being indicated in [317]. It is interesting to note that the results obtained clearly indicate (see Figs. 3 and 4 in [317]) that not only the m value depends on ξ, but the value of K also changes considerably with ξ. This means that Eq. (3.4) is not adequate as a general description of superplastic flow (see discussion in Sect. 1.3. and [96] as well). The procedure suggested in [317] has been generalised for the case of n independent variables and later applied in [318] to take into account the effect of temperature as well

$$m = F_1(\xi,d,T) \approx \sum_{r,s=0}^{r+s+p=N_1} A_{rsp}x^r y^s z^p$$

$$K = F_2(\xi, d, T) \approx \sum_{r,s=0}^{r+s+p=N_2} B_{rsp} x^r y^s z^p \tag{3.6a}$$

where $z = (T\text{–}273)/100$, A_{rsp}, B_{rsp} are numerical coefficients. These equations have been used to describe the experimental data for the tin–lead eutectic alloy of Zehr and Backofen [319]. One sees clearly, e.g., Tables III, V in [318], that the values of m and K depend on the strain rate and so Eq. (3.4) is not adequate for accounting for superplastic flow in tin–lead eutectic alloy.

Experimental data pertaining to 27 systems drawn from metals and alloys, superalloys, ceramics, glass ceramics, metal-matrix composites and an intermetallic were analysed in [320] using the standard power law, Eq. (3.4). It was demonstrated that a satisfactory agreement between the theoretical predictions and the experimental data considered exists.

In summary, it is stated that the standard power relation (3.1) is written in a closed but not a mathematically correct form. It breaks down as $\xi \to 0$ (since $\sigma \to 0$). Thus, it describes a liquid rather than a solid (if $m \neq 0$). Using this equation, material constants can not be usually determined unambiguously and reproducibility of results is not high. Thus, the equation is valid only for uniaxial steady state loading over narrow strain rate ranges.

Some ways of improving Eq. (3.1) have been suggested recently [131]. As mentioned in Sect. 1.4.3, the mathematical form of Eq. (3.4) is not quite correct and it is more appropriate to use the mathematical form given in Eq. (1.12). Then, the following generalisation can be made.

$$\sigma = \sigma_q(\varepsilon_f) \cdot \left(\frac{\xi}{\xi_q} \right)^m \tag{3.7}$$

where $\sigma(\varepsilon_f)$ is the stress at a fixed strain corresponding to $\xi = \xi_q$ = constant. If m on the right hand side of Eq. (3.7) is assumed to be constant, the equation should only be used in a narrow strain rate domain around ξ_q. Eq. (3.7) places the following limits on the shape of the experimental curves: the curves should be similar for different constant strain rates: $\sigma^{(1)}/\sigma^{(2)}$ = constant; $\sigma^{(i)} = \sigma(\varepsilon_f)$ at ξ_i = constant (i = 1, 2). A further generalisation would be

$$\sigma = f(\varepsilon)\, \varphi(\xi) \tag{3.8}$$

It is easy to see that both Eq. (3.8) and Eq. (3.7) require the stress–strain curves for different ξ to be similar. In particular, this generalisation has been used in [92] to develop a new method of evaluating m. Another generalisation which can not be reduced to Eq. (3.8), is

$$\sigma = f(\varepsilon)\left(\frac{\xi}{\xi_q}\right)^{m(\varepsilon)} \tag{3.9}$$

This does not require the $\sigma - \xi$ diagrams for different strain rates to be similar. Also, the equation can be used not only for liquids but also solids if $m(\varepsilon) = 0$ when $0 \leq \varepsilon \leq \varepsilon*$. Then, the initial part of the $\sigma - \varepsilon$ diagram can be described correctly. One can show that the material functions $f(\varepsilon)$ and $m(\varepsilon)$ can be determined unambiguously from uniaxial tensile tests.

3.2.2 Polynomial Models

If a large number of experimental points were availiable, then a suitable mathematical function can be selected to describe the sigmoidal $\sigma - \xi$ curve. Then, the constitutive equations can be represented as

$$\sigma = \varphi(\xi, T) \quad or \quad \xi = \psi(\sigma, T) \tag{3.10a,b}$$

where T is the absolute temperature, φ and ψ are single-valued functions of their arguments. One can decompose them in a Taylor series as

$$\sigma = \sum_{i=0}^{N} A_i \xi^i, \quad \xi = \sum_{k=0}^{N} B_k \sigma^k, \quad T = \text{constant} \tag{3.11a,b}$$

For example, Panchenko [321] has suggested the use of a third order polynomial for the sigmoidal superplastic curve. Then, the slope of the sigmoidal curve M will be given by

$$M = \frac{\sum_{i=0}^{N} iA_i \xi^i}{\sum_{i=0}^{N} A_i \xi^i} \quad or \quad M = \frac{\sum_{k=0}^{N} B_k \sigma^k}{\sum_{k=0}^{N} kB_k \sigma^k} \tag{3.12a, b}$$

Let the pairs of experimentally measured values of σ and ξ be $N_{\exp}$, i.e.,

$$\sigma_i, \xi_i \left(I = 1,\ 2, \ldots, N_{\exp}\right)$$

In this case, the coefficients A_i and B_k on the right hand side of Eqs. (3.11) can be found by a linear regression analysis. Evidently, $N+1 \leq N_{\exp}$. If $N+1 = N_{\exp}$, there will be an exact fit, an undesirable situation as then no allowance is made for experimental errors. It is of interest to determine the parameters at the point of inflection in the sigmoidal $\log\sigma$–$\log\xi$ curve: M_{max}, σ_{opt} and ξ_{opt} (Fig. 1.3). Use of Eqs. (3.11) and (3.12) requires the solution of transcendental equations. It is not very convenient. Bhattacharya and Padmanabhan [322] have used the following representation in logarithmic co-ordinates.

$$\log\sigma = a_0 + a_1 \log\xi + a_2 \log^2 \xi + a_3 \log^3 \xi \tag{3.13}$$

Material constants a_0, a_1, a_2 and a_3 can be found by a regression analysis $N_{exp} \geq 4$. Then,

$$M_{max} = a_1 - \frac{a_2^2}{3a_3};$$

$$\log\xi_{opt} = -\frac{a_2}{3a_3}; \log\sigma_{opt} = a_0 - \frac{a_1 a_2}{3a_3} + \frac{2a_2^3}{27a_3^2} \tag{3.14}$$

Ghosh and Hamilton [282] have suggested that

$$\sigma = \sigma_0 + B\varepsilon \tag{3.15}$$

with

$$\log\sigma_0 = a_0 + a_1 \log\xi + a_2 \log^2 \xi + a_3 \log^3 \xi + a_4 \log^4 \xi$$

The same relation has been considered in [10]. Similar to Eq. (3.13), one can consider the following relation

$$\log\xi = b_0 + b_1 \log\sigma + b_2 \log^2 \sigma + b_3 \log^3 \sigma \tag{3.16}$$

In this case also the values of b_0, b_1, b_2 and b_3 can be found by a regression analysis. The following relations are valid.

$$M_{max} = \frac{3b_3}{3b_1 b_3 - b_2^2};$$

$$\log\sigma_{opt} = -\frac{b_2}{3b_3}; \ \log\xi_{opt} = b_0 - \frac{b_1 b_2}{3b_3} + \frac{2b_2^3}{27b_3^2} \tag{3.17}$$

Obviously, other functions, e.g., the one that involves a Fourier analysis, can also be used to define the CE Eq.(3.10).

The above models are phenomenological. They can be easily used because standard procedures of linear regression analysis are available in many softwares including the very simple ones. It is well-known that linear methods provide unambiguous and stable results. But the range of applicability of these models is yet to be investigated. Generalisation to cover non-uniform stress–strain states also has not been attempted.

3.2.3 Mechanical Modelling

In this section, the usefulness of mechanical modelling (Sect. 2.4.2) for describing isothermal superplastic flow is briefly considered.

3.2.3.1 Generalised Maxwell Body

The sigmoidal $\sigma - \xi$ plot is valid only for the steady state region of superplastic flow. In technological processes, mechanical response in the transient regions of flow is also important. For example, Eq.(3.10) would require the stress change to be instantaneous (solid lines in Fig. 3.1). In reality, the stress changes on changing the strain rate will be gradual (shown by dashed lines in Fig. 3.1). Therefore, Eq. (3.10) does not apply to the transient region of flow and can not account for the extensive relaxation phenomenon present during superplastic deformation.

In the transient region, $d\sigma/dt \neq 0$. Hence, the elastic strain rate ξ^e will not be equal to zero. The total strain rate ξ may be taken as the sum of the elastic strain rate, ξ^e, and the inelastic strain rate, ξ^i determined by Eq. (3.10). If

$$\xi^i = \Psi(\sigma, \varepsilon^i, T) \tag{3.18}$$

where ε^i is the inelastic strain, the equation of mechanical state well-known in the theory of creep is obtained.

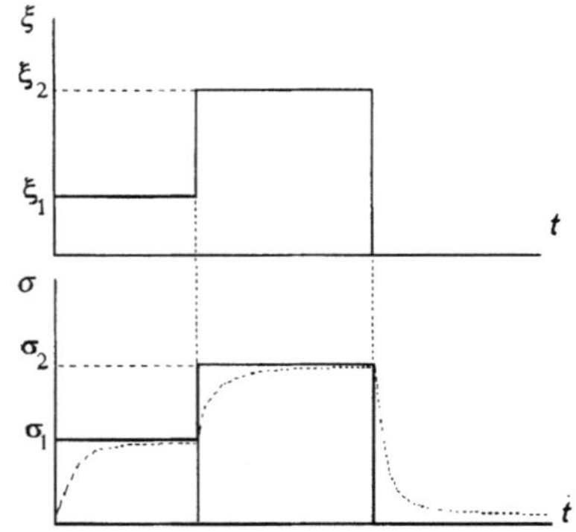

Fig. 3.1. Ideal step strain rate loading **a** and the corresponding mechanical response of a superplastic material **b** Solid lines – theoretical predictions according to Eq. (3.10); dotted lines – the observed response

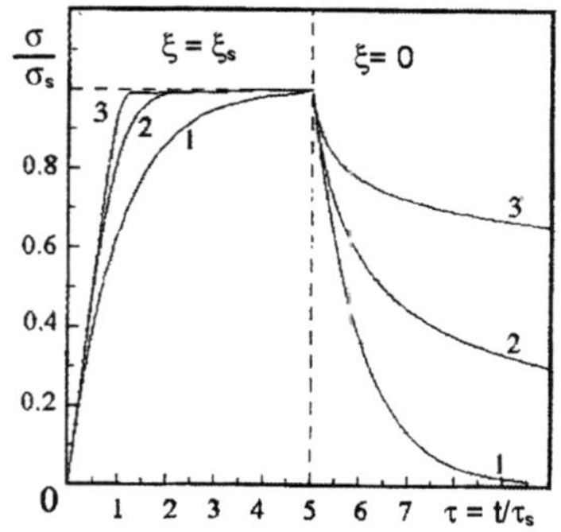

Fig. 3.2. Theoretical time dependences of the flow stress σ for the generalised Maxwell body in a constant strain rate test, cal-culated using Eq.(3.20) and Eq.(3.22) with different m: m = 1(1); m = 0.3(2); m = 0.1(3)

The simplest way of taking both inelastic and elastic strain rates into account is to use the power law Eq.(1.3) for inelastic strain rate to describe the behaviour of a superplastic under transient regimes of loading. Then,

$$\xi = \left(\frac{1}{E}\right)\frac{d\sigma}{dt} + C\sigma^n \tag{3.19}$$

Eq. (3.19) is a result of combination of the elastic (E) and the inelastic (m) elements in series (Fig. 2.5a) – the so-called Maxwell body when $m = 1$. The material obeying Eq.(3.19) will pertain to a generalised Maxwell body. The influence of the finite stiffness of a testing machine is introduced by icluding one more elastic element E_0 in series with the generalised Maxwell body, see. Fig.2.9a. If the loading is determined by a known function $\sigma(t)$ (e.g., σ = constant), the material response will be determined directly by Eq. (3.19). If the loading conditions are determined by a known function $\xi(t)$, then the ordinary differential Eq. (3.19) should be solved for the unknown function $\sigma(t)$.

Let us consider a constant strain rate test $\xi = \xi_s$ = constant. Let σ_s be the corresponding steady state flow stress ($d\sigma/dt = 0$). According to Eq. (3.19)

$$\xi_s = C(\sigma_s)^n$$

The solution for Eq. (3.19), satisfying the initial condition $\sigma(0) = 0$ may then be written as

$$\frac{t}{\tau_s} = \int_0^{\frac{\sigma}{\sigma_s}} \frac{dx}{1 - x^n} \tag{3.20}$$

where τ_s is the characteristic time defined as

$$\tau_s = \frac{\sigma_s}{E\xi_s}$$

In a load relaxation test, usually, the specimen is loaded up to some value of the axial force $P = P_k$. Then the cross head is arrested and the time dependence of P is recorded. Let σ_k be the value of the stress at the moment of cross head arrest ($t = t_k$). Then, Eq. (3.19) may be written as

$$\frac{d\sigma}{dt} = -EC\sigma^n, t \geq t_k \tag{3.21}$$

Integrating Eq. (3.21) and using the initial condition $\sigma(t_k) = \sigma_k$ one obtains

$$\sigma(t)=\frac{\sigma_k}{\left[1+EC(n-1)(\sigma_k)^{n-1}(t-t_k)\right]^{\frac{1}{n-1}}},\ t\geq t_k \tag{3.22a}$$

Denoting $\xi_k = C(\sigma_k)^n$ and $\tau_k = \sigma_k / E\,\xi$, Eq. (3.22a) can be rewritten as

$$\sigma(t)=\frac{\sigma_k}{\left[1+\frac{(n-1)(t-t_k)}{\tau_k}\right]^{\frac{1}{n-1}}},t\geq t_k \tag{3.22b}$$

It should be mentioned that when the steady state flow stress is reached during active loading (at a constant strain rate), $d\sigma/dt$ becomes negligible and the following relations are obtained: $\sigma_k \cong \sigma_s$ and $\tau_k \cong \tau_s$.

The time dependences of the flow stress calculated using Eqs. (3.20) and (3.22) for different values of the strain rate sensitivity index m are represented in Fig. 3.2 in normalised co-ordinates σ/σ_s versus t/τ_s. One can see that as m becomes smaller, the initial part of the $\sigma - t$ plots becomes steeper while the stress decay at larger time is less.

It is worthwhile to examine if Fig. 3.2 is of relevance to superplastic flow. A simple method [147] has been suggested for a generalised Maxwell body. But, m in superplastic flow can be found in many ways, e.g., in a constant cross head velocity test, step strain rate test, load relaxation test, creep test, using the analytical solutions of boundary value problems or results obtained with a technological probe, etc. It is not yet clear which method is to be preferred. However, it has to be mentioned that if it is necessary to choose a suitable method, this problem has to be solved in the following way. If different methods (different types of tests) lead to similar values of constant m, then the method which gives the most stable results should be preferred. One is faced with a more difficult problem when different methods (different types of tests) lead to different values of m. In this case it is not correct to discuss which method is better or worse. If different values of m are obtained, it means that the model cannot be used for describing flow in the given range. Then, the range of applicability of a model and the test that is valid for that range has to be identified or another model that can cover the entire range of the tests will have to be chosen.

The following quantitative characteristics are introduced in respect of the curves presented in Fig. 3.2: i) $\tau_{1/2}$ is the time interval during which the value of the stress increases from its initial value $\sigma(0) = 0$ to $\sigma_s/2$, where σ_s is the steady state flow stress corresponding to ξ_s; ii) $\tau_{1/2R}$ is the time interval during which the value of stress decreases from σ_k to $\sigma_k/2$. These definitions are illustrated in Fig. 3.3, with the assumption that $\sigma_k = \sigma_s$. It is easy to show that for a Maxwell body ($m = n = 1$) the following relations are valid: $\tau_{1/2} = \tau_{1/2\mathrm{R}} = \tau_s \ln 2 \cong 0.7\tau_s$, One can also see that as m decreases, $\tau_{1/2}$ decreases, but $\tau_{1/2R}$ increases. The following relations are easily obtained from Eqs. (3.20) and (3.22).

$$\tau_{\frac{1}{2}} = \tau_s \int_0^{0.5} \frac{dx}{1-x^n} \cong \frac{\tau_s}{2}\left[1+\frac{1}{n+1}\frac{1}{2^n}\right] \tag{3.23}$$

$$\tau_{\frac{1}{2R}} = \tau_s \frac{2^{n-1}-1}{n-1} \tag{3.24}$$

$\tau_{1/2}$ depends only weakly on m; as m increases from 0 to 1, $\tau_{1/2}/\tau_s$ increases from 0.5 to $\cong$ 0.7. But, the dependence of $\tau_{1/2R}(m)$ has quite a different character: an abrupt change in $\tau_{1/2R}$ is present for a non-superplastic material ($m < 0.3$) – see Table 3.1. When the value of m decreases from 0.07 to 0.04 in steps of 0.01, $\tau_{1/2R}$ increases by an order of magnitude for each interval. Thus, superplastics are characterised by a relatively small time of relaxation. This result is in good agreement with the well-known empirical technique of determining the optimal temperature interval for superplastic deformation by measuring the relaxation time at various temperatures.

Yadava and Padmanabhan [323] have studied the effect of strain rate, grain size and test temperature on the shape of the stress–strain curves. The time required for the attainment of steady state superplastic flow in a hot rolled and annealed material has been predicted and experimentally verified [323]. The following theoretical prediction has been made:

$$\frac{d}{2} = \frac{DaC}{kT} \int_0^{t_s} \tau dt \tag{3.25}$$

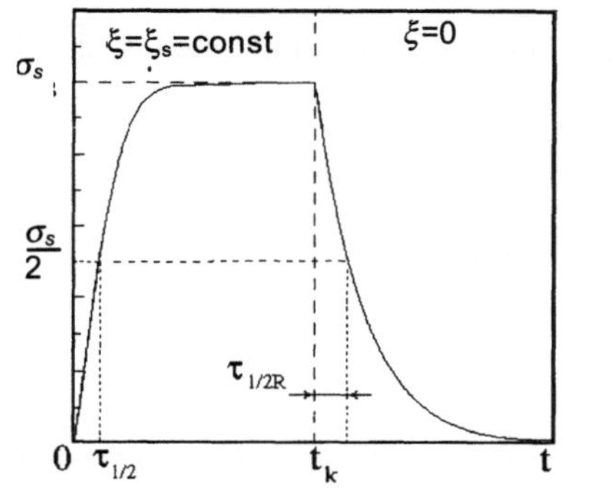

Fig. 3.3. Schematic diagram illustrating the definitions of characteristic time intervals $\tau_{1/2}$ and $\tau_{1/2R}$ at constant strain rate (it is assumed that $\sigma_k = \sigma_s$.)

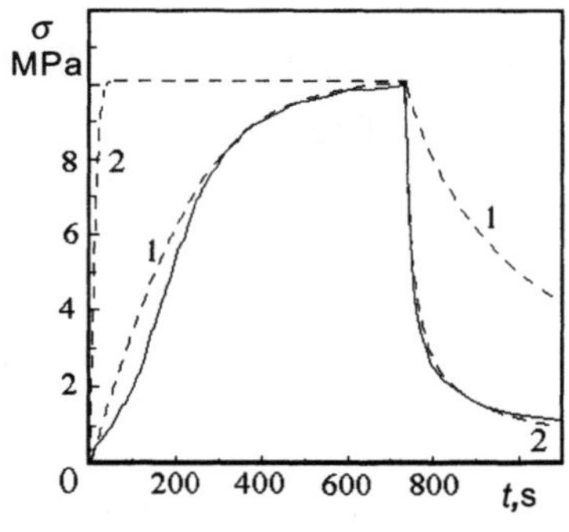

Fig. 3.4. Time dependence of the stress for Wood's alloy at room temperature (solid line). Dotted lines–theoretical curves, calculated with :1 – m = 0.4, E = 4300 MPa; 2 – m = 0.5, E = 280 MPa.

where d is the average grain size, D the diffusivity, a the cross-sectional area of an atom, C the concentration of vacancies, τ the applied shear stress, k the Boltzmann constant and T is the absolute temperature of deformation. This equation may be used to correlate the results of the mechanical tests (by measuring the value of $t_s = 2\tau_{1/2}$) with the physical nature of superplastic flow.

Let the ratio $k = \tau_{1/2}/\tau_{1/2R}$ be fixed for superplastic flow ($m > 0.3$). Using Eqs. (3.23) and (3.24) it can be shown that

$$k \cong m \tag{3.26}$$

That is, the ratio of the characteristic time interval k ($= \tau_{1/2}/\tau_{1/2R}$) for a generalised Maxwell body is approximately equal to the value of m in the superplastic flow region. Thus, the typical curve $\sigma(t)$ recorded in a load relaxation test (Fig.3.4) may be characterised by the following parameters: ξ_s the strain rate during active loading; $\sigma_k = \sigma_s$ the steady state stress; $\tau_{1/2}$ and $\tau_{1/2R}$ which characterise the rates of change in the value of stress during active and passive loading respectively. The special feature of a polycrystalline material in the superplastic state is that its characteristic relaxation time $\tau_{1/2R}$ is (in theory) considerably smaller than that for the same material in the normal state. The ratio $k = \tau_{1/2}/\tau_{1/2R}$ may be used as a normalised characteristic of any material under consideration. In accordance with Eq. (3.26), k should be approximately equal to m. However, this is not borne out by experiments (see below column 6 of Table 3.2).

Consider a constant cross head velocity test (v = constant). The value of the strain rate is given by $\xi = v/(L_0 + vt)$ where L_0 is the initial specimen length. Here, it is necessary to solve the ordinary differential equation Eq.(3.19) which may be rewritten as

$$\bar{\xi} = \tau_s \dot{\bar{\sigma}} + \bar{\sigma}^n \tag{3.27}$$

where τ_s is from (3.20), $\bar{\sigma} = \sigma / \sigma_s$, $\bar{\xi} = \xi / \xi_s$. For a constant strain rate test $\bar{\xi} = 1$ and $\xi_0 = v/L_0$. Therefore,

$$\bar{\xi} = \frac{L_0}{L} = \frac{1}{\left(1 + \dfrac{vt}{L_0}\right)} = \frac{1}{(1 + \eta\tau)}$$

Table 3.1 Dependence on m of characteristic time of relaxation $\tau_{1/2R}$ for the generalised Maxwell body at low values of m

m	0.30	0.25	0.20	0.15	0.10	0.08	0.07	0.06	0.05	0.04
$\tau_{1/2R}/\tau_s$	1.73	2.33	3.75	8.79	56.8	252	752	3320	27600	699000

where $\tau = t/\tau_s$ is normalised time, while $\eta = v\tau_s/L_0$ is a numerical parameter. Using a standard procedure (e.g., Runge–Kutta method of fourth order) to solve Eq. (3.27) numerically, it was found that the discrepancy between the theoretical and the experimental $\sigma - t$ curves was negligible if the absolute value of η did not exceed ~ 0.01. Otherwise, the stress does not reach steady state because the strain rate is not constant with time, e.g., in a tension test the value of the strain rate $\xi = v/L$ decreases; consequently, the value of σ also decreases. In contrast, $\tau_{1/2}$ depends only weakly on the kind of active loading: a constant cross head velocity tension or compression test and a constant strain rate test are characterised by nearly the same value of $\tau_{1/2}$.

Compression tests were done on Wood's alloy of a fine microstructure at 20°C at constant cross head velocities in the range of 0.01 to 1.0 mm/min. The value of m determined by standard procedures was 0.5 ± 0.1. The experimental σ–ε curves were typical of superplastic materials (Fig. 1.2) and load relaxation curves were recorded.

These results also are included in Table 3.2 (ξ_s is the initial strain rate). The decrease in specimen height led to an increase in strain rate during active loading (before cross head arrest). Columns 4 and 5 display the measured values of $\tau_{1/2}$ and $\tau_{1/2R}$ respectively. The last three columns present the values of the elastic modulus E measured in three different ways: from the initial slope of the $\sigma - \varepsilon$ curve (column 7) and calculated on the basis of Eq. (3.21) (columns 8 and 9 respectively). (The following simplifications were made: as $\tau_{1/2} = (0.5–0.7)\tau_s$, it was assumed that $\tau_{1/2} = 0.6\tau_s$ for calculating E. One can find from Eq. (3.24) that $\tau_{1/2R} = \tau_s$ at m = 0.5.)

It is seen from Table 3.2 that the experimental results differ significantly from the theoretical prediction based on Eq. (3.26): a similar comment concerns the

Table 3.2 Experimental and theoretical results for relaxation experiments on Wood's alloy (Bi–2.5% Pb–12.5% Sn–12.5% Cd) at room temperature

σ_k (MPa)	ξ_s (s^{-1})	ξ_k (s^{-1})	$\tau_{1/2}$ (s)	$\tau_{1/2R}$ (s)	k	$d\sigma/d\varepsilon$ (MPa)	$E_{1/2}$ (MPa)	$E_{1/2R}$ (MPa)
1	2	3	4	5	6	7	8	9
10.2	$1.28\cdot10^{-4}$	$1.40\cdot10^{-4}$	180	17	10.6	282	270	4300
13.2	$2.27\cdot10^{-4}$	$2.70\cdot10^{-4}$	75	11	6.8	408	460	4400
14.8	$2.61\cdot10^{-4}$	$2.80\cdot10^{-4}$	96	13	7.4	368	350	4100
17.1	$3.90\cdot10^{-4}$	$5.48\cdot10^{-4}$	60	8	7.5	440	440	3900
20.0	$4.86\cdot10^{-4}$	$7.10\cdot10^{-4}$	36	7	5.1	582	690	4000
21.0	$6.34\cdot10^{-4}$	$1.07\cdot10^{-3}$	50	7	7.1	550	400	2800

Remarks: 1. $E_{1/2}$ was calculated according to the relation $E_{1/2} = 0.6\sigma_s/(\xi_s\tau_{1/2})$.
2. $E_{1/2R}$ was calculated according to the equation $E_{1/2R} = \sigma_s/(\xi_k\tau_{1/2R})$.

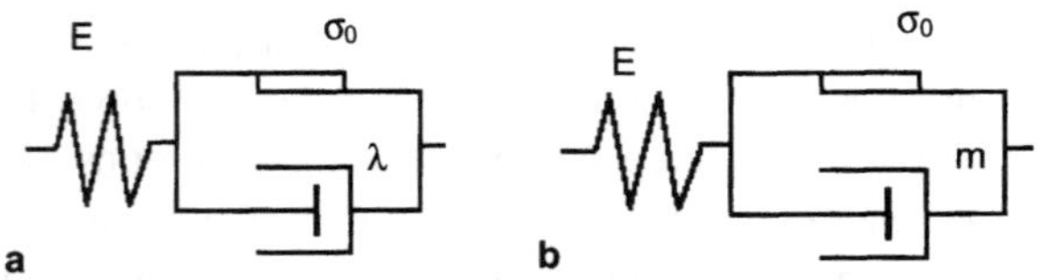

Fig. 3.5. Mechanical analogues of **a** Bingham body and **b** generalised Bingham body

value of E calculated from the initial slope of the $\sigma - \varepsilon$ curve and from Eq. (3.26). Such a large disagreement results from Eq.(3.19) being inadequate for both the active and the passive regimes of loading when the same set of material constants are given a single set of fixed values.

3.2.3.2 Generalised Bingham Body

A Bingham body is introduced in rheology as linear newtonian and Saint-Venant elements in parallel and connected in series to a Hookian element (Fig. 3.5a). If the linear viscous element is replaced by a non-newtonian (as in superplasticity) element, a generalised Bingham body is obtained (Fig. 3.5b). In this section, the relevance of these models for describing superplastic flow is briefly considered.

Two models considered earlier, viz., the three parameter power laws– Karim's model and the generalized Maxwell solid – can be obtained from Fig. 3.5b, by setting $E \to \infty$ and $\sigma_o \to 0$ respectively. A Bingham body, shown in Fig.3.5a, has been used [147, 283] to describe the mechanical behaviour of a lead–tin eutectic alloy. The CE for this model can be written as

$$\xi = \frac{\dot{\sigma}}{E} + \begin{cases} 0, & \sigma < \sigma_0 \\ \dfrac{(\sigma - \sigma_0)}{\lambda}, & \sigma \geq \sigma_0 \end{cases} \tag{3.28}$$

where σ_0, λ and E are material constants. Here, Hooke's law $\sigma = E\varepsilon$ is valid when $\sigma \leq \sigma_0$. If the loading conditions are given by a known function $\sigma(t)$, then the mechanical response $\xi(t)$ is found directly from Eq. (3.28). In particular, in a creep test $\sigma(t) = \sigma_c =$ constant ($\sigma_c > \sigma_0$). The strain rate will then be equal to $\xi_c = (\sigma_c - \sigma_0)/\lambda =$ constant and $\lambda = (\sigma_c - \sigma_0)/\xi_c$.

When kinematic loading is considered, the function $\xi = \xi(t)$ is given. The mechanical response $\sigma(t)$ is found. In this case Eq. (3.28) represents an ordinary differential equation of the first order with respect to the unknown function $\sigma(t)$. The solution of this equation, satisfying the initial condition $\sigma(0) = \sigma_k$, can be presented as

$$\sigma(t) = \sigma_0 + (\sigma_k - \sigma_0)\exp\left(-\frac{t-\tau}{\tau_p}\right) + E\int_0^t \xi(\tau)\exp\left(-\frac{t-\tau}{\tau_p}\right)d\tau \tag{3.29}$$

where $\tau_p = \lambda/E$, $\sigma_k > \sigma_0$. For a load relaxation test $\xi = 0$ and so from Eq. (3.29)

$$\sigma(t) = \sigma_0 + (\sigma_k - \sigma_0)\exp\left(-\frac{t}{\tau_p}\right), \sigma_k > \sigma_0 \tag{3.30}$$

Thus, τ_p is a characteristic relaxation time for a Bingham body. For constant strain rate ($\xi = \xi_k$ = constant) one obtains from Eq.(3.29)

$$\sigma(t) = \sigma_0 + \lambda\xi_k + (\sigma_k - \sigma_0 - \lambda\xi_k)\exp\left(-\frac{t}{\tau_p}\right), \sigma_0 > \sigma_s \tag{3.31}$$

In a constant cross head velocity test, the integral on the right hand side of Eq.(3.29) can be calculated numerically. For a constant load test, there is an analytical solution [283]. The influence of the compliance of the testing machine and its inertia have also been studied [283]. In particular, the following system of equations have been derived.

$$\frac{d\sigma}{dt} = \frac{\dfrac{Ev}{L} - H(\sigma - \sigma_0)\cdot\dfrac{\sigma - \sigma_0}{\tau_p}\cdot\left(1 - \dfrac{\sigma v}{K_m L^2}\right)}{1 + \dfrac{(E - \sigma)v}{K_m L^2}} \tag{3.32a}$$

$$\frac{dL}{dt} = \frac{v(t) + H(\sigma - \sigma_0)\cdot\dfrac{v}{K_m L}\cdot\dfrac{\sigma - \sigma_0}{\tau_p}}{1 + \dfrac{(E - \sigma)v}{K_m L^2}} \tag{3.32b}$$

where $L = L(t)$ is the current length of specimen, K_m is the rigidity of the testing machine, $v = v(t)$ is the cross head velocity (a given function), $H(x)$ is the Heaviside step function ($H(x) = 0$ for $x \le 0$, while $H(x) = 1$ for $x > 0$). Initial conditions for Eq. (3.32) are

$$\sigma(0) = \sigma_k, \qquad L(0) = L_0 \tag{3.32c}$$

The system of Eq. (3.32) has been solved numerically using the Runge–Kutta method after appropriate normalisation. The results are presented in Figs. 3.6. and 3.7. The results for cyclic loading ($v = v_0\cos\omega t$) are presented in Fig. 3.6. (It was assumed that R_0 = 5 mm, L_0 = 10 mm, $\sigma(0) = \sigma_k = 0$, E = 24 GPa, σ_0 = 7 MPa, λ = 16.4 GPa·s, $v = v_0 \cos(2\pi t/\tau_p)$, v_0 = 1mm/min, τ_p = 50 s. Time dependences shown in Fig. 3.7. were calculated with $\sigma(0) = \sigma_k = 2\sigma_0$, R_0 = 5 mm, L_0 = 10mm,

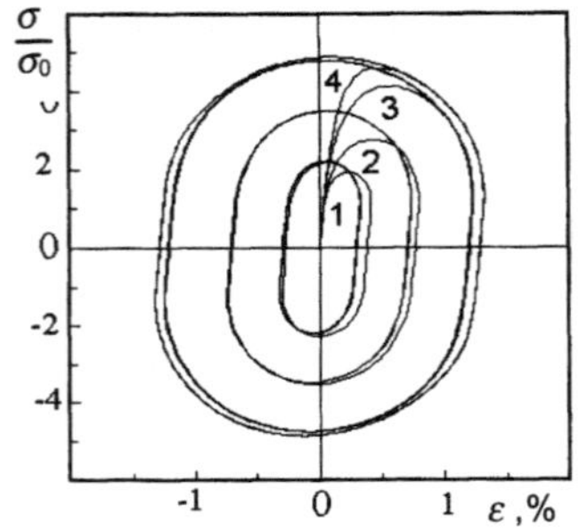

Fig. 3.6. Cyclic stress–strain curves for a Bingham body calculated using Eq. 3.32 and for loading with different values for the rigidity of the testing machine K_m (kN/mm): 1–10, 2–20 , 3–100 , 4–1000

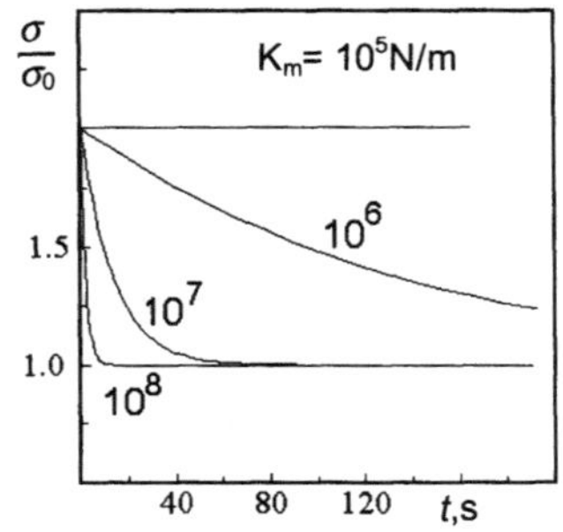

Fig. 3.7. Time dependences of stress, σ, for a Bingham body, calculated using different values for the rigidity of the testing machine (indicated by numbers)

$E = 24$ GPa, $\sigma_0 = 7$ MPa, $\lambda = 16.4$ GPa·s, $v = v_0 \cos(2\pi t/\tau_p)$, $v_0 = 0.1$mm/min, $\tau_p =$ 50 s. The material constants σ_0, λ and E have been determined in [283] in the following way. Several attempts by numerical methods to determine σ_0, λ and E gave unstable results. Then, with the aid of Eq. (3.30), the following goal function was considered.

$$\Phi(\sigma_0, \tau_p) = \sum_{i=1}^{N} \left\{ \frac{P_i}{P_{max}} - \frac{\sigma_0}{\sigma_k} - \left[1 - \frac{\sigma_0}{\sigma_k}\right] exp\left(-\frac{t_i}{\tau_p}\right) \right\}^2 \rightarrow min \tag{3.33}$$

where $P_i = P_i(t_i)$, $i = 0, 1, 2, \ldots, N$ are the experimental data points (from a load relaxation curve), $P_{\max} = P(0)$, $\sigma_k = \sigma(0)$, $\tau_p = \lambda/E$ and σ_0 are material constants to be determined. The values of τ_p and σ_0 were found successfully by a non-linear regression analysis.

Constant load tests were also used to determine the same material constants. Here, the following goal function was examined.

$$\Psi(\lambda, \sigma_0, E) = \sum_{i=1}^{N} \left\{ \lambda \left[\frac{v_i}{L_i} - \frac{\sigma_k v_i}{E L_0} \right] + \sigma_k \frac{L_i}{L_0} - \sigma_0 \right\}^2 \rightarrow \min \tag{3.34}$$

The value of E was assumed while the other two constants were found from the conditions $\partial\Psi/\partial\lambda = 0$ and $\partial\Psi/\partial\sigma_0 = 0$.

It was seen that the values of the constants obtained from a constant load test did not agree with those of a stress relaxation test. The theoretical curves calculated with the first set of constants were in good agreement with the results of strain rate jump tests, while those calculated with the second set vastly differed from the experimental curves. The conclusion reached was that the Bingham

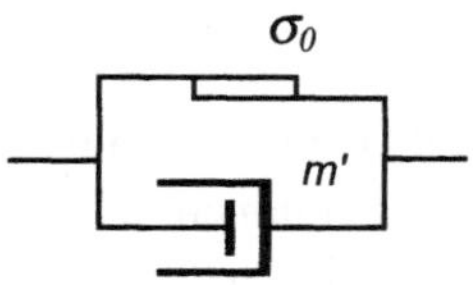

Fig. 3.8. Mechanical analogue for Karim's model

model does not describe both the passive and the active loading regions with the aid of the same set of material constants. Thus, the situation is analogous to that for the generalised Maxwell body.

3.2.3.3
Mechanical Threshold: Analyses of Karim and Murty

The models for the Maxwell and Bingham solids have the following features. In the first model the viscous element starts to operate under any arbitrarily small stress, while in the second only when the stress exceeds σ_0. (But in both the models, deformation commences in a material under any arbitrarily small stress.) If in the general Bingham model, the elastic element is considered to be absolutely rigid ($E \to \infty$), then it will be known as a rigid-visco-plastic material. In this model the role of the parameter σ_0 can be clearly seen – it will be the control value. If $0 < \sigma < \sigma_0$, there will be no deformation in the material. When $E \to \infty$ the general Bingham model leads to the model proposed by Karim Eq. (1.15a) (discussed in Chap. 1). It can be obtained from Fig. 3.5b. when $E \to \infty$; see Fig. 3.8. Therefore,

$$\sigma = \sigma_0 + K'\xi^{m'}, \quad \sigma \geq \sigma_0 \tag{3.35}$$

Eq. (3.35) has a superior capability to describe the experimental data than Eq.(3.4) since it contains three material constants. For this equation

$$\begin{aligned} M &= \frac{d\log\sigma}{d\log\xi} = m'\frac{K'\xi^{m'}}{\sigma_o + K'\xi^{m'}} = m'\frac{\sigma - \sigma_0}{\sigma}; \\ \frac{dM}{d\ln\xi} &= \frac{dM}{d\ln\sigma}\frac{d\ln\sigma}{d\ln\xi} = M\sigma\frac{dM}{d\sigma} = m'\frac{\sigma_0}{\sigma}\cdot\left(1 - \frac{\sigma_0}{\sigma}\right) \end{aligned} \tag{3.36}$$

Thus, the inequality $dM/d\ln\xi > 0$ is realised, since $\sigma_0 \leq \sigma$. Consequently, M monotonically increases with log ξ; the curve log σ – log ξ does not have a point of inflection. So, this equation can not be used to determine the point of inflection on this curve (M_{max}, σ_{opt} and ξ_{opt}). Even when the values M_{max}, σ_{opt} and ξ_{opt} are known, it is not possible to use them to derive the values of σ_o, K' and m' in Eq. (3.35). Independent methods have to be developed for this purpose. The constant

σ_0 (threshold stress) has a physical meaning. So, it is desirable to have a reliable procedure for its determination.

In the literature, a procedure suggested by Mohammed [148] is often used to estimate the values of σ_0 and m' in Eq. (3.35). However, as mentioned in Sect. 1.4.4, this procedure is erroneous. But, a simple procedure can be devised to estimate σ_0, K' and m' [157]. (This will be called Procedure A). The following goal function is considered.

$$\Phi(\sigma_0, K', m') = \sum_{i=1}^{N_{\exp}} \left[\sigma_i - \sigma_0 - K'\xi_i^{m'}\right]^2 \to \min \tag{3.37}$$

Minimisation of Φ with respect to σ_0, K', m' results in a system of 3 non-linear equations in the unknowns σ_0, K', m'. Choosing an arbitrary value of m' ($0 \le m' \le 1$), the values of σ_0 and K' and hence, Φ can be estimated. For different values of m', Φ is then obtained. From this set of values, the combination σ_0, K', m' for which Φ is minimum is chosen as the final answer. An alternative procedure (Procedure B) uses a reference point σ_r, ξ_r. Let $\bar{\sigma} = \sigma/\sigma_r$, $\bar{\sigma}_0 = \sigma_0/\sigma_r$, $\bar{\xi}_i = \xi/\xi_r$. Using the relationship Eq. (3.35)

$$K'\xi^{m'} = K'\bar{\xi}^{m'}\xi_r^{m'} = \bar{\xi}^{m'}(\sigma_r - \sigma_o)$$

One can obtain

$$\bar{\sigma} = \bar{\sigma}_0 + (1 - \bar{\sigma}_0)\bar{\xi}^{m'} \tag{3.38}$$

If now one starts with Eq. (3.38), the goal function will be

$$\Lambda(\bar{\sigma}_0, m') = \sum_{i=1}^{N_{Exp}} \left[\bar{\sigma}_i - \bar{\sigma}_0 - (1 - \bar{\sigma}_0)\bar{\xi}_i^{m'}\right]^2 \to \min \tag{3.39}$$

As before, the best estimates of $\bar{\sigma}_o$ are obtained by an appropriate choice of m'. Let $m' = 0.01$, then one can find the value of $\bar{\sigma}_0$ from the condition $\partial\Lambda/\partial\bar{\sigma}_0 = 0$ as follows.

$$\bar{\sigma}_0 = \sum_{i=1}^{N_{\exp}} \left(\bar{\sigma}_i - \bar{\xi}_i^{m'}\right)\cdot\left(1 - \bar{\xi}_i^{m'}\right)\left[\sum_{i=1}^{N_{\exp}} \left(1 - \bar{\xi}_i^{m'}\right)^2\right]^{-1}$$

In the next step, one has to assume $m' = 0.02$ and the calculations repeated. As a result, the function $\vartheta(m') = \Lambda(\bar{\sigma}_0(m'), m')$ is obtained. In order to satisfy approximately the condition $\partial\Lambda/\partial m' = 0$ one has to find the minimum of $\vartheta(m')$ with respect to all the values of m which were used in the calculations. Let $m' = m'^{**}$ give a minimum in $\vartheta(m')$. The value of K' and σ_o are found as follows:

$$m' = m'^{**} \sigma_0(m'^{**}) = \sigma_r \bar{\sigma}_0(m'^{**}), \quad K' = \frac{\sigma_r - \sigma_0(m'^{**})}{\xi_r^{m'^{**}}}$$

Computer programs needed to execute the above procedures A and B have been developed and tested. The feasibility of using the procedures developed to analyse the experimental data has been demonstrated. (Experimental stress–strain rate data pertaining to Al–12Si [324], Al–33Cu [325], and Al–33Cu–0.4Zr [326] systems and the commercial alloy Supral–100 (Al–6Cu–0.4Zr) [158] were used to validate the procedures developed.)

Initially Procedure 2 of Mohamed [148] (see Chap. 1) was used to determine the values of the threshold stress. The main problem was the need to determine slope M of the sigmoidal curve in Region II. Mohamed [148] has not described the method of estimating the value of M from the experimental data. Here, this value has been estimated in two ways. Firstly, M was obtained as the strain rate sensitivity index m using the set of experimental data. Then, σ_0 was determined by linear extrapolation to zero strain rate. In that method σ_0 turned out to be negative for all the cases considered. The following conclusion could be reached: the value of m assumed was lower than the slope of the sigmoidal plot in Region II. Therefore, M was taken as equal to M_{max}, the maximum value of M at the point of inflection. (The values of the parameters at the point of inflection in the sigmoidal curve can be obtained, for example, within the framework of the Padmanabhan–Schlipf model (see below)). The threshold stress was then calculated for all materials with $M = M_{max}$. Such an approach allowed one to obtain more reasonable values for the threshold stress. However, it is necessary to know the value of M_{max} *a priori*. To determine experimentally the value of M_{max} is difficult. Thus, it is not surprising that in [146, 151] the values of σ_0 have been calculated arbitrarily for $n' = 1/m' = 1$, 2, 3, 4, 5. Presently, the results obtained using Procedures A and B will be described.

The results obtained are summarized in Table 3.3, where they are also compared with those found independently within the framework of the Padmanabhan–Schlipf model. The values of the material constants σ_0, K' and m', found by Procedure A (see Table 3.3), were used to calculate the theoretical $\sigma(\xi)$ curves in accordance with Eq. (3.25). These curves are presented in Fig. 3.9 by solid lines. One can see from Fig. 3.9 and Table 3.3 (columns 6 and 10) that a good agreement between the theoretical curves and the corresponding experimental values was present in all cases. The theoretical predictions for Procedure B were practically the same as shown in Fig. 3.9 and so they are not shown here.

The stability of Procedure A with respect to the number of experimental points used in the calculations has also been considered in [157]. In particular, it is shown that Procedure A gives rise to reasonably stable results even for a minimum number of experimental data ($N_{min} = 3$). In addition, the influence of the choice of the

Table 3.3 Best fit values of the material constants σ_0, K' and m' in the Eq.(3.35)

T (K)	Procedure A						Procedure B			[30]
	N	m'	σ_0 (MPas$^{m'}$)	K' (MPa)	$\Delta\sigma$ (MPa)	m'	σ_0 (MPa)	K' (MPa.s$^{m'}$)	$\Delta\sigma$ (MPa)	σ_0 (MPa)
1	2	3	4	5	6	7	8	9	10	11
					Al–12Si					
831	12	0.88	0.260	2590	0.014	0.84	0.251	1630	0.015	0.267
811	14	0.75	0.311	862	0.016	0.76	0.315	942	0.016	0.350
791	14	0.68	0.375	592	0.018	0.68	0.375	591	0.018	0.430
763	15	0.63	0.636	524	0.026	0.59	0.604	361	0.030	0.705
					Al–33Cu					
793	25	0.73	0.064	46.2	0.007	0.78	0.073	63.5	0.008	0.090
753	22	0.57	0.201	38.8	0.017	0.57	0.205	38.4	0.018	0.306
713	17	0.59	0.212	112	0.010	0.57	0.191	97.5	0.011	0.306
633	10	0.58	0.739	649	0.047	0.51	0.530	357	0.052	0.844
					Al–33Cu–0.4Zr					
793	23	0.84	0.369	2610	0.033	0.86	0.390	3060	0.034	0.450
753	22	0.85	0.627	8150	0.035	0.86	0.639	8940	0.035	0.725
713	11	0.75	1.308	6410	0.103	0.91	1.65	29100	0.121	1.689
					Supral					
763	20	0.63	0.741	472	0.042	0.63	0.740	473	0.042	1.158
743	20	0.63	1.078	601	0.068	0.67	1.187	807	0.080	1.512
723	15	0.58	1.207	490	0.061	0.59	1271	522	0.061	1.856

Remarks: **a** $\Delta\sigma$ was evaluated by standard procedures as mean square deviation of the experimental data from the corresponding theoretical values obtained using Eq. (3.35); **b** the pairs σ_N, ξ_N were always used with Procedure B.

reference point for Procedure B has been studied [157]. It is shown that a change in the magnitude of the reference point does not lead to very large changes in the values of the material constants determined using Procedure B.

The difference between Procedures A and B is as follows. Procedure A gives some average values of σ_0, K' and m', which ensure 'best fit' for the given set of experimental data. Procedure B 'attaches' the theoretical curve Eq.(3.35) to the reference point σ_r, ξ_r. In other words, Procedure B forces the approximation curve described by Eq. (3.35) to pass exactly through the reference point. This restricts pairs σ_K, ξ_K in terms of Eq. (3.35). One can see that the approximation error (to some extent the possibility of finding the very best fit for the set of experimental data) for Procedure B (column 10 of Table 3.3) is somewhat larger than that for Procedure A (column 6 of Table 3.3). But, Procedure B allows one to eliminate the experimental points with large errors (see [157]). Thus, the methods suggested facilitate the unambiguous determination of the values of the material constants σ_0, K' and m' in Eq. (3.35) and can be recommended for analysing the mechanical

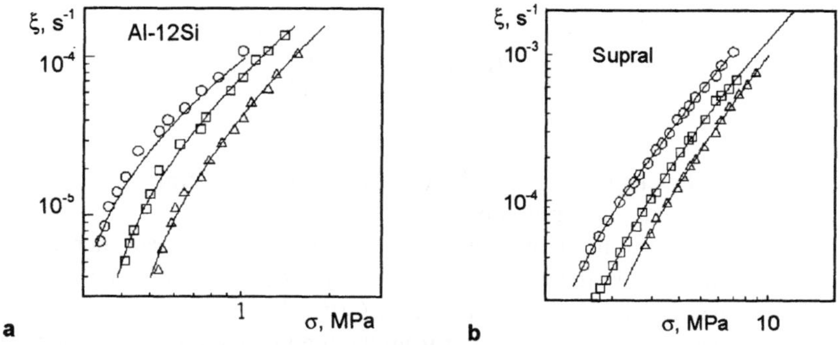

Fig. 3.9. Comparison between the predicted (solid lines) and the experimentally measured values of the strain rate as a function of the applied stress for different alloys (log–log plots): **a** Al–12Si alloy [324]: ooo–831 K, □□□–811 K, ΔΔΔ – 791K; **b** Supral alloy [158]: ooo – 763 K, □□□ – 743 K, ΔΔΔ – 723K.

response of superplastic materials.

The choice of Procedure (A or B) will depend on the task to be performed: if one seeks the very best fit values for K', m', one may use Procedure A. If it is necessary to 'attach' the approximation curve to any point σ_r, ξ_r, or to analyse the set of experimental data with a view to excluding unreliable observations, Procedure B should be used. Murty [69] has examined the stress relaxation behaviour of a superplastic material where the above analysis applies. He has modified Karim's model Eq.(3.35) as

$$\sigma - \sigma_0 = K' \xi_p^{m'} \tag{3.40}$$

where ξ_p is the plastic (inelastic) strain rate and σ_0, K', m' are materials constants. Generally speaking, these material constants are not the same as the ones in Eq. (3.35). One can see that Eq. (3.40) describes the behaviour of a generalised Bingham body.

$$\xi = \frac{\dot{\sigma}}{E} + \begin{cases} 0, & \sigma < \sigma_0 \\ C'(\sigma - \sigma_0)^{n'}, & \sigma \geq \sigma_0 \end{cases} \tag{3.40a}$$

where $n' = 1/m'$, $C' = 1/(K')^{n'}$. Murty has used Eq. (3.40a) to analyse load relaxation in uniaxial tension. For this case,

$$C'(\sigma - \sigma_0)^{n'} = -F\dot{\sigma} \tag{3.40b}$$

where $F = 1/E + A/\chi L$, χ is the compliance of the testing machine, E the Young's modulus, A and L cross-sectional area and gauge length, respectively, of a specimen. Integration of this equation gives the stress relaxation curve

$$\sigma(t)=\sigma_0+\frac{\sigma_k-\sigma_0}{\left[1-\frac{(n-1)C't}{F}(\sigma_k-\sigma_0)^{n'-1}\right]^{\frac{1}{(n'-1)}}} \tag{3.41}$$

where $\sigma_k = \sigma(0)$. In Chap. 1, some methods of evaluating the strain rate sensitivity index from load relaxation curves were given.

The following algorithm to analyse the load relaxation curves can then be suggested: (i) plot the asymptote for $\sigma(t)$ as $t \to \infty$; (ii) this asymptote is considered as the new zero level for the load relaxation curve; (iii) this curve is examined using the equations, available for evaluating m from the load relaxation curve (see Chap. 1). It is important that such a procedure should exclude the influence of the practically uncontrollable shift in the zero level of the machine used to record load relaxation. From a mathematical point of view, all results obtained in this Section are applicable for a generalised Bingham body if σ is replaced by $(\sigma - \sigma_0)$.

3.2.3.4
Smirnov's Mechanical Analogue

Smirnov [2] has suggested two different rheological models for superplastic materials–Fig. 3.10. (The general Smirnov model is discussed later and only the mechanical analogues used by Smirnov are examined here.)

The first analogue (Fig. 3.10a) is practically the same as Shvedov's body if $\sigma_s >> \sigma_0$, which is different from a Bingham body (Fig. 3.5a). The difference lies in the presence of a second Saint-Venant element σ_s. This element restricts the upper value of stress.

In the second model the initial elastic part of the stress–strain diagram is absent and strain will be zero when $\sigma \leq \sigma_0$. When $\sigma > \sigma_0$ the elastic and superplastic elements will operate in series. The upper limit of stress is restricted in this case by the value $\sigma_0 + \sigma_s$. It is evident that Smirnov has assumed that $\sigma_0 \leq \sigma_s$.

Recently, Galimov [327] has discussed the possibilities of the models shown in Fig. 3.10 for describing the sigmoidal variation of the logarithmic flow stress with the logarithmic strain rate.

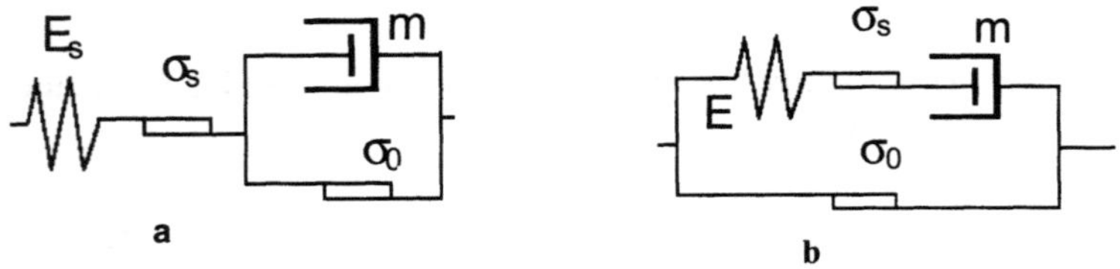

Fig. 3.10. Mechanical analogues for the model of Smirnov [2]

3.2.3.5
Models of Murty–Banerjee and Zehr–Backofen

Murty and Banerjee [153] have suggested the following modificaion to Karim's model.

$$\sigma = K_0 \xi^{m_0} + K\xi^{m} \tag{3.42}$$

where K_0, m_0, K and m are material constants. For the calculations, it was assumed that $m = 0.5$ and $m_0 = 0.1$. This modification, in effect, takes into account the strain rate dependence of the threshold stress. But this is a model for a fluid because in Eq. (3.42), $\sigma = 0$, when $\xi = 0$. Equation (3.42) represents superplastic elements joined in parallel (Fig. 3.11a). The properties of such a combination are considered below.

Zehr and Backofen [319] have suggested the mechanical model shown in Fig. 3.11b. They have developed graphical methods for estimating the material constants K_i, m_i (i = 1, 2, 3). (No analytical procedures have been suggested.) It is emphasised that Zehr and Backofen have indicated that the mixed combination shown in 3.11 can describe the sigmoidal variation of the logarithmic flow stress with the logarithmic strain rate. But, their analysis is not mathematically correct.

3.2.3.6
Combinations of Non-Linear Viscous Elements

Physical models of superplasticity consider the physics of the problem. The following physical processes are often invoked: grain boundary sliding (GBS), diffusion creep (DC) and intragranular slip (IS) (see, e.g., [1, 3, 5, 328, 329]). In some cases, the total strain rate is written as

$$\xi^{p} = \xi_{GBS} + \xi_{DC} + \xi_{IS} \tag{3.43}$$

The contribution of each micromechanism, e.g., GBS, is defined as follows: $\beta_{GBS} = \xi_{GBS}/\xi^{p}$ [1, 3, 5, 328, 329]. It is often assumed that each mechanism is character ised by its own value of the strain rate sensitivity index m, in the relation Eq.(3.4), e.g., for GBS it is assumed that $m_{GBS} \approx 0.5$, for DC $m_{DC} \approx 1$, while for IS $m_{IS} \approx 0.1$. The resulting value of the strain rate sensitivity index, m, is believed to be determined by the interactions among these mechanisms.

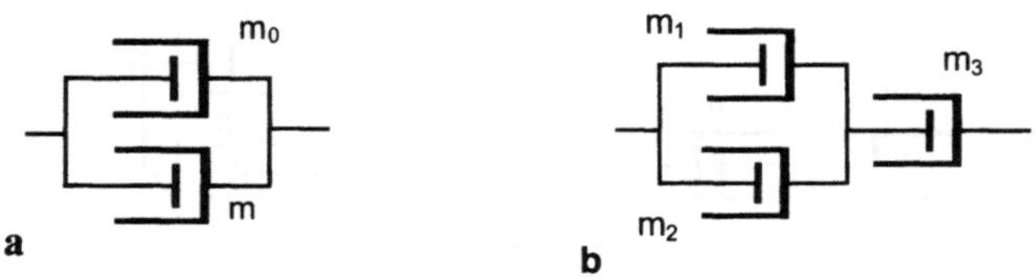

Fig. 3.11. Mechanical analogues of the model of **a** Murty–Banerjee, and **b** Zehr–Backofen

It is often assumed that the contribution of GBS is maximum in the vicinity of the point of inflection of the sigmoidal curve. Some reports have questioned this view, see, e.g. [330–335]. From a mechanical point of view, Eq. (3.43) may be considered as a combination of three non-linear viscous elements (dashpots) connected in series [336]. The properties of such a combination may be described by the equation.

$$\xi^p = C_{GBS}\sigma^{n_{GBS}} + C_{DC}\sigma^{n_{DC}} + C_{IS}\sigma^{n_{IS}} \tag{3.44}$$

where $n_{GBS} = 1/m_{GBS}$; $n_{DC} = 1/m_{DC}$; $n_{IS} = 1/m_{IS}$; C_{GBS}, C_{DC} and C_{IS} are material constants. The question as to whether the sigmoidal curve of superplasticity can be described by Eq. (3.44) has remained unattended to. If this is not possible, the constants in Eq. (3.44) should be so chosen as to achieve this. With this in view, the mechanical behaviour of various combinations of the non-linear viscous elements are considered below.

It is assumed that the properties of the *i*-th element can be described by the following power relation

$$\sigma_i = K_i \xi_i^{m_i}, \quad i = 1, 2, \ldots, N \quad (\textit{no summation over } i) \tag{3.45}$$

where σ_i is the stress in the *i*-th element; ξ_i is its strain rate; K_i and m_i are material constants characterising the properties of the *i*-th element ($0 \le m_i \le 1$). The sequential (Fig. 3.12a), parallel (Fig. 3.12b) and mixed (Fig. 3.11b) combinations are considered below. The capability of such combinations to describe the superplasticity curves has received the main consideration.

Consider the series joining of elements. Many physical models of superplasticity can be regarded as combinations of non-linear viscous elements, e.g., [329]. In this case the total strain rate is equal to the sum of the strain rates produced by different processes – see also Sect.3.2.2.; Eq. (3.11b).

$$\xi = \xi_1 + \xi_2 + \ldots + \xi_k = \sum_{i=1}^{k} C_i \sigma^{n_i} \equiv \beta_1 \xi + \beta_2 \xi + \ldots + \beta_k \xi \tag{3.46}$$

where $\beta_i = \xi_i/\xi$ is the contribution of the *i*-th element. It is evident that $\beta_1 + \beta_2 + \ldots + \beta_k = 1$; $0 \le \beta_i \le 1$ ($i = 1, 2, \ldots, k$).

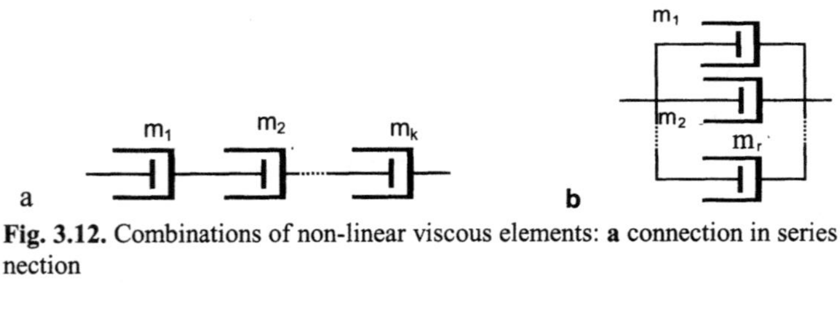

Fig. 3.12. Combinations of non-linear viscous elements: **a** connection in series **b** parallel connection

Let $\sigma_s = \sigma(\xi_s)$ be a reference stress, e.g., σ_s, ξ_s may coincide with the co-ordinates of the point of inflection on the sigmoidal curve, σ_{opt}, ξ_{opt} (see Fig. 1.3.a). Let the contributions of the various elements at $\xi = \xi_s$ be denoted by β_{1s}, β_{2s},...,β_{ks}

$$\beta_{iq}\xi_q = C_i\sigma_q^{n_i} \quad (i = 1,\ 2,...,k;\ \beta_{1s} + \beta_{2s} + ... + \beta_{ks} = 1)$$

Then, one can replace C_i by means of β_{is} in Eq. (3.46).

$$\bar{\xi} = \beta_{1s}\bar{\sigma}^{n_1} + \beta_{2s}\bar{\sigma}^{n_2} + ... + \beta_{ks}\bar{\sigma}^{n_k} \equiv \bar{\sigma}^{n_c} \tag{3.46a}$$

where $\bar{\sigma} = \sigma/\sigma_s$ and $\bar{\xi} = \xi/\xi_s$ are the normalised stress and strain rate respectively, $n_c = 1/m_c$. For the chain under consideration m_c is the effective strain rate sensitivity index. In general, $m_c \neq$ constant. It should be mentioned that the contribution β_{is} of the i-th element at $\xi = \xi_s$ is not equal to that at $\xi \neq \xi_s$. It can be shown that

$$\beta_i = \beta_{is}\frac{\bar{\sigma}^{n_i}}{\bar{\xi}} \qquad (i = 1,2,...,k) \tag{3.47}$$

The introduction of β_{is} ensures that C_i are unambiguously determined by the relationships

$$C_i = \frac{\beta_{iq}\xi_q}{\sigma_q^{n_i}} \qquad (i = 1,\ 2,...,k)$$

In turn, the clear definition of C_i determines the values of β_{is} through the relations

$$\beta_{iq} = \frac{C_i\sigma_q^{n_i}}{\xi_q}$$

The slope of the sigmoidal curve is defined as $1/M = d\log\xi/d\log\sigma = d\log\bar{\xi}/d\log\bar{\sigma}$. It is easy to show that

$$M_c(\xi) = \frac{\beta_{1s}\bar{\sigma}^{n_1} + \beta_{2s}\bar{\sigma}^{n_2} + ... + \beta_{ks}\bar{\sigma}^{n_k}}{\beta_{1s}n_1\bar{\sigma}^{n_1} + \beta_{2s}n_2\bar{\sigma}^{n_2} + ... + \beta_{ks}n_k\bar{\sigma}^{n_k}} \tag{3.48}$$

where $\bar{\sigma} = \bar{\sigma}(\bar{\xi}) = \bar{\xi}^{m_c}$ is the solution for the transcendental Eq. (3.46). It can be shown that

$$\lim_{\xi \to 0} M_c(\xi) = \lim_{\xi \to 0} m_c(\xi) = m_{\max} \equiv \max_{i=1,2,\dots k} m_i; \qquad \lim_{\xi \to \infty} M_c(\xi) = \lim_{\xi \to \infty} m_c(\xi) = m_{\min} \equiv \min_{i=1,2,\dots,k} m_i \tag{3.49a,b}$$

In Fig. 3.13, the strain rate dependences for $\bar{\sigma}$, M_c, m_c and contributions β_i (i = 1, 2, 3) are presented. One can see in Fig. 3.13a that the curve $\log \bar{\sigma} - \log \bar{\xi}$ is convex. The values of M_c and m_c decrease monotonically with increasing ξ (Fig. 3.13b). At low ξ, the element with maximum m ($m_{\max} = m_1 = 1$) dominates, so that $\beta_1 \to 1$ as $\xi \to 0$ (Fig. 3.13c). At large ξ, the element with minimum m ($m_3 = 0.1$), makes the largest contribution, so that $\beta_3 \to 1$ as $\xi \to \infty$ (Fig. 3.13c). The contribution of the non-linear viscous element $m_2 = 0.5$ is maximum at intermediate strain rates. For example, even when the number of non-linear viscous elements were increased (two more mechanisms in addition to the three specified), a sigmoidal $\sigma - \xi$ as well as a realistic m–log ξ (or M–log ξ) relationships could not be obtained. Hence, *physical models based on the hypothesis that their effects are additive* (similar to Eq. 3.43) can not account for *the sigmoidal curve seen during superplastic flow*. So, other possible ways of combining the elements must be considered.

Consider a parallel combination of the elements. In this case each element makes a contribution to the total stress. For example, when two non-linear viscous elements are connected in parallel, one obtains the model of Murty and Banerjee (Fig. 3.11a). Expressions similar to Eq. (3.42) are widely used in mechanics of compounds; they may be used to simulate the rheological behaviour of two-phase titanium alloys. For the case where r non-linear viscous elements are connected in parallel (see also Sect. 3.2.2., Eq. (3.11a))

$$\sigma = \sum_{i=1}^{r} K_i \xi^{m_i} \equiv \alpha_1 \sigma + \alpha_2 \sigma + \dots + \alpha_r \sigma \tag{3.50}$$

where $\alpha_i = \sigma_i / \sigma$ is the contribution of the i-th element to the total stress. Let for some reference point $\sigma_q = \sigma\ (\xi_q)$ the contributions of individual elements be equal to $\alpha_{1q}, \alpha_{2q}, \dots, \alpha_{rq}$. Then, Eq. (3.50) can be rewritten as

$$\bar{\sigma} = \sum_{i=1}^{r} \alpha_{iq} \bar{\xi}^{m_i} \overset{def}{=} \bar{\xi}^{m_p} \tag{3.51}$$

where $\bar{\sigma} = \sigma / \sigma_q$, $\bar{\xi} = \xi / \xi_q$, m_p is an effective value of the strain rate sensitivity index, m. The contributions of elements depend on the strain rate. One can obtain that

$$\alpha_i = \frac{\sigma_i}{\sigma} = \alpha_{iq} \frac{\bar{\xi}^{m_i}}{\bar{\sigma}}, \quad i = 1, 2, \dots, r \tag{3.52}$$

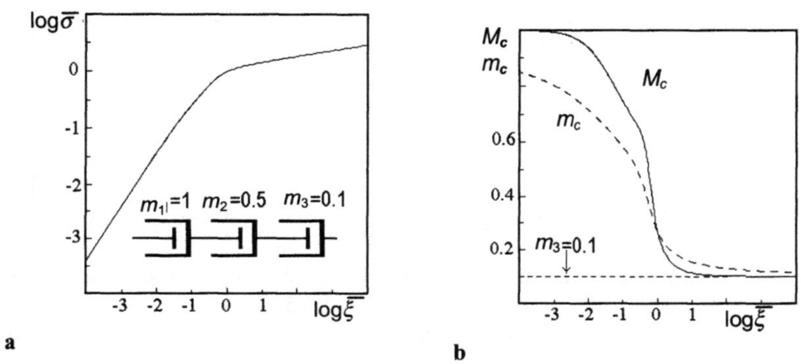

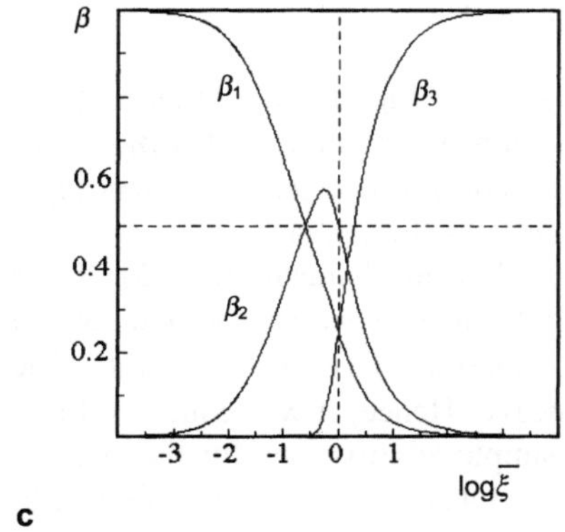

Fig. 3.13. Theoretical dependences calculated for the chain joined in series with k = 3, m_1 =1, m_2 = 0.5, m_3 = 0.1, β_{1s} = 0.25, β_{2s} = 0.5, β_{3s} = 0.25:**a** stress – strain rate curve (Eq. 3.46); **b** $M_c(\xi)$ curve (solid line – Eq. 3.48) and $m_c(\xi)$ curve (dashed line – Eq. 3.46) **c** contributions β_i (Eq. 3.47).

where $\overline{\xi} = \overline{\xi}(\overline{\sigma})$ is the solution of the transcendental Eq. (3.51). In this case, the slope of the superplastic curve is

$$M_p = \frac{\alpha_{1q} m_1 \overline{\xi}^{m_1} + \alpha_{2q} m_2 \overline{\xi}^{m_2} + \ldots + \alpha_{rq} m_r \overline{\xi}^{m_r}}{\alpha_{1q} \overline{\xi}^{m_1} + \alpha_{2q} \overline{\xi}^{m_2} + \ldots + \alpha_{rq} \overline{\xi}^{m_r}} \tag{3.53}$$

It can be shown that

$$\begin{aligned} &\lim_{\xi \to 0} M_p(\xi) = \lim_{\xi \to 0} m_p(\xi) = m_{min} \equiv \min_{i=1,2,\ldots r} m_i; \\ &\lim_{\xi \to \infty} M_p(\xi) = \lim_{\xi \to \infty} m_p(\xi) = m_{max} \equiv \max_{i=1,2,\ldots r} m_i \end{aligned} \tag{3.54a,b}$$

In Fig. 3.14, the theoretical strain rate dependences of $\overline{\sigma}$, M_p and m_p values calculated for 3 elements connected in parallel are presented. One can see from Fig. 3.14a that the curve $\log\overline{\sigma} - \log\overline{\xi}$ is concave. The values of M_p and m_p increase monotonically with ξ (Fig. 3.14b). At low ξ, the element with minimum

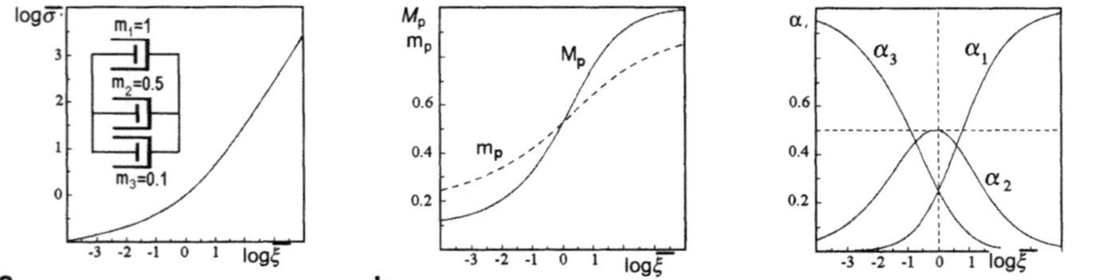

a b c

Fig. 3.14. Theoretical dependences calculated for a parallel combination of non-linear viscous elements with $r = 3$, $m_1 = 1$, $m_2 = 0.5$, $m_3 = 0.1$, $\alpha_{1q} = 0.25$, $\alpha_{2q} = 0.5$, $\alpha_{3q} = 0.25$:
a stress – strain rate curve (Eq. 3.51); **b** $M_p(\xi)$ curve (solid line– Eq. 3.53) and $m_p(\xi)$ curve (dashed line – Eq.3.51); **c** contributions α (Eq. 3.52)

m (= $m_3 = 0.1$) dominates, so that $\beta_3 \to 1$ as $\xi \to 0$ (see Fig. 3.14c). At large ξ, the element with maximum m (= $m_1 = 1$) makes the main contribution so that $\beta_1 \to 1$ as $\xi \to \infty$ (see Fig. 3.14c). The contribution of the non-linear viscous element $m_2 = 0.5$ has its maximum value for intermediate strain rates (Fig. 3.14c).

Numerical calculations were done for up to five mechanisms ($r = 5$). Their relative contributions α_{iq} and strain rate sensitivity indexes m_i were varied. But in all cases considered, $M_p(\xi)$ and $m_p(\xi)$ increased monotonically with strain rate. The shape of the $\log\sigma - \log\xi$ curve was always concave. Hence, it was concluded that physical models based on the hypothesis of a simple additive law for the stresses arising from different mechanisms (similar to Eq. 3.42) were incapable of describing superplastic flow.

Now, the case of mixed combinations of elements is addressed. When Figs. 3.13b and 3.14b are compared, it is seen that series and parallel combinations lead to antipodal properties. Hence, one may expect that mixed combinations will allow non-monotonic dependences for $m(\log\xi)$ and $M(\log\xi)$. There are a few examples in the literature where mixed combinations have been suggested. Zehr and Backofen (Sect. 3.2.3.5) have considered a mixed combination to describe the mechanical response of a superplastic lead–tin eutectic alloy. Recently, Perevezentsev et al. have suggested a physical model [277] which accounts for the sigmoidal variation of $\log\sigma$ with $\log\xi$. This model is rather cumbersome and so it is not easy to handle. However, based on their many publications in Russian, it is concluded that they use a mixed combination while assessing the contributions of different micromechanisms. Such an approach enables them to describe the experimental sigmoidal curve for the titanium alloy Ti–6Al–4V, reported by Ghosh [337], Ghosh and Hamilton [338]. So, a consideration of mixed combinations of nonlinear elements is of practical interest.

The mixed combination of three elements (Fig. 3.11b) leads to the equation

$$\sigma = K_1\xi_1^{m_1} + K_2\xi_2^{m_2} + K_3\xi_3^{m_3},\ \xi_1 = \xi_2,\ \xi = \xi_1 + \xi_3 \qquad (3.55)$$

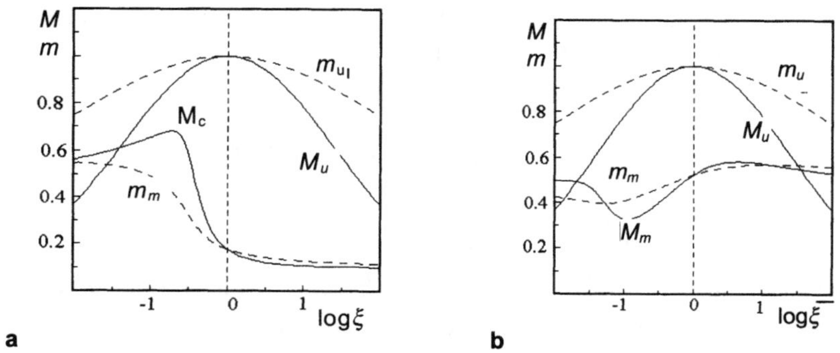

Fig. 3.15. Strain rate dependences of the M values (solid lines) and the m values (dashed lines) calculated according to Eqs. (1.6), (1.7), (3.57), (3.58) with two different sets of material constants: **a** $m_1 = 1$; $m_2 = 0.5$; $m_3 = 0.1$; $\alpha_{10} = 0.5$; $\beta_{10} = 0.5$ (M_m– and m_m–curves); **b** $m_1 = 1$; $m_2 = 0.1$; $m_3 = 0.5$; $\alpha_{10} = 0.5$; $\beta_{10} = 0.5$ (M_m –and m_m–curves).M_u and m_u-curves have been calculated with $a = 0.5$ and $M_{\max} = 1$.

where ξ_i (i = 1, 2, 3) is the strain rate of the i-th element; K_i, m_i (i = 1, 2, 3) are material constants. For some reference point $\sigma = \sigma(\xi_0)$

$$\begin{aligned}\sigma_0 &= K_1\xi_{10}^{m_1} + K_2\xi_{20}^{m_2} = \alpha_{10}\sigma_0 + \alpha_{20}\sigma_0 \\ \xi_0 &= \xi_{10} + \xi_{30} = \beta_{10}\xi_0 + \beta_{20}\xi_0\end{aligned} \tag{3.56}$$

Then,

$$\bar{\sigma} = \frac{\alpha_{10}}{\beta_{10}^{m_1}}\left(\bar{\xi} - \bar{\xi}_3\right)^{m_1} + \frac{\alpha_{20}}{\beta_{20}^{m_2}}\left(\bar{\xi} - \bar{\xi}_3\right)^{m_2} = \frac{1}{\beta_{30}^{m_3}}\bar{\xi}_3^{m_3} \overset{def}{=} \bar{\xi}^{m_m} \tag{3.57}$$

where $\bar{\sigma} = \sigma / \sigma_0$, $\bar{\xi} = \xi / \xi_0$, $\beta_{20} \equiv \beta_{10}$, $\bar{\xi}_i = \xi_i / \xi_0$, m_{m} is an effective value of the strain rate sensitivity index; symbol 'm' being used to indicate a mixed combination. It is easy to see that for a given strain rate ξ, the middle part of Eq. (3.57) represents a transcendental equation for the unknown function $\bar{\xi}_3$. The value of M for this mixed model may be found as

$$\frac{1}{M_m(\xi)} = \frac{\bar{\xi}_3}{m_3\bar{\xi}} + \frac{\bar{\xi}_3^{m_3}}{\beta_{30}^{m_3}} \frac{1 - \dfrac{\bar{\xi}_3}{\bar{\xi}}}{\bar{\alpha}_{10} m_1\left(\bar{\xi} - \bar{\xi}_3\right)^{m_1} + \bar{\alpha}_{20} m_2\left(\bar{\xi} - \bar{\xi}_3\right)^{m_2}} \tag{3.58}$$

where $\bar{\alpha}_{i0} = \alpha_{i0} / \beta_{i0}^{m_i}$ (i = 1, 2), and $\bar{\xi}_3\left(\bar{\xi}\right)$ is the solution of Eq. (3.57). One can show that $\lim_{\alpha \to 0} M_m(\bar{\xi}) = M_c$; $\lim_{\beta_3 \to 0} M_m(\bar{\xi}) = M_p$; $\lim_{\xi_3 \to \xi} M_m(\bar{\xi}) = m_3$.

Also, it is possible to derive the following relation for $M_m(\xi)$ directly from Eq.(3.55) without introducing a reference point.

$$M_m = \frac{m_m \xi L(\xi_1)}{K_1\xi_1^{m_1}(m_3 - m_1) + K_2\xi_1^{m_2}(m_3 - m_2) + \xi L(\xi_1)} \quad (3.59)$$

where $L(\xi_1) = K_1 m_1 \xi_1^{m_1} + K_2 m_2 \xi_2^{m_2}$ *and* $\xi_1(\xi)$ can be found from the following trancendental equation: $K_1\xi_1^{m_1} + K_2\xi_1^{m_2} = K_3(\xi - \xi_1)^{m_3}$. The results of calculations based on Eqs. (3.57) and (3.58) are displayed in Fig. 3.15. One can see that the $M_m(\log\xi)$ dependences are not monotonic. However, there is a significant difference between the $M_m(\log\xi)$ plots and the dependence $M_u(\log\xi)$ for the universal curve shown in Fig. 1.6b. But, good agreement between $M_m(\log\xi)$ and $M_u(\log\xi)$ curves as also between $m_m(\log\xi)$ and $m_u(\log\xi)$ curves could be found using appropriate values of m_i, α_{10} and β_{10} (see Fig. 3.16). It should be noted that the set of constants that enables one to describe adequately the universal superplasticity curve is not unique. Many other sets are possible, see, e.g., Table 3.4. The last column in that table gives the standard mean square deviation Δ for the results of the calculation from the universal superplasticity curve shown in Fig.1.6b. It is evident that one can find many sets of material constants to describe the universal superplasticity curve with reasonable accuracy.

It is pertinent to note that when all non-linear viscous elements contained in a mechanical model have the same strain rate sensitivity index (that is, $m_1 = m_2 = \ldots = m_0$), then independent of the type of connection $M = m = m_0$. This fact is a consequence of Eqs. (3.48), (3.53) and (3.59).

In conclusion, it can be said that when the mechanical elements, regardless of their number, are connected in series or in parallel it is not possible to obtain the sigmoidal variation of logarithmic flow stress with logarithmic strain rate. In con-

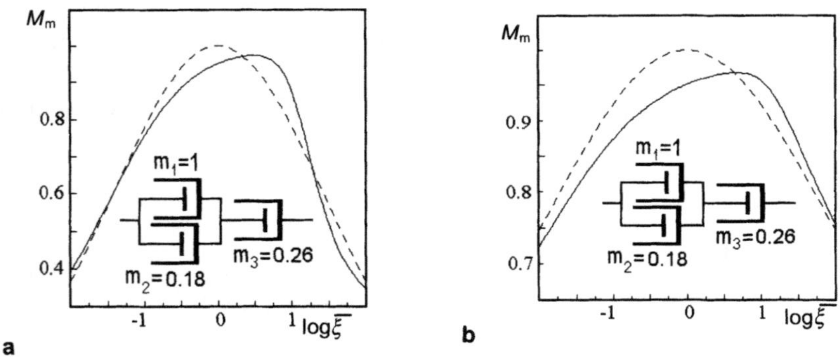

Fig. 3.16. Strain rate dependences of the M–value **a** and the *m*–value **b** for a mixed combination, calculated using Eqs. (3.57) and (3.58) with $m_1 = 1$; $m_2 = 0.18$; $m_3 = 0.26$; $\alpha_{10} = 0.94$; $\beta_{10} = 0.9999$. For comparison the corresponding universal curves calculated in terms of Eqs. (1.6) and (1.7) with $M_{max} = 1$ are presented by dashed lines

Table 3.4 Sets of material constants for Eqs. (3.57) and (3.58) useful in describing the universal superplasticity curve shown in Fig. 1.6b

m_1	m_2	m_3	α_{10}	α_{20}	β_{10}	β_{20}	Δ
1	0.18	0.26	0.94	0.06	0.9999	0.0001	0.05
1	0.1	0.46	0.96	0.04	0.97	0.03	0.05
1	0.4	0.25	0.87	0.13	0.9999	0.0001	0.09
1	0.3	0.25	0.90	0.10	0.9999	0.0001	0.06
1	0.33	0.25	0.88	0.12	0.9999	0.0001	0.07
1	0.2	0.5	0.94	0.06	0.92	0.08	0.08

trast, when mixed combinations are used, the sigmoidal curve can be produced with the required level of accuracy. The results of this section are discussed later while considering the model of Padmanabhan and Schlipf [138].

3.2.4 Smirnov's Model

Smirnov [2, 339] has considered a superplastic to be an elasto-viscoplastic medium. The CE is written as

$$\sigma = \sigma_s \frac{\sigma_0 + K_v \xi^{m_v}}{\sigma_s + K_v \xi^{m_v}}, \quad \xi \neq 0, \tag{3.60}$$

where σ_0, σ_s, K_v and m_v are material constants. Eq. (3.60) can be rewritten as

$$\lim_{\xi \to 0} \sigma = \sigma_0, \quad \lim_{\xi \to \infty} \sigma = \sigma_s \tag{3.61}$$

$$\xi = \begin{cases} 0, & \sigma \le \sigma_0 \\ \left[\dfrac{\sigma_s}{K_v} \dfrac{\sigma - \sigma_0}{\sigma_s - \sigma} \right]^{\frac{1}{m_v}}, & \sigma > \sigma_0 \end{cases} \tag{3.60a}$$

Smirnov has suggested the following methods for determining the values of σ_0, σ_s, K_v and m_v. At first one has to find graphically the point of inflection on the sigmoidal curve and measure the maximum slope M_{max}. Then, the value K_v is determined as $K_v = \sigma_{opt} / (\xi_{opt})^{M_{max}}$. The values of σ_0 and σ_s are found by extrapolation using Eq. (3.61). Finally, the value of m_v is calculated as

$$m_v = M_{max} \frac{\sqrt{\sigma_s} + \sqrt{\sigma_0}}{\sqrt{\sigma_s} - \sqrt{\sigma_0}} \tag{3.62}$$

It can be shown that Smirnov's model allows an accurate description of the universal superplasticity curve. Equation (3.60) describes the steady state flow stress and does not account for the initial part of the stress–strain diagram and the load relaxation curve. As a rule, Eq. (3.60) refers to the creep strain rate rather than to the total strain rate, which is equal to the sum of the elastic and creep strain rates. To use it in the transient region of loading, the equation is modified as

$$\xi = \xi_e + \xi_c = \frac{\dot{\sigma}}{E} + \left[\frac{\sigma_s}{K_v} \frac{(\sigma - \sigma_0)}{(\sigma_s - \sigma)} \right]^{\frac{1}{m_v}} ; \sigma > \sigma_0 \tag{3.63}$$

Careful experimentation at different temperatures and strain rates followed by an analysis based on Eq. (3.63) revealed that the equation is inadequate. The conclusion became inevitable that the entire $\sigma - \xi$ curve including the initial transient part and the relaxation curve can not be described within the framework of Smirnov's model by using the same set of material constants. This result is to be kept in mind for practical applications because in metalworking processes the loading conditions are non-monotonic.

Smirnov's model was later extended to include structural parameters [276, 340–344]. This model can be represented as

$$\xi = \xi_{eq} \cdot \exp\left[K\Omega^{\gamma}\left(\sigma - \sigma_{eq}\right)\right]\left\{\frac{\sigma - \sigma_0}{\sigma_s - \sigma}\right\}^{n_v} ; \qquad \text{(3.64a, b)}$$

$$\dot{\Omega} = \Omega^{-q}\left(1 + p\xi^{g}\right)$$

where ξ_{eq}, K, γ, σ_{eq}, σ_0, σ_s, q, n_v, p, g are material constants, Ω is a generlised structural parameter defined as

$$\Omega = \sqrt{L_{\acute{a}} L_{\hat{a}}} \cdot \frac{L_{\acute{a}}}{L_{\hat{a}}}$$

where L_α, L_β are the average grain sizes in the α- and the β-phases respectively.

One can find other variants of the Smirnov model also in the literature. All of them can be reduced to Eq. (3.60) or ones similar to it. Smirnov's model does not describe the intial part of the stress–strain diagram and the load relaxation curves. It can be generalised for the case of non-uniform stress–strain state (see Chap. 4) and is used in Russia for practical calculations. The number of empirical constants in the analysis, however, is rather large.

3.2.5 Anelasticity

There are many reports in the literature [81–84] where the transient behaviour of superplastic materials is described on the assumption that the total strain rate consists of three parts, viz.,

$$\xi = \xi^e + \xi^p + \xi^a \tag{3.65}$$

where $\xi^e = \dot{\sigma} / E$ is the elastic strain rate (E is Young's modulus), ξ^p is the plastic strain rate, and ξ^a is the anelastic strain rate. It is assumed that the plastic strain rate depends on the applied stress as $\xi^p = C\sigma^n$, where C and n are constants[1]. The anelastic strain ε^a is introduced to describe adequately the elastic after-effect present in superplastics. The value of E available in the literature has been used to estimate the elastic compliance J_e [81–84]. It is shown that the anelastic compliance $J_a = -\varepsilon^a/\sigma$, for a superplastic eutectic Sn–Pb alloy was more than two orders of magnitude greater than the elastic compliance, J_e ($J_a/J_e = 130$). This is referred to in [82] as 'superanelasticity'. It has been shown that including anelasticity allows a more adequate description of the initial part of the stress–strain curve.

The mechanical model suggested in [81] to describe the anelastic behaviour is shown in Fig. 3.17. It is evident that the model includes an elastic element J_e, a non-linear viscous element m and three Voigt elements J_{0i} connected in series. Thus, in [81–84] plastic strain is described by a non-linear viscous element while the anelastic strain is described by viscoelastic Voigt elements.The following relationships have been derived.

$$\varepsilon = \varepsilon^e + \varepsilon^{a_1} + \varepsilon^{a_2} + \varepsilon^{a_3} + \varepsilon^p \tag{3.66}$$

$$\sigma = E\varepsilon_e = \frac{\varepsilon_e}{J_e} = K(\xi_p)^m = E_{0i}\varepsilon^{a_i} + \eta_i\xi_i \tag{3.67}$$

Then, from $\sigma = K(\xi^p)^m$ one obtains after some transformations the relation

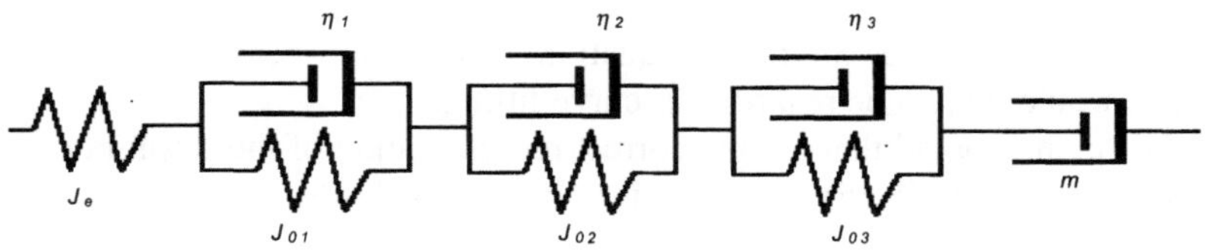

Fig. 3.17. Schneibel's mechanical analogue [81]

[1] It is to be mentioned that in mechanics of solids such a material property is termed as creep rather than plasticity.

$$\begin{aligned}&a_0\sigma^n + a_1\sigma' + a_2\sigma'' + a_3\sigma''' + a_3\sigma'''' + a_5\sigma^{n-1}\sigma' + a_6\sigma^{n-2}\sigma'^2 + \\ &a_7\sigma^{n-3}\sigma'^3 + a_8\sigma^{n-1}\sigma'' + a_9\sigma^{n-2}\sigma'\sigma'' + a_{10}\sigma^{n-1}\sigma''' \\ &\quad = \xi + b_1\xi' + b_2\xi'' + b_3\xi'''\end{aligned} \tag{3.68}$$

where

$$n = \frac{1}{m},\ a_0 = K^{-\frac{1}{m}},\ b_1 = \tau_1 + \tau_2 + \tau_3,\ b_2 = \tau_1\tau_2 + \tau_1\tau_3 + \tau_3\tau_1,\ b_3 = \tau_1\tau_2\tau_3\ ; \tag{3.69}$$

$a_1 = J_e + J_{01} + J_{02} + J_{03}$,
$a_2 = \tau_3(J_e + J_{01} + J_{02} + J_{03}) + \tau_2(J_e + J_{01}) + \tau_1(J_e + J_{02}) + (\tau_1 + \tau_2)J_{03}$,
$a_3 = \tau_1\tau_3(J_e + J_{02}) + \tau_2\tau_3(J_e + J_{01}) + \tau_1\tau_2(J_e + J_{03})$,
$a_4 = J_e\tau_1\tau_2\tau_3$, $a_5 = a_0b_1n$, $a_6 = a_0b_2n(n-1)$, $a_7 = a_0b_3n(n-1)(n-2)$,
$a_8 = a_0b_2n$, $a_9 = 3a_0b_3n(n-1)$, $a_{10} = a_0b_3n$

Schneibel has given an algorithm to solve Eq. (3.68). It is to be mentioned that one has to set up 4 initial conditions (for σ, σ', σ'', σ''') if the loading conditions are determined from the known function $\xi = \xi(t)$. If the loading conditions are given by the known function $\sigma = \sigma(t)$ one has to set up 3 initial conditions (for ξ, ξ', ξ'').

Other generalisations of Eq. (3.68) including that suggested in [81] are possible. It is necessary to note that Eqs.(3.68) and (3.69) are very complicated and involve many material constants. It is clear that such equations can be used to describe a wide range of natural phenomena, including the mechanical response of superplastic materials. However, the procedure for determining the material constants in such models is very complicated. Thus, it is not surprising that two *different* approximations for *the same* experimental curve are suggested (see the caption of Fig. 2 in [81]). Therefore, the procedures suggested do not permit an unequivocal determination of the material constants. Also, these models do not allow a description of the universal superplasticity curve. They also do not account for the load relaxation curve adequately (e.g., see Fig. 7 in [81]). In contrast, if two sets of material constants are permitted – one for the initial part of the $\sigma - \xi$ curve and another for load relaxation – it is possile to account for the results using either the generalised Maxwell solid or Smirnov's equation. (From a physical point of view, however, this would merely be regarded as curve fitting.)

In conclusion, it is noted that a more correct representation of the strain rate using an equation similar to (3.65) has been given by Odqvist[236].

$$\varepsilon = \varepsilon^e + \varepsilon^p + \varepsilon^c \tag{3.70}$$

where ε^e is the elastic strain, ε^p is the plastic strain and ε^c is the creep strain. For active loading conditions one can assume $\varepsilon^p = g(\sigma)$; the creep strain rate depends

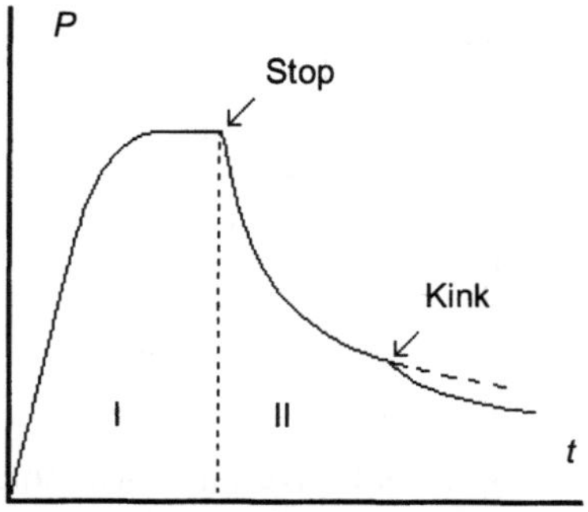

Fig. 3.18. Load relaxation curve with a 'kink': I – Region of plastic deformation; II – Region of load relaxation

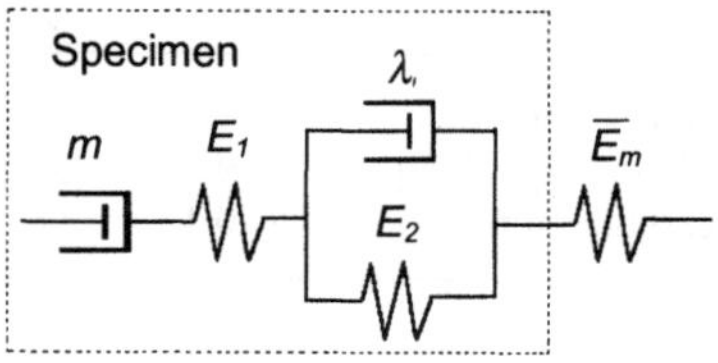

Fig. 3.19. Mechanical analogue for the model of Arrowood and Mukherjee

on stress and temperature as $\xi^c = f(\sigma, T)$. Then;

$$\xi = \frac{\dot{\sigma}}{E} + kg'(\sigma)\,\dot{\sigma} + f(\sigma, T) \tag{3.71}$$

Here, k is of the order of unity for active loading conditions; otherwise $k = 0$.

3.2.6 Kinks on the Load Relaxation Curves

Arrowood and Mukherjee [70] have reported the presence of an unusual 'kink' in the relaxation curve of a Sn–Pb eutectic alloy (Fig. 3.18). (Its genuineness was established by careful experimentation.) To account for this behaviour, the mechanical analogue shown in Fig 3.19 is suggested. The testing machine is modelled by an elastic element E_m, and the specimen by a series connection of a non-linear viscous element m, an elastic element E_1 and a Voigt element (a linear viscous element λ connected in parallel to an elastic element E_2).

For this combination of 'specimen–testing machine' the following relations can be derived.

$$\xi = \frac{\dot{\sigma}}{E_m} + \frac{\dot{\sigma}}{E_1} + C\sigma^n + \xi_V, \quad \sigma = E_2\varepsilon_V + \lambda\xi_V \tag{3.72}$$

where ε_V, ξ_V are respectively the strain and strain rate for the Voigt element. ξ is an apparent strain rate which is equal to (see also Chap. 2; Eqs. (1.18) and (2.72))

$$\xi = \left[\frac{v}{L} - \frac{\sigma A}{XL}\right] \cdot \left[1 - \frac{\sigma A}{XL}\right]^{-1} \tag{3.73}$$

Table 3.5 Characteristics of superplastic flow micromechanisms

Micromechanism	m_i	Q_i	Q_i, kJ/mol for Pb–Sn [5]	β_{iq}
Diffusion creep	1	Q_{DC}	104	0.25
Grain boundary sliding	0.5	Q_{GBS}	40	0.50
Intragranular slip	0.1	Q_{IS}	104	0.25

In general, the system of Eqs. (3.72) and (3.73), do not have analytical solutions. Therefore, it has been solved using numerical methods. When σ = constant (a creep test), this system of equations has the following solution satisfying the initial condition $\xi(0) = 0$, viz.,

$$\xi = \xi_0 \left[1 - \exp\left(-\frac{t}{\tau_V} \right) \right] \tag{3.74}$$

where $\tau_v = \lambda / E_2$ is the characteristic time for the Voigt element and $\xi_0 = C\sigma^n$ is the strain rate in the stable region of deformation. Thus, the strain rate will be constant in the constant stress regime of loading; the strain rate being defined by the non-linear viscous element. One can also show that the system of equations has an analytical solution for an arbitrary regime of loading $\sigma = \sigma(t)$. However, when the kinematics of the process is given, the response can be found by numerical procedures only.

Arrowood and Mukherjee [70] explain the kink in the stress relaxation curve by arbitrarily changing n from 2 to 3 at the kink. Earlier experiments by Murty [69] and Geckinli and Barrett [145] did not reveal evidence for the presence of a kink in a stress relaxation test. So, it is debatable if the observation was merely an artefact. In careful relaxation experiments on the Supral-100 alloy (Al–6 wt.% Cu–0.4 wt.% Zr), the present authors observed perceptible kinks in 4 tests, mild kinks in 4 others and no kink at all in 7 other experiments.

The following explanation is offered. It is not necessary to invoke such a complicated model like the one used by Arrowood and Mukherjee [70] if one can change the values of the material constants depending on the loading conditions. Even a simple equation like Eq. (3.19) can explain the kink in the stress relaxation curve. In this case, at the kink it would be necessary to change the value of E. In an Al-alloy like Supral-100, the presence of solute elements like Cu and Mg gives rise to serrated flow arising from the phenomenon of dynamic strain ageing [345]. This is a mathematically well treated problem in dislocation dynamics. When the impurities are anchored to dislocations the force required to detach the dislocations for causing plastic flow would be greater than when they are free of the solute atmosphere. Hence, the change in E and the kink in the relaxation curve. In fact, Haasen [345] has noted that the phenomenon has a strong effect on anelasticity. In this view, the Pb–Sn alloy used by Arrowood and Mukherjee [70] contained some solute elements responsible for dynamic strain ageing, whereas the Pb–Sn alloy used by Murty [69] and Geckinli and Barrett [145] was free of such impurities. In the experiments of the present authors also in the specimens in which the kink was sharp, the solute elements concentration was more than in those in which the kinks were milder. Further experiments are necessary for unambiguous conclusion.

3.2.7 Mechanistic Model

The phenomenology of grain boundary sliding controlled optimal structural superplastic flow [161, 162] leads to the following equation

$$\ln \xi = C + A \ln \sigma - B\sigma \tag{3.75}$$

where A, B, C are constants. Regression procedures have been developed in [161] to calculate A, B, C for a number of superplastic systems. It is noted that Eq. (3.75) can be considered as a specific form of Odqvist's Eq. (3.71). One has to assume that $\xi \equiv \xi^c$ (*i.e.*, $\xi^e = \xi^p \equiv 0$). Then, one can predict the strain rates of deformation accurately. Some of the predictions of the analysis have been compared with those of the more detailed physical model due to Padmanabhan and Schlipf [138], discussed later. This phenomenological analysis also should be examined from the viewpoint of mechanics as was done recently in case of the physical model [105].

3.2.8 Activation Energies

A number of physical models of superplasticity (see Sect. 3.3) assume that the total strain rate can be expressed as a sum of the strain rates due to different micromechanisms (see Sect. 3.2.3.6, Eq. (3.43)). Specific activation energy (AE) and strain rate sensitivity index values are assigned to each micromechanism (see Table 3.5). From the viewpoint of mechanics, Eq. (3.44) can be considered as a series connection of three non-linear viscous elements, as shown in Fig. 3.13a. The mechanical properties of such combinations were considered in Sect. 3.2.3.6 It is of interest to calculate the resulting activation energy for such a chain. The definitions of activation energy were presented in Chap. 1.

For a series connection of the viscous elements shown in Fig.3.13a the properties of the i-th element are described by the following equation

$$\xi_i = \hat{C}_i \sigma^{n_i} \cdot \exp\left(-\frac{Q_i^\sigma}{RT}\right) \tag{3.76}$$

Then, Eq. (3.44) can be rewritten as

$$\xi = \xi_1 + \xi_2 + ... + \xi_k = \sum_{i=1}^{k} C_i \sigma^{n_i} = \beta_1 \xi + \beta_2 \xi + ... + \beta_k \xi \tag{3.77}$$

where $\beta_i = \xi_i/\xi$ is the relative contribution of the i-th mechanism, $i = 1, 2, 3$; indexes 1, 2, 3 refer to diffusion creep (DC), grain boundary sliding (GBS) and intragranular slip (IS) respectively.

Let $\sigma_q = \sigma(\xi_q)$ be a typical value of the flow stress, at a temperature of deformation $T = T_q$. Let σ_q, ξ_q, T_q be the *reference values of the three variables.* Dividing (3.77) by ξ_q one obtains

$$\frac{\xi}{\xi_q} = \sum_{i=1}^{k} \frac{\hat{C}_i \sigma^{n_i}}{\xi_q} \exp\left(-\frac{Q_i^\sigma}{RT}\right) = \sum_{i=1}^{3} \frac{\hat{C}_i \sigma_q^{n_i} \exp\left(-\frac{Q_i^\sigma}{RT_q}\right)}{\xi_q} \left(\frac{\sigma}{\sigma_q}\right)^{n_i} \exp\left(-\frac{Q_i^\sigma}{R\theta}\right) \tag{3.78}$$

Introducing the notations

$$\bar{\sigma} = \frac{\sigma}{\sigma_q}, \; \bar{\xi} = \frac{\xi}{\xi_q}, \; \frac{1}{\theta} = \frac{1}{T} - \frac{1}{T_q}, \; \beta_{iq} = \frac{\xi_{iq}}{\xi_q}$$

one can rewrite Eq. (3.78) as

$$\bar{\xi} = \sum_{i=1}^{3} \beta_{iq} \bar{\sigma}^{n_i} \exp\left(-\frac{Q_i^\sigma}{R\theta}\right) \tag{3.78a}$$

The relative contribution of the i-th mechanism depends on T, ξ and σ. One can then obtain that

$$\beta_i = \frac{\xi_i}{\xi} = \beta_{iq} \frac{\bar{\sigma}^{n_i}}{\bar{\xi}} \exp\left(-\frac{Q_i^\sigma}{R\theta}\right) \tag{3.79}$$

Regime of loading $\sigma = \sigma_q =$ *constant*: In this case $\bar{\sigma} \equiv 1$. One can find from Eq. (3.78a) that

$$\bar{\xi} = \sum_{i=1}^{3} \beta_{iq} \exp\left(-\frac{Q_i^\sigma}{R\theta}\right) \stackrel{def}{=} \exp\left(-\frac{Q_f^\sigma}{R\theta}\right)$$

Then,

$$Q_f^\sigma = -R\theta \ln\left[\sum_{i=1}^{3} \beta_{iq} \exp\left(-\frac{Q_i^\sigma}{R\theta}\right)\right] \tag{3.80}$$

The value of Q_a^σ can be found by substituting from Eq. (3.78).

$$Q_a^\sigma = \frac{\sum_{i=1}^{3} \beta_{iq} Q_i^\sigma \exp\left(-\frac{Q_i^\sigma}{R\theta}\right)}{\sum_{i=1}^{3} \beta_{iq} \exp\left(-\frac{Q_i^\sigma}{R\theta}\right)} \tag{3.81}$$

Contributions $\beta_i = \xi_i/\xi$ are determined by Eq. (3.79), with $\bar{\sigma} \equiv 1$.
Regime of loading $\xi = \xi_q$ *constant*: In this case $\bar{\xi} = 1$ and Eq. (3.78a) becomes

$$\sum_{i=1}^{3} \beta_{iq} \bar{\sigma}^{n_i} \exp\left(-\frac{Q_i^\sigma}{R\theta}\right) \equiv 1 \tag{3.82}$$

This equation gives implicitly the value of $\bar{\sigma}$ which can be used to find Q_a^ξ and Q_f^ξ. The value of Q_a^ξ can be found (Chap.1) as

$$Q_a^\xi = \left[\sum_{i=1}^{3} \beta_{iq} \bar{\sigma}^{n_i} Q_i^\sigma \, exp\left(-\frac{Q_i^\sigma}{R\theta}\right)\right]\left[\sum_{i=1}^{3} \beta_{iq} \bar{\sigma}^{n_i} n_i \, exp\left(-\frac{Q_i^\sigma}{R\theta}\right)\right]^{-1} \tag{3.83}$$

where $\bar{\sigma}$ is from Eq.(3.82). From Sect.1.4. $\bar{\sigma} = \exp\left(Q_f^\xi / R\theta\right)$ and substituting $\bar{\sigma}$ in (3.82) the following transcendental equation as to the unknown Q_f^ξ is obtained.

$$\sum_{i=1}^{3} \beta_{iq} \exp\left(\frac{n_i Q_f^\xi - Q_i^\sigma}{R\theta}\right) = 1 \tag{3.84}$$

Contributions $\beta_i = \xi_i/\xi$ are determined by Eq. (3.79), where $\bar{\xi} = 1$. For practical calculations the lead–tin eutectic alloy was selected. In this case, according to [5] the activation energies for lattice and grain boundary diffusion are as follows: Q_L = 104 kJmol^{-1}, Q_{gb} = 40 kJmol^{-1}. The relative contributions of the different mechanisms were taken as $\beta_{1q} = 0.25$, $\beta_{2q} = 0.5$, $\beta_{3q} = 0.25$, with $T = T_q = 300$ K. The values of m_i, Q_i and β_{iq} used in the calculations are given in Table 3.5. The results are presented in Figs 3.20–3.23. Theoretical curves $\log\bar{\sigma} - \log\bar{\xi}$ and the dependence of their slope $M = \partial\log\bar{\sigma} / \partial\log\bar{\xi}$ on the normalized strain rate $\bar{\xi}$ were

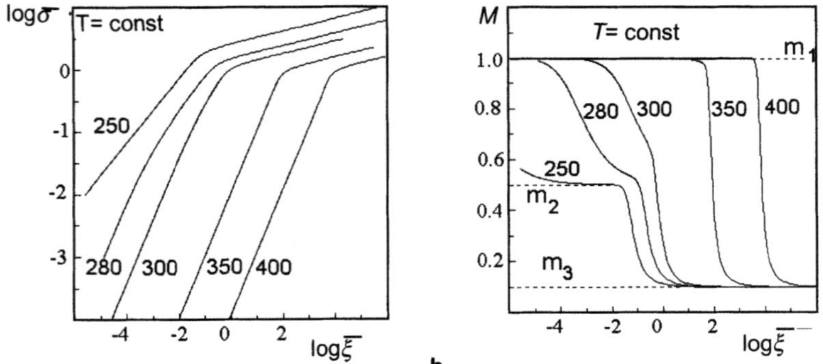

a b

Fig. 3.20. **a** log $\bar{\sigma}$ –log $\bar{\xi}$ plots and **b** strain rate dependence of M, calculated using Eq. (3.78a) for a eutectic Pb–Sn alloy tested at various temperatures (indicated by the numbers near the curves)

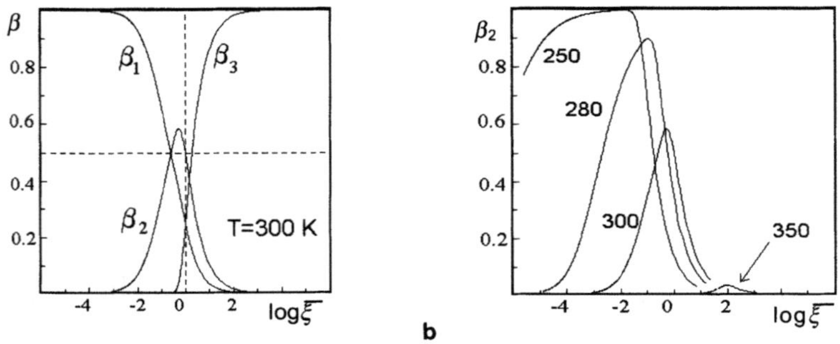

a b

Fig. 3.21. **a** Contributions of β_i vs. normalized strain rate $\bar{\xi}$, calculated according to Eq. (3.79) at $T = 300$ K; **b** for different temperatures T = constant (indicated near the curves)

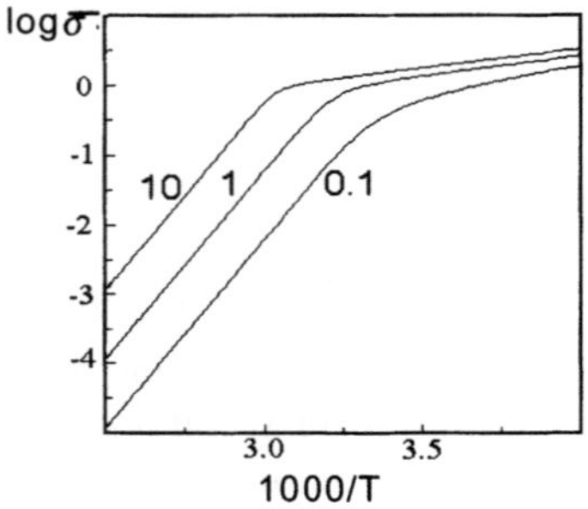

Fig. 3.22. Curves calclulated according to Eq. (3.82) with various $\bar{\xi}$ = constant values

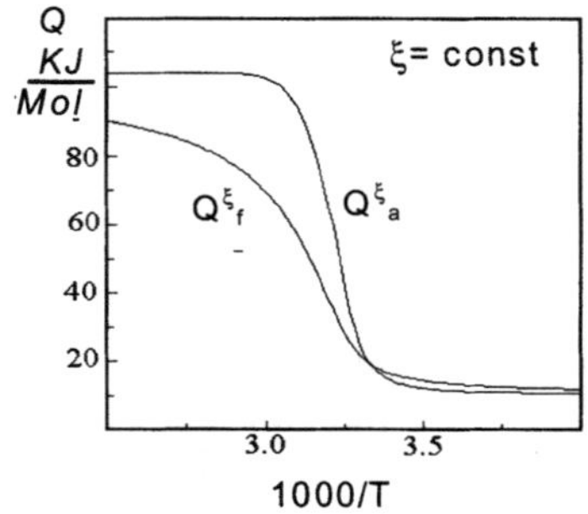

Fig. 3.23. Temperature dependence of activation energies (Pb–Sn eutectic alloy) calcluated according to Eqs. (3.83) and (3.84)

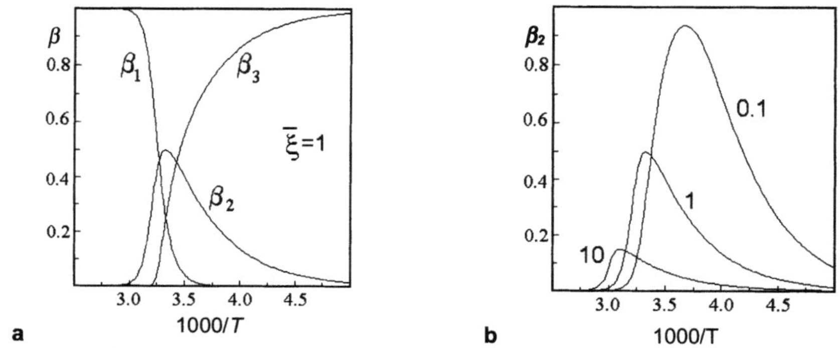

Fig. 3.24. Temperature dependence of β_i, calculated according to Eq. (3.79) **a** with $\bar{\xi} \equiv 1$ **b** with different $\bar{\xi}$ (indicated near the curves)

calculated in accordance with Eq. (3.78a), see Fig. 3.20. The curves $\log\bar{\sigma} - \log\bar{\xi}$ are convex and M decreases with increasing $\bar{\xi}$. Thus, as was already mentioned, a model based on summing the contributions of micro-mechanisms to obtain the total strain rate, according to Eq. (3.44), does not describe the sigmoidal variation of the log flow stress with log strain rate.

The strain rate dependences of the contributions β_i are presented in Fig. 3.21. It is clear that DC dominates at low ξ while IS makes its main contribution at large ξ. The contribution of GBS is maximum at intermediate strain rates. It appears from Fig. 3.21 that the contribution from GBS reaches its maximum value at a lower strain rate as the temperature is lowered. Temperature dependence of flow stress for constant strain rate loading is given in Fig.3.22. The log $\bar{\sigma} - 1/T$ curves shift towards higher stresses with increasing strain rate. Another important feature is that these curves are convex. Temperature dependence of the activation energies Q_a^{ξ} and Q_f^{ξ} calculated for the Pb–Sn eutectic alloy, with $\bar{\xi} \equiv 1$, are presented in Fig. 3.23. The apparent activation energy at constant strain rate decreases with decreasing temperature (in the range considered).

Temperature dependence of the contributions β_i is presented in Fig. 3.24 ($\bar{\xi} \equiv 1$). DC dominates at high temperatures, while IS dominates at low T. The contribution from GBS is maximum at intermediate temperatures. Temperature dependence of GBS contribution β_2 is presented in Fig. 3.24, for different constant strain rates. In the range studied β_2 increases with decreasing ξ. Temperature dependence of strain rate at constant stress is presented in Fig. 3.25. $\mathrm{Log}\,\bar{\xi} - 1/T$ curves are concave and they are shifted to higher strain rates with increasing stress. Temperature dependence of activation energies Q_a^{σ} and Q_f^{σ} are presented in Fig. 3.26. The apparent activation energies at constant stress decrease with decreasing temperature. Temperature dependence of the contributions β_i are presented in

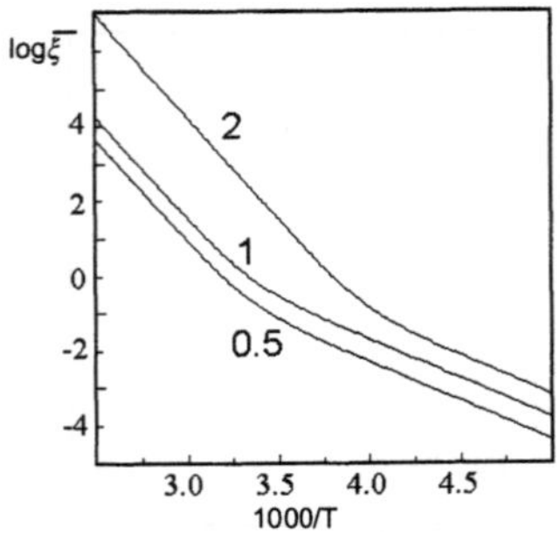

Fig. 3.25. Curves calculated according to Eq. (3.78) with different $\bar{\sigma}$ = constant (numbers near the curves)

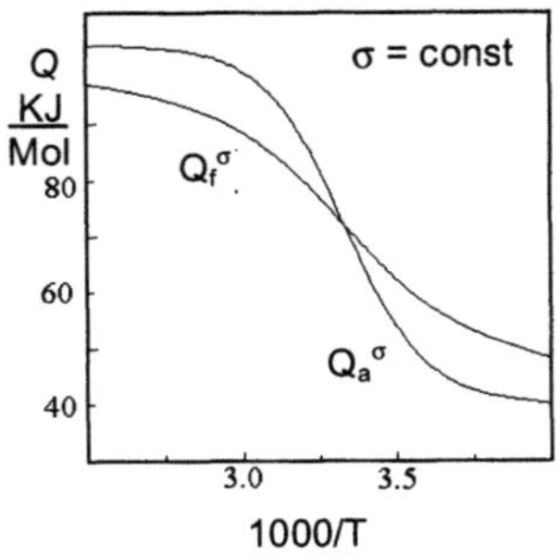

Fig. 3.26. Temperature dependence of activation energies for the lead–tin eutectic alloy, calculated according to Eqs. (3.80) and (3.81) with $\bar{\sigma} \equiv 1$

Fig. 3.27(a). Contributions from DC and IS coincide, while the GBS contribution increases with decreasing temperature (as ξ at a given stress would also decrease). Temperature dependence of GBS contribution β_2 is presented in Fig.3.27(b). β_2 decreases with increasing temperature at a given stress and this observation has no experimental support.

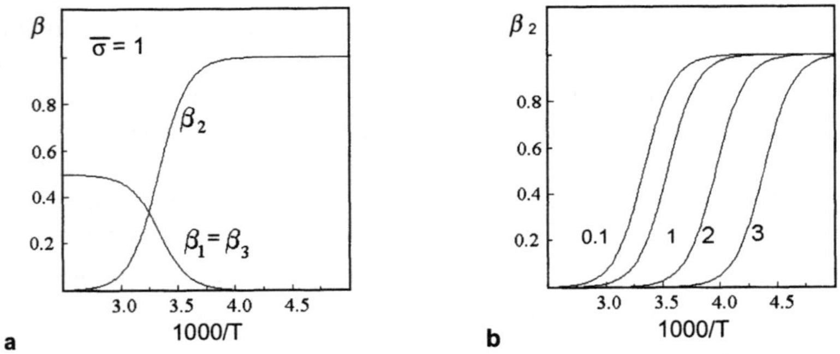

Fig. 3.27. Temperature dependences β_i, calculated according to equation (3.79) with **a** $\bar{\sigma}$ =1 **b** with different $\bar{\sigma}$ = constant (numbers near the curves)

Table 3.6 Qualitative character of different dependences in case of series and parallel combinations of non-linear viscous elements [346] with the other variable(s) constant

Connection	$\log\bar{\sigma} - \log\bar{\xi}$	$\mathrm{Log}\,\bar{\xi} - 1/T$	$\log\bar{\sigma} - 1/T$	$Q - 1/T$	M–$\log\bar{\xi}$
Series	Convex	Concave	Convex	Decreasing	Decreasing
Parallel	Concave	Convex	Concave	Increasing	Increasing

It is important to note that the activation energy obtained in the analysis is merely a phenomenological parameter and should not be identified with any single physical process (as many mechanisms have been combined to write down the constitutive equation). As was pointed out earlier, such a series combination of micromechanisms can not also account for the sigmoidal $\log\sigma$–$\log\xi$ curve.

If the constitutive equation relating the steady state stress to the strain rate and temperature can be written as $\Phi(\sigma,\xi,T)=0$, sigmoidal superplastic curves $\log\sigma$–$\log\xi$ can be obtained as intersections of the plane T = constant. Then, the concept of temperature sensitivity can be introduced. The plane σ = constant gives rise to the $\xi - T$ curve, while ξ = constant plane is responsible for the $\sigma - T$ curve. Thus, *two* independent temperature sensitivity values can be introduced for a superplastic material. One can then introduce different parameters characterising temperature sensitivity (see Sect. 1.4.4). It is emphasised that two different parameters can be introduced at any given point on the surface $\Phi = 0$. For example, one can use apparent activation energies Q_a^{ξ} and Q_a^{σ} to do this. The difference between these two parameters can be quite significant and so it should be considered while analyzing the mechanical response of superplastic materials. Physical models of superplasticity which are based on Eq. (3.44) have been considered above in detail. Temperature dependence of mechanical properties arising from parallel and mixed connections of non-linear viscous elements can also be examined. The analysis is rather extended and so it is not presented here. The main results are given in Table 3.6. Evidently, mixed combinations will give intermediate results.

The foregoing, however, should serve a note of caution to the physicists/metallurgists about multi-mechanisms. By suitably combining these in series, parallel or mixed modes, many properties of superplastic alloys can be simulated. This does not necessarily mean that those mechanisms are present in reality. Experimental verification is essential to validate each of the micromechanisms assumed.

3.3 Physical Constitutive Equations

In this section, physical models of superplastic flow, which include microstructural parameters (e.g., Burgers vector, grain boundary sliding displacement) and introduce them directly into the CE are considered. A discussion of the validity and/or the meaning of the physical parameters is beyond the scope of this book. The present treatment is based on the common requirements of the general theory of constitutive equations presented in Sect. 3.1. Initially, the models that do not include kinetic equations for the structural parameters are examined. Subsequently, the physical models involving kinetic equations for the physical parameters are discussed.

3.3.1 Classical Models

Many early models of superplasticity can be reduced to the following form.

$$\xi = A\frac{(\sigma - \sigma_0)^{n'}}{d^q}\exp\left(-\frac{Q}{RT}\right) \tag{3.85}$$

where Q is an activation energy, R the gas constant, d the average grain size, A a temperature dependent material function, σ_0 a threshold stress, and n' and q are constants.

Ashby and Verrall [347], Alden [348], Mukherjee [349], Gifkins [350], Shorshorov et al. [351], Hayden et al. [352], Coble [353], Nabarro [354], Herring [355] Valiev–Kaibyshev [5, 329], Padmanabhan [162], and others have presented physical models that fall in this category. Detailed reviews of these physical models are available in the literature, see, e.g., [1, 3, 6]. From a mechanical point of view, all these models can be analysed in terms of what is given in Sect. 3.2.3, with m being assumed as constant. There are a few other physical models that cannot be reduced to Eq.(3.85). Avery and Backofen [356] have suggested a model that leads to a CE

$$\xi = \frac{A\sigma}{d^2} + B\, sinh(\beta\sigma) \tag{3.86}$$

where $A,\ B,\ \beta$ are material constants. Hence,

$$M_{AB} = \frac{1}{\dfrac{d\ln\xi}{d\ln\sigma}} = \frac{\left(\dfrac{A}{d^2}\right)\sigma + B\sinh(\beta\sigma)}{\sigma\left[\left(\dfrac{A}{d^2}\right) + \beta B\cosh(\beta\sigma)\right]} \tag{3.87}$$

Packer and Sherby [357] have criticized this model and have suggested an alternative CE.

$$\xi = \frac{A'\sigma^2}{d^3} + B'\sigma^2\, sinh\left(\beta'\sigma^{2.5}\right) \tag{3.88}$$

where A', B', β' are material constants. Subsequently this CE was modified to use d^2 instead of d^3 on the right hand side of Eq. (3.88). The slope of the sigmoidal curve in accordance with Eq. (3.88) is

$$M = \frac{\left(\frac{A'}{d^3}\right)\sigma^2 + B'\sigma^2 \sinh\left(\beta'\sigma^{2.5}\right)}{2\left(\frac{A'}{d^3}\right)\sigma^2 + 2.5\beta'B'\sigma^{4.5}\cosh\left(\beta'\sigma^{2.5}\right) + 2B'\sigma^2 \sinh\left(\beta'\sigma^{2.5}\right)} \tag{3.89}$$

It was found [357] that Eq. (3.89) was better than Eq. (3.87). But the general usefulness of this model is very limited and the analysis also is not rigorous.

Valiev and Kaibyshev [5, 329] have suggested a CE which can be written as

$$\xi = \frac{\dot{\sigma}}{E^*} + \frac{\Omega D}{16kTG} \cdot \left(\frac{\sigma - \sigma_0}{d}\right)^2, \quad \sigma > \sigma_0 \tag{3.90}$$

where σ_0 is a threshold stress, Ω the volume of a vacancy, k the Boltzmann constant, G the shear modulus, D the coeficient of grain boundary diffusion and E^* is the effective modulus of the system 'specimen–testing machine'[2]. For steady state flow, Eq. (3.90) is written as

$$\sigma = \sigma_0 + d\sqrt{\frac{16kTG\xi}{\Omega D}} \tag{3.91}$$

and so

$$M_{VK} = \frac{\partial \log \sigma}{\partial \log \xi} = \frac{\sigma - \sigma_0}{2\sigma} \tag{3.92}$$

From Eq. (3.92) it follows that M increases monotonically with σ from 0 (when $\sigma = \sigma_0$) to 0.5 (as $\sigma \to \infty$). (It is interesting to note that as early as 1934, Cook [358] showed that the experiemental results of Pearson [34] can be described by the equation

$$\sigma = \sigma_0 + C\sqrt{\xi} \tag{3.93}$$

where σ_0 and C are constants. Equation (3.91) can be obtained from Eq. (3.93) if one assumes that $c = d\sqrt{16kTG/\Omega D}$, d, T and Ω are constants.)

For conventional deformation, the well-known Hall–Petch relationship is valid.

$$\sigma_d = \sigma_0 + Kd^{-\frac{1}{2}} \tag{3.94}$$

[2] It should be noted that it is not quite correct to use a characteristic of a testing machine in a CE. Material properties should not depend on the testing machine used. This comment is valid for some other models also.

Equations (3.91) and (3.94) have been used to estimate the critical grain size d_{cr} at which a transition from conventional to superplastic flow is present, so that.

$$d_{cr} \approx \sqrt[3]{K^2 \frac{D\Omega}{16\xi GkT}} \tag{3.95}$$

Numerical estimate for the Zn–0.4%Al alloy, assuming suitable values for K, D, Ω, G, k and T gave a value of d_{cr} = 5–12 μm which is in good agreement with experimental results.

This method has two limitations: (a) the variation of m with logξ is not predicted correctly; (b) the method of evaluating the constants is not given so that it is impossible to verify the stability and reliability of the calculations. If d is treated as a constant[3], the following phenomenological equation will correspond to the Valiev–Kaibyshev model.

$$\xi = \frac{\dot{\sigma}}{E^*} + \mathrm{A} \cdot (\sigma - \sigma_0)^2, \quad \sigma > \sigma_0 \tag{3.96}$$

where E^*, A and σ_0 are material constants. Equation (3.96) can be considered as a particular case of Eq. (3.40).

3.3.2 Modern Theories

In this section physical models for structural superplasticity proposed within the last 10 years will be considered.

3.3.2.1 Model of Ghosh

Ghosh [359] has suggested a physical model that extends the 'core and mantle' concept. The CE can be written as

$$\xi = \xi_m + \xi_c = \frac{A}{d^2}(\sigma - \sigma_0)^{1+q} + \left(1 + \frac{A_1}{d^3}\right)\sigma^n \tag{3.97}$$

$$d = d_0 + \alpha t^p + \beta \varepsilon_m \tag{3.98}$$

where d_0 is the initial grain size, σ_0, A, q, A_1, n, α, p and β are material constants, suffixes 'm' and 'c' refer to mantle and core respectively. The five constants in Eq. (3.97) are evaluated from macro-experiments (($\sigma - \xi$) plots). The 3 constants of

[3] Otherwise one obtains a CE, which is not closed

Eq. (3.98) are calculated from micro-experiments (d–t–ξ relationship). With such an over-prescription, it was easy to get a good fit with the experimental data concerning two Al–alloys.

Assuming $\varepsilon_m = \xi_m t$, Eq. (3.98) can be written as $d = F_1(t, \sigma)$ for ξ = constant. Substituting $d = F_1(t, \sigma)$ on the right hand side of Eq. (3.97), one obtains an expression $\xi = F_2(\sigma, t)$ containing 8 independent constants plus the initial grain size d_0. Thus, from the viewpoint of mechanics, Ghosh's model is a specific form of the equation $\xi = F(\sigma, t)$, where F is a function (not a functional) that includes the constants. The difference between this model and those described by Eqs. (3.10) is that time t is included here implicitly. The relationship $\xi = F(\sigma, t)$ is widely used in the theory of creep (see Chap. 2).

It is not surprising that a model containing 9 constants can describe satisfactorily a restricted set of experimental data (8 experimental points for alloy 1 and 6 for alloy 2 were used). Methods for evaluating the constants in a general way are not given in [359], i.e., no crucial experiments have been suggested. So, the range of applicability of the equations is undefined.

3.3.2.2 Model of Hamilton

Hamilton [360] has suggested a model where the CE has the form

$$\xi = \frac{K_{II}(\sigma - \sigma_0)^{\frac{1}{m}} \exp\left(-\frac{Q_s}{RT}\right)}{d^p kT} + \frac{K_{III}\sigma^n \exp\left(-\frac{Q_c}{RT}\right)}{kT} \tag{3.99}$$

with Q_s the activation energy for superplastic flow, Q_c the activation energy for creep, σ_0 a threshold stress, n, p, m, K_{II} and K_{III} material constants, k the Boltzmann constant and R the gas constant. To describe grain size evolution, Hamilton has used the results of Wilkinson and Caceres [361, 362], viz.,

$$\dot{d} = \dot{d}_a + \frac{\lambda d \xi \dot{d}_u}{\lambda d \xi + \dot{d}_u} \tag{3.100}$$

where $\dot{d}_u$ is the upper limit of the rate of grain growth, λ is a numerical coefficient ($\lambda \approx 1$)and $\dot{d}_a$ is the rate of static grain growth. After some transforms Hamilton obtains the following relation

$$\dot{d} = d \cdot \left[\lambda \xi + \frac{C}{q}\left(d_0^q + Ct\right)^{-1}\right] \tag{3.101}$$

where d_o is the initial grain size. Integration leads to

$$d(t) = \exp\left[\frac{C}{q}\ln\left(d_0^q + Ct\right)\right] + \int_0^t \lambda\xi(t)dt \tag{3.102}$$

where C, λ, q are material constants. It is to be noted that Eq. (3.102) does not satisfy the initial condition $d(0) = d_0$ if $C \neq 1$.

Combining Eqs. (3.99) and (3.101) one can obtain a relation between σ and ξ.

$$\xi = \frac{K_{II}(\sigma - \sigma_0)^{\frac{1}{m}} exp\left(-\frac{Q_s}{RT}\right)}{kT\left\{exp\left[\frac{C}{q} ln\left(d_0^q + Ct\right) + \int_0^t \lambda\xi dt\right]\right\}^p} + \frac{K_{III}\sigma^n exp\left(-\frac{Q_c}{RT}\right)}{kT} \tag{3.103}$$

In a later modification based on finite strain behaviour [279], the following form has been suggested.

$$\xi = \frac{K_{II}(\sigma - \sigma_0)^{\frac{1}{m}} exp\left(-\frac{Q_s}{RT}\right)}{d^a} + K_{III}\sigma^n exp\left(-\frac{Q_c}{RT}\right) \tag{3.104}$$

In Eq. (3.99) or in Eqs. (3.104) and (3.101) there are 8 constants to be obtained from macro-experiments and 3 from micro-experiments. No general methods (crucial experiments) for evaluating the constants have been given. The range of applicability of the equations is not spelt out. Fitting experimenal data in a limited range of experiments using 11 constants is of limited significance.

3.3.2.3
The Model of Pschenichniuk–Astanin–Kaibyshev

A model for superplasticity in terms of grain boundary sliding stimulated by intragranular slip has been suggested recently [363–365]. This model uses the concept of cooperative grain boundary sliding [366–370]. This model will be presently considered with the intention of using the CE for practical calculations.

It is assumed that the total strain rate equals the sum of diffusion creep rate (ξ_1), co-operative grain boundary sliding (CGBS) rate (ξ_{CGBS}) and intra-granular slip rate *(*ξ_2).

$$\xi = \xi_1 + \xi_{CGBS} + \xi_2 \tag{3.105}$$

where

$$\xi_1 = 7\xi_0\left(\frac{b_L}{d}\right)^2 \frac{\tau}{\mu_0}\left(\frac{D_L}{D_{GB}} + \pi\frac{b_L}{d}\right), \quad \xi_0 = \mu b_L \frac{D_{GB}}{kT} \tag{3.106}$$

$$\xi_2 = C\mu_0 b_L \frac{D_L}{kT}\left(\frac{\tau}{\mu_0}\right)^n P_D(\tau,d) \tag{3.107}$$

$$\frac{\xi_{CGBS}}{\xi_0} = \frac{\langle M\rangle}{M_0}\cdot\frac{C_G}{4}\frac{b_B}{b_L}\left(\frac{\tau}{\mu_0}\right)^2\left[\frac{b_L}{d}p_B(\tau,d)+\frac{\tau}{\mu_0}e^{D^2}p_D(\tau,d)\right] \tag{3.108}$$

$$\frac{\langle M\rangle}{M_0} \cong \frac{1}{2}\left\{1-\Phi\left[\frac{\sqrt{n_1}(p_c-p_b)}{\sqrt{2p_b(-p_b)}}\right]\right\}, \quad \Phi(x)=\frac{2}{\sqrt{\pi}}\int_0^x \exp\left(-y^2\right)dy$$

$$C_G = \frac{1}{10}\left(\frac{b_L}{b_B}\right)^3 \frac{\sin\varphi}{\cos^2\varphi}, \qquad D^2 = ln\frac{\langle d^2\rangle}{\langle d\rangle^2};$$

$$p_B(\tau,d)=\frac{1}{2}\left[\Phi\left(\frac{1}{\sqrt{2D}}\ln\frac{L_D\exp\left(\frac{D^2}{2}\right)}{d}\right)-\Phi\left(\frac{1}{\sqrt{2D}}\ln\frac{L_{LM}\exp\left(\frac{D^2}{2}\right)}{d}\right)\right]$$

$$p_D(\tau,d)=\frac{1}{2}\left[1-\Phi\left(\frac{1}{\sqrt{2D}}\ln\frac{L_D(\tau)\exp\left(D^2/2\right)}{d}\right)\right]$$

Here b_L, b_B are, respectively, the absolute values of the Burgers vectors of the lattice and grain boundary dislocation, D_L, D_{GB} the coefficients of lattice and grain boundary diffusion, $\mu_o = \mu/2\pi(1-\nu)$ with μ the shear modulus, d the average grain size, τ the shear stress, C a constant, k the Boltzmann constant, L_{LM} and L_D the upper and the lower limits of grain size when local migration is present, $< M >$ is the average number of CGBS bands, $M_0 = \max < M >$, n_1 the number of grain boundaries per CGBS band and p_c, p_b are constants ($0 < p_b < 1$, $0 < p_c < 1$).

The following remarks can be made with respect to the form of Eq. (3.105). If one were to assume all the coefficients, parameters and the average grain size, d, to be of constant value, then this system of equations would be closed and can be reduced to a particular case of Eq. (3.10). However, it should be stated that the microstructure changes in the transient region; $d(t)$ is not defined. Grain size

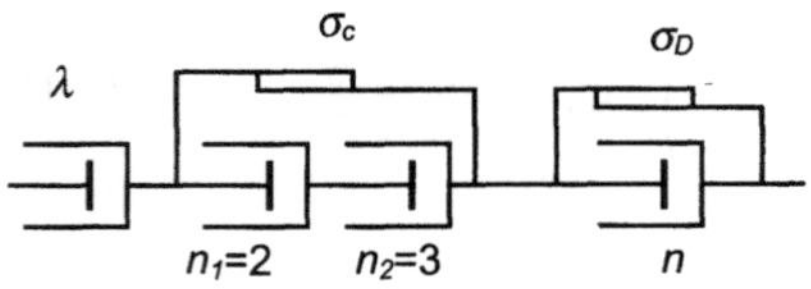

Fig. 3.28. Mechanical analogue for the Pschenichniuk–Astanin–Kaibyshev model

distribution is characterised by the values of d, $< d >$ and $< d^2 >$. Need for calculation of D becomes unnecessary if D is taken to be a constant, as is done in the numerical verification of the model.

Methods for evaluating the physical constants (either experimental or analytical) are yet to be suggested. As the constants for commercial alloys are not likely to be readily available in the literature, the application of the model in metal forming will not be possible. The range of applicability of the equations can not also be defined. Multi-mechanisms like Eq. (3.105), where the magnitude of one mechanism depends on that of the other, should be validated by physical experiments. This is yet to be done.

An analysis of the system of equations (3.105) enables one to suggest the mechanical analogue presented in Fig. 3.28. At low stresses only the element λ will operate (diffusion creep). When $\sigma > \sigma_C$, CGBS starts to operate. Finally, when $\sigma > \sigma_D$ intragranular slip becomes increasingly dominant. Therefore, the CE becomes ($\sigma_c < \sigma_D$)

$$\xi = \begin{cases} \sigma / \lambda, & \sigma < \sigma_C \\ \sigma / \lambda + C_1(\sigma - \sigma_c)^2 + C_2(\sigma - \sigma_c)^3, & \sigma_C \le \sigma < \sigma_D \\ \sigma / \lambda + C_1(\sigma - \sigma_c)^2 + C_2(\sigma - \sigma_c)^3 + C(\sigma - \sigma_D)^n, & \sigma \ge \sigma_D \end{cases} \tag{3.109}$$

where λ, C, n, C_1, n_1, C_2, n_2, σ_C and σ_D are material constants. Thus, even the simplified form of the model consists of 9 material constants. Also, it is a model of a liquid, as $\sigma(0) = 0$. This model can be transformed into a model for a solid if the element λ (Fig. 3.28) is excluded from it. Equation (3.109) belongs to the type of Eqs. (3.10) and so it can not be used to describe the transient loading regimes.

3.3.2.4
The Model of Perevezentsev et al.

Perevezentsev et al. [277] have developed the following relationships based on physical considerations. The total strain rate is assumed to be the sum of intragranular strain rate ξ_v, grain boundary strain rate ξ_b and diffusion creep strain rate ξ_{cr}.

$$\xi = \xi_v + \xi_{cr} + \xi_b \tag{3.110}$$

From physical considerations the following equations are additionally derived.

$$\xi_b = \frac{\xi_s \xi_a}{(\xi_s + \xi_a)}, \quad \xi_a = \frac{A' D_b \sigma^2}{d^2}; \frac{1}{\xi_s} = \frac{(r_d \rho_b)^2}{\xi^1} + \frac{(1 - r_d \rho_b)^2}{\xi^p}$$

$$\xi_v = B' D_b (\sigma - \sigma_0^v)^4; \xi_{cr} = \frac{A_{cr} D_b}{a^3 (\sigma - \sigma_0^{cr})}; \xi^1 = \left[\frac{r_d \rho_b \delta}{\mu_1 d}\right] (\sigma^1 - \sigma_0^1)$$

$$\xi^p = \left[\frac{(1 - r_d \rho_b)\delta}{\mu_p d}\right] (\sigma^p - \sigma_0^p); \rho_b = \left(\frac{\xi \xi_v t_d}{b}\right)\left[1 + \frac{\xi \xi_v t_d r_d}{b}\right]; t_d = \frac{t_d^0 d}{\sqrt{\xi D_b}}$$

$$\sigma = \sigma^1 r_d \rho_b + \sigma^p (1 - r_d \rho_b) \ ; \ \xi_v = \rho_b \psi \xi d \left(\frac{\xi^1}{r_d \rho_b} - \frac{\xi^p}{1 - r_d \rho_b}\right) \tag{3.111}$$

The system of equations includes the following 11 material constants: μ_1, μ_p, $\sigma^1{}_0$, $\sigma^p{}_0$, $\sigma^v{}_0$, $\sigma^{cr}{}_0$, A', B', A_{cr}, D_b, $t^0{}_d$. The method used to determine these constants are outlined in the paper (non-linear regression analysis). Apart from the 11 constants listed above, 5 more constants are assumed. (r_d, a, δ, b, ψ). Larin et al. [371] have added a kinetic equation for the grain growth rate as well.

$$\dot{d} = \frac{a' \xi_v \xi \sigma}{1 + b' \xi_v \xi} \tag{3.112}$$

where a', b' are material constants. (Thus, the model has a total of 13 constants and 5 more constant values are assumed). Bylia [372] has analysed the system of equations (excluding Eq. (3.112)). She has shown that taking into consideration the relative values of the various parameters one can reduce it to

$$\xi = \frac{a_1 \sigma^4 + a_2 \sigma^3 + a_3 \sigma + a_4}{a_5 \sigma + 1} \tag{3.113}$$

where a_i ($i = 1, 2, 3, 4, 5$) are material constants. Hence, Perevezentsev's model belongs to type Eq. (3.10). Therefore, it can not describe the behaviour of superplastics under transient regimes of loading, i.e., it is a model for the steady state.

3.4 Construction of Constitutive Equations

3.4.1 Common Scheme

In this section, a formal scheme useful in constructing constitutive equations suitable for practical applications is presented. These CEs should satisfy the requirements stated in Sect. 3.1. This approach leads to a way of developing phenomenological constitutive equations from physical models and thus helps in bringing together the efforts of physicists/materials scientists and specialists in mechanics of solids.

At first, the physical equations describing one or more mechanisms of deformation are used to establish the relationship among stress, strain and strain rate. These equations will include a set of constants. They may be either fundamental physical constants or 'microconstants' characterising the properties of a material at micro- or meso-levels. Some of them can be measured only at microlevel. The second step is to forget the physical meaning of the material constants and consider them as fitting coefficients of the equations under consideration. This step is based on the conviction that the physical equations to be used in practical calculations have to involve only those material constants that are important in understanding the *macrobehaviour* of a material. If this is so, then one can, in principle, determine the values of these constants from macro-experiments. If the equations include some physical parameters which can only be determined from micro-scale experiments (e.g., absolute value of the Burgers vector) but can not be related to the macroconstants of a CE, then such equations are likely to be of only limited use in practical situations. Moreover, from the viewpoint of mechanics, if one can not determine *all* the material constants of a system of equations from macro-experiments only, then the CE will be of no use. Once these conditions are satisfied, basic experiments for evaluating the material constants and crucial experiments to define the area of applicability of the CE should be performed.

The above proposed scheme is illustrated below using the recently proposed model of Padmanabhan and Schlipf as an example.

3.4.2 Model of Padmanabhan and Schlipf

In this physical model for optimal (defined as the region of flow starting from the lowest strain rates till the point of inflection in the $\log\sigma - \log\xi$ curve) structural superplasticity, grain/interphase boundary sliding controls the rate of flow [138–141]. The physical features of this model will not be discussed here. Discussion will centre around the constitutive equation resulting from the analysis which, for the isostructural isothermal case can be written as

$$\xi = \left(\frac{C_1}{\sigma_m}\right)\left\{(\sigma - \sigma_0)\cdot[\sigma_m + \alpha(\sigma - \sigma_0 - 2\sigma_m)] + \alpha\sigma_m^2 \exp\left(-\frac{1}{\alpha}\right)\right\} \quad (3.114)$$

where C_1, σ_0, σ_m and α are material constants (defined in [138–141]). According to the common scheme outlined in the preceding sections, we shall set aside the physical meaning of these material constants and consider them as fitting coefficients. Eq. (3.114) can be rewritten as

$$\xi = a\sigma^2 + b\sigma + c \tag{3.115}$$

where

$$a = \frac{\alpha C_1}{\sigma_m}, b = \left(\frac{C_1}{\sigma_m}\right)\cdot\left[\sigma_m - 2\alpha\sigma_0 - 2\alpha\sigma_m\right];$$
$$c = \left(\frac{C_1}{\sigma_m}\right)\cdot\left[\alpha\sigma_0^2 + 2\alpha\sigma_0\sigma_m - \sigma_0\sigma_m + \alpha\sigma_m^2 \exp\left(-\frac{1}{\alpha}\right)\right] \tag{3.116}$$

The 'best fit estimates' of the constants a, b, c in Eq. (3.115) can be determined unambiguously by a least squares analysis. However, from a physical point of view, it is desirable to know the four material constants of Eq. (3.114). For this, an additional independent condition will have to be added as follows. It follows from the physical model that the constants C_1, σ_0, σ_m and α are all positive [138–141]. Therefore, from the first of Eq. (3.116), a is always greater than zero; b and c, in general, can be either positive or negative. Consider the equation for the mechanical threshold stress, σ_{th}.

$$\xi = a\sigma^2 + b\sigma + c = 0 \tag{3.117}$$

If the discriminant D (= $b^2 - 4ac$) is greater than zero, Eq. (3.117) will have two roots:

$$\sigma_{th1,2} = -\frac{b}{2a} \pm \frac{\sqrt{D}}{2a}, D = b^2 - 4ac \tag{3.118}$$

All possible combinations of the independent constants a, b, c and the derived constant D are listed in Table 3.7 and the corresponding $\sigma - \xi$ plots for the first quadrant of (σ, ξ) plane are presented in Fig. 3.29. Of these, curves 1, 5 and 6 are physically unrealistic as they suggest definite flow rates even when $\sigma = 0$. Curve 4 is physically unrealistic because according to this curve at low strain rates, the

Table 3.7 Combinations of the signs of the constants a, b, c

	a	b	c	D	σ_{th1}	σ_{th2}
1	+	+	+	+	–	–
2	+	+	–	+	+	–
3	+	–	–	+	+	–
4	+	–	+	+	+	+
5	+	+	+	–	*	*
6	+	–	+	–	*	*

* indicates that the value is non-existent

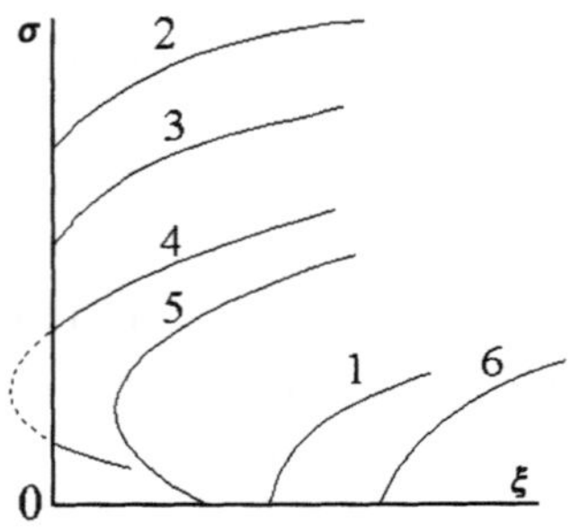

Fig. 3.29. Stress–strain rate dependences according to Eq. (3.115) for various values of a, b, c (schematic)

same strain rate can arise from two different stresses. Therefore, only those combinations of a, b and c corresponding to curves 2 and 3 or the upper portion of curve 4 in Fig. 3.29 need further consideration. From Fig. 3.29 and Eq. (3.118) the mechanical threshold σ_{th} can be written as

$$\sigma_{th} = \frac{\left(\sqrt{D} - b\right)}{2a} \tag{3.119}$$

(The other root is negative and, hence, is physically meaningless.) Then, Eq. (3.115) becomes (when $D \geq 0$)

$$\sigma = \frac{\sqrt{D + 4a\xi} - b}{2a}, D > 0 \tag{3.115a}$$

From this expression one obtains

$$M(\xi) = \frac{\partial \log \sigma}{\partial \log \xi} = \frac{2a\xi}{\sqrt{D + 4a\xi}} \cdot \frac{1}{\sqrt{D + 4a\xi} - b} \tag{3.120}$$

The values of ξ_{opt} and σ_{opt} can be found from the condition for maximum M, viz.,

$$\left. \frac{dM}{d \log \xi} \right|_{\xi = \xi_{opt}} = 0 \tag{3.121}$$

Then, the value of the optimal strain rate will be

$$\xi_{opt} = \frac{b^2 - 4ac}{b} \cdot \left[-\frac{2c}{b} + \sqrt{\left(\frac{2c}{b}\right)^2 - \frac{c}{a}} \right] \tag{3.122}$$

$$\sigma_{opt} = -\frac{2c}{b} + \sqrt{\left(\frac{2c}{b}\right)^2 - \frac{c}{a}} \tag{3.123}$$

From Equation (3.120)

$$M_{max} = \frac{1}{1 + 2\left(-ac/b^2 - 4ac\right)^{\frac{1}{2}}} \tag{3.124}$$

During superplastic flow, ξ_{opt} exists and is positive. But, if c is positive, ξ_{opt} will be imaginary. Eq. (3.122) and so c has to be negative. Likewise, when b is negative, ξ_{opt} will be negative which is absurd and so it has to be positive. Therefore, for optimal superplasticity, a and b are positive but c is negative. Thus, the following conditions are to be imposed on the constants of the constitutive equation:

$$a > 0, \qquad b > 0, \qquad c < 0, \qquad D > 0$$

Validation of the analysis using experimental data has been carrying out in [105] using experimental strain rate – stress data pertaining to Al–12Si [324], Al–33Cu [325], and Al–33Cu–0.4Zr [326] systems and Supral-100 (Al–6Cu–0.4Zr) [158]. Some of the results obtained are presented in Fig. 3.30 and Table 3.8. It is seen from Fig. 3.30 that the agreement between the predicted and the experimentally measured values is good. (As Eq. (3.115) is quadratic, the constants a, b and c can be calculated unambiguously by the method of least squares. Thus, using the present procedure, the mechanical response of superplastic alloys in the optimal region of steady state flow can be described accurately with the aid of three mate-

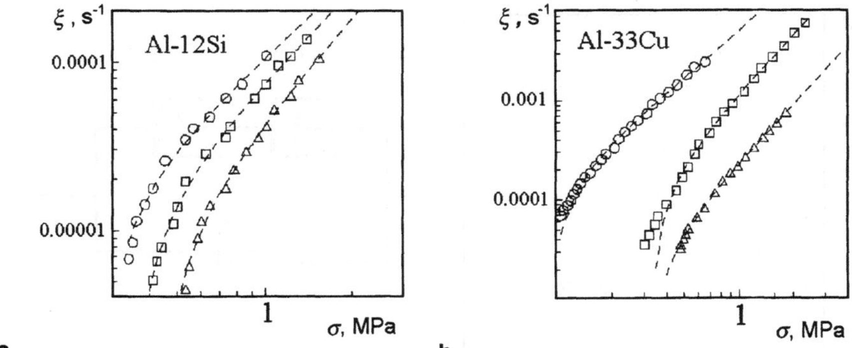

Fig. 3.30. Comparison between the predicted (dashed lines) and the experimentally measured values of the strain rate as a function of the applied stress for two different alloys (log–log plots):**a** Al–12Si alloy: ooo – 831 K, □□□– 811 K, ΔΔΔ – 791K; **b** Al–33Cu alloy: ooo -793K, □□□ – 753 K, ΔΔΔ – 713K

Table 3.8 Material constants *a, b, c* for the Padmanabhan–Schlipf model [105]

T (K)	$a(10^{15}\,s^{-1}Pa^{-2})$	$b(10^{10}\,s^{-1}Pa^{-1})$	$c(10^5\,s^{-1})$	M_{max}	$\xi_{opt}\,(s^{-1})$	σ_o (MPa)
			Al–12Si			
831	0.0372	0.952	−2.8079	0.64	$2.281 \cdot 10^{-3}$	1.640
811	0.0439	0.540	−2.4288	0.57	$2.750 \cdot 10^{-4}$	2.068
791	0.0412	0.141	−1.3702	0.51	$6.894 \cdot 10^{-4}$	3.965
			Al–33Cu			
793	2.782	24.19	−24.105	0.64	$1.957 \cdot 10^{-3}$	0.555
753	1.065	2.224	−16.801	0.51	$1.058 \cdot 10^{-2}$	3.073
713	0.219	0.361	−3.1568	0.51	$2.837 \cdot 10^{-3}$	3.538
			Al–6Cu–0.4Zr			
763	0.01618	0.473	−7.644	0.55	$1.083 \cdot 10^{-3}$	7.131
743	0.01052	0.289	−6.773	0.53	$1.278 \cdot 10^{-3}$	10.02
723	0.00949	0.082	−4.798	0.51	$5.389 \cdot 10^{-3}$	23.50

rial constants.) Similar results could also be obtained for eutectic Pb–Sn and Zn–22%Al eutectoid alloys.

Bricknell and Bentley [158] have reported experimental data over a wide range for the Al–6Cu–0.4Zr alloy. It is of interest to use them for an independent validation of the model. One can see from Table 3.8 that for the Al–6Cu–0.4Zr alloy $M_{max} = 0.55$ at $T = 763K$ and $\xi_{opt}=1.08 \cdot 10^{-3}\ s^{-1}$ which are in good agreement with the corresponding data in [158]. A similar conclusion can be reached by comparing the curves shown in Fig. 3.31 with analogous curves in ref. [158]. Considering that in all calculations a non-zero value for the threshold stress was obtained, Eq. (3.115) can be written as

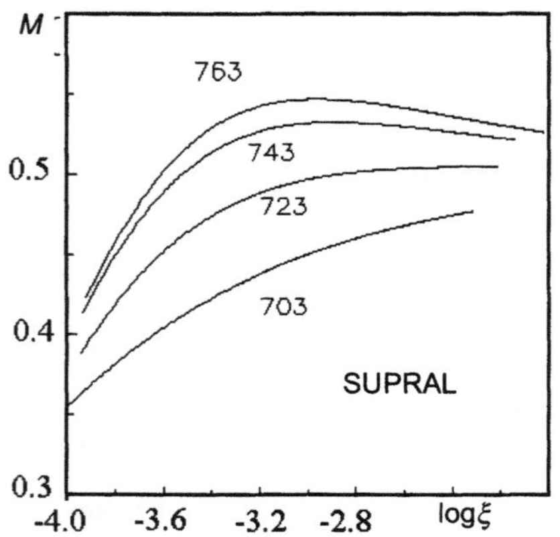

Fig. 3.31. Strain rate dependence of M in case of the Al–6Cu–0.4Zr alloy

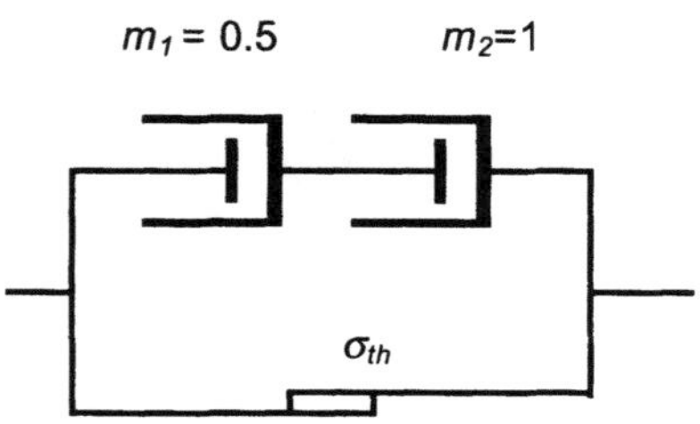

Fig. 3. 32. Mechanical analogue for Eq. (3.115); $a > 0, b > 0, c < 0$.

$$\xi = \begin{cases} 0 & \text{if } \sigma < \sigma_{th}, \\ a \cdot (\sigma - \sigma_{th})^2 + \sqrt{b^2 - 4ac}(\sigma - \sigma_{th}) & \text{if } \sigma \ge \sigma_{th} \end{cases} \tag{3.125}$$

where σ_{th} is from Eq. (3.119). The mechanical analogue for the Padmanabhan–Schlipf model is presented in Fig. 3.32. It is interesting to compare the results of the theoretical predictions made within the framework of the model with the universal sigmoidal curve considered in Chap.1. Dividing Eq. (3.120) by Eq. (3.124) one obtains

$$\frac{M}{M_{max}} = \frac{2a\xi}{\sqrt{D + 4a\xi}} \cdot \frac{1 + \sqrt{-\frac{4ac}{D}}}{\sqrt{D + 4a\xi} - b} \tag{3.126}$$

Let the relative strain rate $\bar{\xi}$ be given by $\bar{\xi} = \xi / \xi_{opt}$. Then, from Eqs. (3.126) and (3.122)

$$\frac{M}{M_{\max}} = \frac{(x+1)\bar{\xi}}{x^2 - x + 2x\bar{\xi} - (x-1)\sqrt{x^2 - 1 + 2\bar{\xi}(x+1)}} \tag{3.127}$$

where

$$x = \sqrt{-\frac{D}{4ac}} = \sqrt{1 - \frac{b^2}{4ac}}$$

Equation (3.127) was studied numerically – Fig. (3.33). It is obvious that a given value of x defines unambiguously the shape of the sigmoidal curve. Taking into account Eq. (3.116) one can obtain

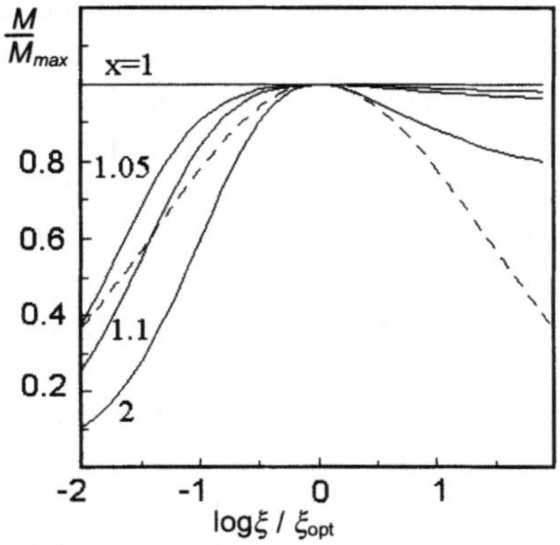

Fig. 3.33. Strain rate dependence of the normalised slope M/M_{max} in Supral calculated in accordance with Eq. (3.127) for different x values (indicated near the curves)

$$x^2 = \frac{[4\alpha^2 \exp(-1/\alpha) - 1 - 4\alpha^2]\sigma_m^2}{4\alpha[\alpha\sigma_0^2 + 2\alpha\sigma_0\sigma_m - \sigma_0\sigma_m + \alpha\sigma_m^2 \exp(-1/\alpha)]} \tag{3.128}$$

This relation reveals that the shape of the sigmoidal curve is determined by the values α and the ratio σ_0/σ_m and it does not depend on C_1.

The question whether a universal curve, as suggested by a cursory empirical examination (Fig. 1.3b) really exists is of interest. The (M/M_{max}) vs. $\log(\xi/\xi_{opt})$ plots for two alloys at different temperatures are presented in Fig. 3.34 using the model constants. It is noted that the plots are temperature dependent – a fact not recognized in [98]. In many systems, both the values of M_{max} and ξ_{opt} are clearly temperature dependent and this would lead to the observed temperature dependence in the (M/M_{max}) vs. $\log(\xi/\xi_{opt})$ plot. At nearly equal (T/T_m) ratios (T = test temperature, K) the relevant curves practically coincide, which is clear indication of the existence of the 'universal curve'. (The original 'universal curve' of [98] is also superimposed in Fig. 3.33–dotted curve.) It is evident that the (T/T_m) ratios for the alloys examined in [98] varied widely. When appropriate (T/T_m) values are chosen that are relevant for Figs. 3.35–3.36, the experimental points considered in [98] also fell on the universal curves of these figures.

When curves similar to Figs. 3.34–3.36 are available, the 'universal curve' at any temperature in the experimental range can be obtained by the method of interpolation by evaluating the temperature dependence of x as $x = E_1 + E_2/T$, where E_1 and E_2 are material constants. The procedure is illustrated in Fig. 3.33. Results pertain to an Al –6Cu–0.4Zr alloy.

Thus, based on the systems analysed to date [105], it is reasonable to suggest that a 'universal curve' for superplastic flow exists *provided* the (M/M_{max}) vs. $\log(\xi/\xi_{opt})$ plot is made at a constant (T/T_m) ratio. Such a plot is material independent. However, if the grain size varies widely there could be an additional effect on the 'universal curve', as both M_{max} and ξ_{opt} are known to be affected by the grain size also. Paucity of experimental data has prevented verifications in this regard. The existence of a universal curve may be interpreted as an indication of a common mechanism of deformation for structural superplasticity.

Figures 3.34–3.36 also reveal that the shape of the universal curve on the left hand side of ξ_{opt} is similar for the different alloys. The curves actually coincide (true universal curves) when the (T/T_m) ratios are nearly equal. But the shapes are different on the right hand side of ξ_{opt}. Thus, based on mechanics also one is able to draw the conclusion drawn earlier from physical considerations [138–141] that the grain boundary sliding controlled flow model should not be used at strain rates greater than ξ_{opt}.

A common requirement in mechanics of solids is that a constitutive law/equation should be stable against small variations in the values of the material constants. To check the reliability of the present method, in Eq. (3.115) data pertaining to Supral alloy (Al–6.5 Cu–0.4Zr) at 763 K were used. For different computations, $\sigma - \xi$ pairs of varying numbers, taken from the optimal range of superplastic flow at the given temperature, were used. The results are presented in

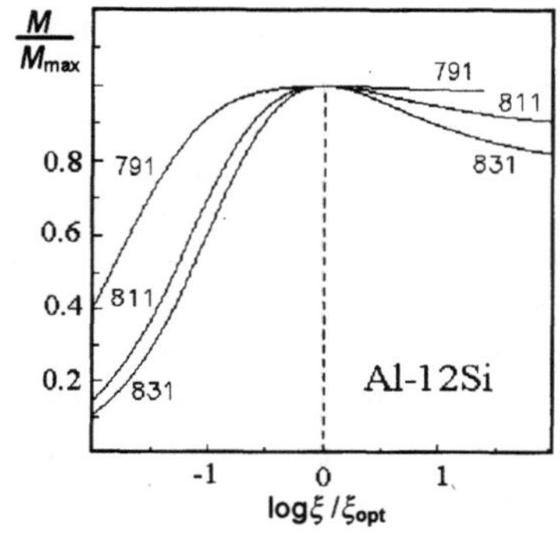

Fig. 3.34. Normalized (M/M_{max}) vs. $\log(\xi/\xi_{opt})$ curves for the Al–12Si alloy at different temperatures (indicated in degrees K near each curve)

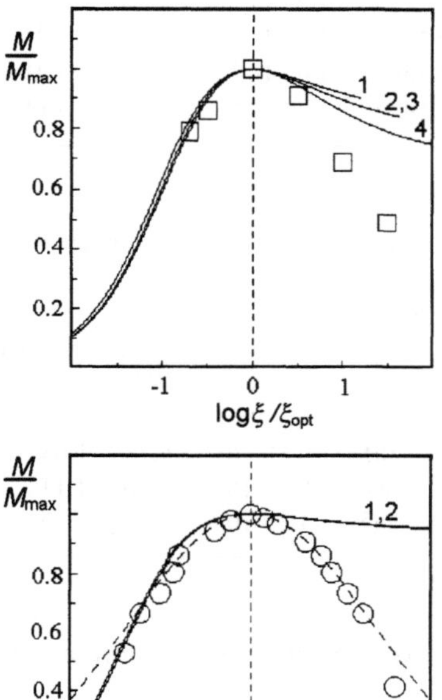

Fig. 3.35. (M/M_{max}) vs. $\log(\xi/\xi_{opt})$ plots at comparable (T/T_m) ratio for the following alloys (solid lines): 1 – Sn–38Pb, 443 K, $(T/T_m) = 0.972$; 2 – Al–12Si, 831 K, $(T/T_m) = 0.985$; 3 – Al–33Cu, 793 K, $(T/T_m) = 0.966$; 4 – Al–33Cu–0.4Zr, 793 K, $(T/T_m) = 0.97$. □□□□ – MA21 alloy deformed at $(T/T_m) = 0.973$

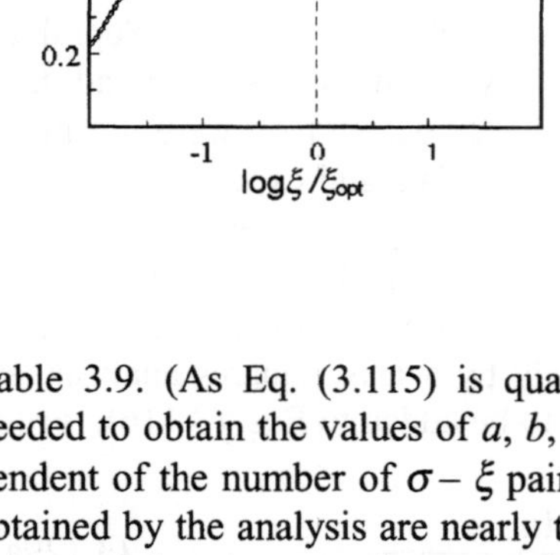

Fig. 3.36. (M/M_{max}) vs. $\log(\xi/\xi_{opt})$ plots at comparable (T/T_m) ratio for the following alloys (solid lines): 1 – Zn–22%Al, 503 K, $(T/T_m) = 0.918$; 2 – Supral alloy, 743 K, $(T/T_m) = 0.911$. OOOO Al5083 deformed at $(T/T_m) = 0.922$. The original 'universal curve' of [98] is also shown by dashed line.

Table 3.9. (As Eq. (3.115) is quadratic, the minimum number of $\sigma - \xi$ pairs needed to obtain the values of a, b, c in the equation is 3.) It is evident that independent of the number of $\sigma - \xi$ pairs chosen, the values of M_{max}, ξ_{opt}, σ_{opt} and σ_0 obtained by the analysis are nearly the same. This establishes the reliability of the values of the constants determined by the present procedure It also leads to the desirable situation in which an accurate idea of the values of the constants of the

Table 3.9 Material constants *a*, *b*, *c* for Supral–100 alloy deformed at 763 K estimated with different number of experimental points *N*

N	$a\times10^{17}$ ($s^{-1}Pa^{-2}$)	$b\times10^{11}$ ($s^{-1}Pa^{-1}$)	$c\times10^{5}$ (s^{-1})	$D\times10^{20}$ ($s^{-2}Pa^{-2}$)	M_{max}	ξ_{opt} (s^{-1})	σ_{opt} (MPa)	σ_0 (MPa)
20	1.618	4.727	–7.644	0.718	0.547	$1.07\cdot10^{-3}$	7.13	1.16
17	1.583	4.910	–7.891	0.741	0.549	$1.07\cdot10^{-3}$	7.13	1.17
15	1.640	4.515	–7.217	0.677	0.545	$1.05\cdot10^{-3}$	7.02	1.13
12	1.629	4.617	–7.453	0.699	0.545	$1.07\cdot10^{-3}$	7.10	1.15
10	1.769	3.841	–6.495	0.607	0.535	$1.15\cdot10^{-3}$	7.27	1.12
7	1.666	4.689	–7.565	0.724	0.545	$1.10\cdot10^{-3}$	7.09	1.15
6	1.579	5.010	–7.905	0.750	0.551	$1.05\cdot10^{-3}$	7.02	1.16
5	1.618	4.722	–7.611	0.716	0.547	$1,07\cdot10^{-3}$	7.11	1.15
4	1.502	5.681	–8.791	0.851	0.559	$1.05\cdot10^{-3}$	7.02	1.18
3	1.681	4.199	–6.467	0.611	0.542	$9.77\cdot10^{-4}$	6.73	1.08

Table 3.10 Parameters M_{max}, ξ_{opt}, σ_{opt} and σ_0 calculated for three different sets of experimental data

σ_1 MPa	ξ_1 s^{-1}	σ_2 MPa	ξ_2 s^{-1}	σ_3 MPa	ξ_3 s^{-1}	M_{max}	ξ_{opt} s^{-1}	σ_{opt} MPa	σ_0 MPa
1.49	$3.52\cdot10^{-5}$	3.10	$2.27\cdot10^{-4}$	6.99	$1.05\cdot10^{-3}$	0.54	$9.77\cdot10^{-4}$	6.73	1.08
1.42	$3.40\cdot10^{-5}$	2.95	$2.32\cdot10^{-4}$	7.35	$1.1\cdot10^{-3}$	0.61	$8.91\cdot10^{-4}$	6.50	1.11
1.65	$3.8\cdot10^{-5}$	2.70	$1.9\cdot10^{-4}$	6.3	$9.3\cdot10^{-4}$	0.59	$1.35\cdot10^{-3}$	7.88	1.35

constitutive equation is obtained with minimum experimentation (here a minimum of 3 pairs of $\sigma-\xi$ values).

In another test (Table 3.10) a set of 3 pairs of σ–ξ data pertaining to Supral alloy deformed at 763 K was deliberately distorted by 5 and 10% (second and third rows) respectively. The resultant values of M_{max}. ξ_{opt}, σ_{opt} and σ_{th} were not much different. This test shows that small variations in the data and experimental errors are not likely to affect the usefulness of the constitutive equation developed.

From a physical point of view, it is of interest to determine the values of α, σ_m, σ_0 and C_1 in Eq. (3.114). For this, apart from the 'best fit estimates' of the constants *a*, *b*, *c* in Eq. (3.115) determined unambiguously by a least squares analysis, an additional independent condition will also have to be added to Eqs. (3.116). From Eq. (3.116)

$$\sigma_m^2 = \frac{1}{4a^2} \cdot \frac{b^2 - 4ac}{\left(\frac{1}{2\alpha} - 1\right)^2 - \exp\left(-\frac{1}{\alpha}\right)} \tag{3.129a}$$

$$\sigma_0 = \sigma_m \left(\frac{1}{2\alpha} - 1 \right) - \frac{b}{2a}, \; C_1 = \frac{\sigma_m a}{\alpha} \tag{3.129b}$$

As $(b^2–4ac) \geq 0$, the denominator on the right hand side of Eq. (3.129) is positive. Then, keeping in mind that $\alpha = 2/\left(\lambda\sqrt{\pi}\right)$ [138, 141], it is easy to show that

$$0 \leq \alpha \leq \alpha_{cr} \cong 0.391 \qquad \lambda_{cr} \cong 2.886 \leq \lambda < \infty \tag{3.130}$$

where λ is the standard deviation and σ_m the mean of the two parameter log-normal internal stress distribution arising from atomic scale boundary sliding. σ_0 the threshold stress that has to be applied to convert atomic scale sliding into a mesoscopic (co-operative) grain/interphase boundary sliding process involving a number of interconnected grain boundaries and C_1 is a grain size and temperature dependent material constant [138, 141].

As four constants have to be determined and there are only three Eq. (3.129), the fourth condition is introduced through the well-known criterion in mathematical statistics of 'minimum variance unbiased estimate'. According to this criterion, λ should be minimum (or α maximum). In addition, due to the physical nature of the problem and definitions the following inequalities have to be satisfied [141]:

$$\sigma_0 < \sigma_{\min} = \min_{i=1,2...N} (\sigma_i); \qquad \sigma_m < \sigma_{\max} = \max_{i=1,2...N} (\sigma_i) \tag{3.131}$$

where σ_i represents the set of experimental data points used to estimate a, b and c. Thus, α is incremented from 0 to its limiting value of 0.391 in steps of 0.001 (say). For each value of α, σ_m, σ_0 and C_1 values are calculated using Eqs. (3.129). The highest value of α for which the inequalities (3.131) are satisfied is accepted as the final answer and the corresponding α, σ_m, σ_0 and C_1 values are taken as the constants of the physical model, Eq. (3.114). (From the closeness between σ_{th} and σ_0 values, the mechanical threshold σ_{th} may be approximately equated to the threshold stress needed to cause mesoscopic boundary sliding, σ_0.) The values of α, σ_m, σ_0 and C_1 obtained by such a rigorous procedure [105] are compared in Table 3.11 with the values got by a trial and error procedure [141]. The values of α, σ_0 and C_1 obtained by the two procedures are in reasonably good agreement but the values of σ_m may differ sometimes significantly. It is of interest to estimate the values of the parameter x on the right hand side of Eq.(3.127) proceeding from the experimental data (Table 3.12).

Equation (3.115) can not be used beyond the point of inflection in the steady state sigmoidal $\log\sigma - \log\xi$ plot. It has been suggested [105] that at those higher strain rates, apart from the mechanism proposed by Padmanabhan and Schlipf, a climb controlled multiple dislocation deformation mechanism (Weertman creep) also will operate. Preliminary unpublished work of Padmanabhan and Daniel indi-

Table 3.11 Comparison of the values of the material constants obtained by the rigorous method [105] and an earlier trial and error procedure [141]

T	C_1 (GPa^{-1} s^{-1})		α		σ_0 (Mpa)		σ_m (Mpa)	
(K)	[105]	[141]	[105]	[141]	[105]	[141]	[105]	[141]
				Al–12Si				
831	0.1858	0.140	0.19	0.1	0.269	0.27	0.948	0.53
811	0.1367	0.120	0.19	0.3	0.351	0.32	0.592	0.40
791	0.0800	0.090	0.19	0.2	0.431	0.43	0.369	0.68
763	0.0423	0.066	0.19	0.2	0.706	0.77	0.300	1.20
				Al–33Cu				
793	6.461	1.60	0.27	0.26	0.099	0.048	0.627	0.05
753	1.935	1.30	0.27	0.33	0.313	0.250	0.490	0.37
713	0.3763	0.40	0.27	0.23	0.313	0.300	0.464	0.57
633	0.0198	0.04	0.27	0.21	0.853	1.200	0.653	1.90
				Al–33Cu–0.4Zr				
793	0.1283	0.090	0.16	0.1	0.451	0.35	2.754	0.8
753	0.0393	0.028	0.16	0.1	0.727	0.63	3.965	1.2
713	0.0182	0.012	0.16	0.3	1.691	1.10	4.679	2.3

Table 3.12 The values of the physical constants σ_0, σ_m, α and parameter x

T (K)	$1/T$ (K^{-1})	σ_0 (MPa)	σ_m (MPa)	σ_0/σ_m	α	x
			Al–12Si			
831	0.001203	0.269	1.015	0.265	0.198	1.781
811	0.001233	0.383	1.401	0.273	0.292	1.298
791	0.001264	0.527	1.527	0.345	0.343	1.044
763	0.001311	0.860	1.672	0.514	0.361	1.114
			Al–33Cu			
793	0.001261	0.101	0.655	0.154	0.275	1.785
753	0.001328	0.321	0.613	0.524	0.295	1.034
713	0.001403	0.475	1.553	0.306	0.368	1.023
633	0.001580	1.424	3.646	0.391	0.382	1.195
			Al–33Cu–0.4Zr			
793	0.001261	0.454	3.350	0.136	0.182	3.595
753	0.001328	0.727	4.038	0.180	0.162	3.464
713	0.001403	1.696	5.843	0.290	0.185	1.794
			Supral			
763	0.001311	1.483	5.976	0.248	0.335	1.205
743	0.001345	1.683	5.028	0.335	0.311	1.137
723	0.001383	3.758	8.719	0.316	0.366	1.019
703	0.001422	3.159	6.695	0.472	0.383	$x^2<0$

cates that this is indeed the case. (Microstructural evidence for both the mechanisms being present at these higher strain rates is available – see, for example [3]). With this additional mechanism included, an analysis for the entire stress–strain rate range should be done in the future as has been done here for the optimal range. Also, Eq. (3.115) does not describe the behaviour of a superplastic material in the transient regimes of loading and so it is necessary to generalise it. Let a simple variant as given below be used.

$$\xi = \frac{\dot{\sigma}}{E} + a\sigma^2 + b\sigma + c \tag{3.132}$$

where E is the effective Young's modulus of the system 'specimen–testing machine'[4]. Then, the time dependence of flow stress for constant strain rate loading ($\xi = \xi_0$ = constant) can be written as

$$\sigma(t) = -\frac{b}{2a} + \frac{\sqrt{D+4a\xi_0}}{2a} \cdot \frac{1+B\exp(-Et\sqrt{D+4a\xi_0})}{1-B\exp(-Et\sqrt{D+4a\xi_0})} \tag{3.133}$$

where
$$B = \frac{b+\sqrt{D+4a\xi_0}}{b-\sqrt{D+4a\xi_0}}$$

It is easy to show that

$$\lim_{t\to\infty} \sigma(t) = \sigma_s = -\frac{b}{2a} + \frac{\sqrt{D+4a\xi_0}}{2a}$$

which is identical to Eq. (3.115a) when $\xi = \xi_0$. One can obtain the following expression for characteristic time $\tau_{1/2}$ (see definition in Sect. 3.2.3.2).

$$\tau_{1/2} = \frac{1}{E\sqrt{D+4a\xi_0}} \cdot \ln \frac{b+3\sqrt{D+4a\xi_0}}{b+\sqrt{D+4a\xi_0}} \tag{3.134}$$

For a load relaxation test the following expression is valid:

$$\sigma(t) = -\frac{b}{2a} + \frac{\sqrt{D}}{2a} \cdot \frac{1+C_0\exp(-E\sqrt{D}t)}{1-C_0\exp(-E\sqrt{D}t)} \tag{3.135}$$

where
$$C_0 = \frac{2a\sigma_k + b - \sqrt{D}}{2a\sigma_k + b + \sqrt{D}}$$

[4] It is pointed out that this is only a crude way of including the transient region of flow. For a constitutive equation to be truly acceptable, it should be independent of machine characteristics.

The characteristic time $\tau_{1/2R}$ will be

$$\tau_{1/2} = \frac{1}{E\sqrt{D}} \cdot \ln \frac{b + 3\sqrt{D}}{b + \sqrt{D}} \tag{3.136}$$

Equations (3.132)–(3.136) can be used to analyse the mechanical response of superplastics under transient regimes of loading.

The following conclusions can be reached based on the above analysis [105]. The sigmoidal $\log\sigma - \log\xi$ plot up to the point of inflection can be accounted for with the help of *three* independent constants. A fit of similar accuracy can also be obtained if one uses the three parameter power law $\xi = C\,(\sigma - \sigma_0)^{n'}$, with n' being treated to be of any arbitrary value, see Sect. 3.2.3). But the three parameter power law is empirical while the constitutive equation resulting from grain/interphase boundary sliding controlled flow [138–141] is based on a detailed physical model. The above analysis is able to relate the three independent 'best fit' constants of the constitutive equation unequivocally with the four material constants of the physical model [138] – see Eqs. (3.129–3.131).

The grain/interphase boundary sliding controlled flow model is able to predict quantitatively the existence of an optimal strain rate at which M is maximum. In contrast, in the three parameter power law, n decreases (or M increases) monotonically with increasing stress. To the best of our knowledge, the model of Padmanabhan and Schlipf [138] is the only analysis that is able to predict the existence of an optimal strain rate at which M has its maximum value, Eqs. (3.124) and (3.126).

The proximity between the values of σ_0 and σ_{th} suggests that for practical purposes the stress needed to cause mesoscopic sliding can be equated with the mechanical threshold, σ_{th}. Thus, the model accurately accounts for steady state experimental data up to the point of inflection in the $\log\sigma$–$\log\xi$ curve with the help of three independent material constants. The stability of obtaining the values of the material constants and the reliability of the constitutive equation are clearly established. The model constants [138–141] can be obtained accurately with minimum experimentation. (Number of experimental $\sigma - \xi$ pairs required ≥ 3).

The constitutive equation based on a grain/interphase boundary sliding controlled flow process should not be used in isolation at strain rates greater than that at the point of inflection in the sigmoidal $\log\sigma - \log\xi$ plot. In that range, this model is to be combined (in an additive manner) with an analysis for dislocation climb controlled creep flow, e.g., the model of Weertman. Here, it has also been shown how this model can be generalised to include the regimes of transient loading invariably present during superplastic flow.

This model for optimal superplasticity has also been used to explain deformation in nanostructured materials [373–375] and superplastic deformation in metallic glasses [376]. These aspects are beyond the scope of this book.

3.5
Constitutive Equations in tensor Form

Construction and analysis of constitutive equations (CEs) is the traditional major problem in mechanics of solids. As mentioned in Sect. 3.1, the problem is well investigated for superplastics in uniaxial loading. However, for multiaxial loading this problem has not even been set up. In particular, the sufficiently general tensor form of the CEs and the hypotheses concerning the mutual orientation of such basis tensors as stress, increment of stress, increment of strain etc. are not known. In addition, one should formulate this problem in such a way that the transition from the non-superplastic to the superplastic state for multiaxial loading in the stress or the strain space is clearly defined. For uniaxial flow, the onset condition for superplastic flow is defined usually as $M \geq 0.3$. It is not clear how this condition is to be generalised for multiaxial loading. Also, the effects of thermomechanical treatments are unknown.

In this section some possible routines for the generalisation of uniaxial CE to include multiaxial loading are discussed along with the attendant problems. These procedures are well-known to specialists in mechanics of solids. However, in the superplasticity literature the bulk of the activity has been concentrated on assigning indices in scalar equations obtained from uniaxial tests (to carry out calculations to determine important technological parameters).

3.5.1
Non-Uniaxial Stress–Strain States

The scalar CE should be extended to cover non-uniaxial stress–strain states also and written in a tensor form. In such a treatment, in addition to uniaxial deformation (tension or compression), multiaxial and complex evaluating conditions should be included. Only a few reports are available in the literature on the behaviour of superplastics under multiaxial loading. There is an erroneous belief that by replacing the vectors one can easily analyse superplastic flow in terms of a rheological model. Another simplistic belief is that the deformation theory of plasticity is sufficient to assign the necessary indices in a scalar CE.

In reality, however, the problem of extending the scalar CE to the 3D situation is extremely dificult and has not been solved for all cases even when the strains are infinitesimally small (the case of finite strains is considered in Chap. 5). This problem is illustrated using the simple CE

$$\sigma = K\xi^m \varepsilon^n \tag{3.137}$$

as an example where ε and ξ are the *total* strain and strain rate, respectively. For definiteness let it be supposed that Eq.(3.137) is derived from tensile tests. As the strains are infinitesimally small, there is no problem in selecting the concrete measures of the stress and the strain states. They are unique. Let $\varepsilon_{ij}, \xi_{ij}$ and σ_{ij} be the components of the strain, strain rate and stress tensors respectively. The problem of generalisation of the 'uniaxial' Eq. (3.137) to a scalar equation that will be suitable for a 3D situation is initially considered. Let Eq. (3.137) be written in an invariant form so that this form reduces to Eq. (3.137) for a tensile test.

$$\sigma_e = K\varepsilon_e^n \xi_e^m \tag{3.138}$$

Although the generalization seems obvious, it places certain restrictions on the material properties, e.g., the assumption that the material has 'a unique curve' involving, σ_e, ε_e, ξ_e for any arbitrary mode of deformation. Existence of 'the unique curve' Eq. (3.138) means, for example, that the CE for a torsion test has the form

$$\tau = K\gamma^n \dot{\gamma}^m \left(\sqrt{3}\right)^{-1-n-m}, \quad \tau \equiv \sigma_{12}, \gamma \equiv 2\varepsilon_{12},$$

Then, in this case, the CE for a tension test has the form of Eq. (3.137). Using another invariant of stress, strain, strain rate tensors one can arrive at another 'unique curve'. It should be noted that any such generalisation of Eq. (3.137) is a hypothesis; it is not based on mathematical rules or common concepts used in mechanics of solids only. The correctness and the domain of validity of the hypothesis should be established by special experiments.

To obtain a generalisation of Eq. (3.137) (e.g., Eq. (3.138)), it is restricted to a particular class of deformation processes, e.g., simple loading. A general proposition concerning the existence of a universal relation between the invariants of stress, strain and strain rate tensors, generally speaking, implies the absence of arbitrary histories of trajectories of deformation. But, it is possible to do more accurate and more interesting calculations for practical applications: a formal presentation of Eq. (3.137) in an invariant scalar form does not create a problem. The problem lies in finding its range of validity (i.e., agreement with experimental data for each specified variant of the invariant present in the equation). The last statement applies also to Eq. (3.137) in tensor form (that is, invariant form). But the scalar Eq. (3.137) can be written in a tensor form using different variants, i.e. Eq.(3.137) is only an example. Consider the equations

$$S_{ij} = (2K/3)\varepsilon_e^{n-1}\xi_e^m \varepsilon_{ij}, \quad S_{ij} = (2K/3)\varepsilon_e^n \xi_e^{m-1}\xi_{ij} \tag{3.139a,b}$$

and in a more complex form

$$S_{ij} = A_1\varepsilon_{ij} + B_1\xi_{ij}, \quad S_{ij} = A_2\xi_{ij} + B_2\, \xi_{ip}\xi_{pj}, \quad S_{ij} = \Lambda\varepsilon_{ip}\xi_{pj} \tag{3.140a,b,c}$$

Equations (3.139) and (3.140a) are the so called linear tensor equations while Eq.(3.140b,c) are non-linear tensor equations. Of course, it is possible to suggest other forms. It is important to note that each such representation has its own physical (mechanical) significance and range of applicability, e.g., Eqs. (3.139) and

(3.140) are two variants. One can show that each of these equations leads to the same Eq. (3.137) for a tensile test. It can also be shown that a contraction of tensors in Eqs. (3.139) leads always to Eq. (3.138) but with Eqs. (3.140) it is strictly not so. Now some particular cases of application of the above generalisations will be considered. (i) A strain rate insensitive material (m = 0): In this case Eq. (3.137) transforms into the well-known strain hardening equation $\sigma = K\varepsilon^n$. Then, one can see that Eq. (3.139a) corresponds to the deformation theory of plasticity while Eq. (3.139b) corresponds to the Levy–Mises theory of plastic flow. According to the theorem on simple loading due to Iljushin [169] the results obtained by the use of these two extensions of the strain hardening equation $\sigma = K\varepsilon^n$ coincide for the case of simple loading only. (ii) Non-linear viscous material: In this case Eq. (3. 137) transforms into the well-known relation $\sigma = K\xi^m$. Then one can see that (3.139b) corresponds to viscous flow. Equation (3.139a) corresponds to an idea suggested in 1943 [377], viz., in creep stress and strain are related by finite relationships that are fully analogous to the relationships in the deformation theory of plasticity. This idea was proposed with the objective of constructing a convenient technical theory of creep and was later used in many variants of the theory. Numerous experiments have confirmed that for creep (and hence for superplasticity), tensorial character of Eq. (3.139b) is more realistic than that of Eq.(3.139a). (iii) To make clear the differences in the meaning of Eqs. (3.139) and (3.140) let the following deformation programme (stress–strain states are assumed to be homogeneous) be considered: a) first step, OA is elongation up to $\varepsilon = \varepsilon^0$ = constant with $\xi_v = \xi^0$ = constant; b) second step, AB is a very small shear with $\xi_v = \xi^0$ = constant. This programme is the so called two branch deformation trajectory with an orthogonal break, in Iljushin's theory. This programme (A) is the same for both the variants (3.139), (3.140) because deformation on the first branch is simple. Since $\varepsilon_{ij}(A) \approx \varepsilon_{ij}(B)$ (second step is small and from Eq. (3.139a) it follows that, $S_{ij(A)} \cong S_{ij(B)}$, that is, the stress state at the end of the programme approximately equals the one at point A. But according to Eq. (3.139b) we have $S_{ij(A)} \neq 0$, i = j; $S_{ij(A)} = 0$, $i \neq j$; $S_{ij(B)} = 0$, $i = j$ and $|S_{12(B)}|$ has the order of $|S_{11(A)}|$. On the other hand, from Eq. (3.140c) it follows that an analagous analysis may be performed.

$$|S_{12(B)}| >> |S_{11(B)}|, \quad |S_{12(B)}| >> |S_{22(B)}| \; and \; S_{33(B)} = 0$$

for Eq. (3.140a and b) also.

Thus, it has to be concluded that there is no unique way of extending CE (3.137) for the 3D case even when the strains are infinitesimally small. Consequently even if one has the scalar CE (which may be absolutely adequate for uniaxial loading), one may not yet be able to solve the boundary value problems. It is necessary to select a concrete variant to expand to the 3D case. If one considers the case of finite strains (which is pertinent for analysing superplastic flow), additional problems will arise. In particular, for the same specific method of tensor extension of the same scalar CE, one has to select a specific pair of measures of stress and strain state; this selection can not be made unambiguously.

As mentioned earlier, in most technological processes the stress–strain state is not uniform. Laboratory tests under complex stress conditions present serious technical difficulties; also decoding the results is difficult. Therefore, most of the available experimental data pertain to simple cases of loading (usually, uniaxial tension or compression tests). It is *emphasised* that the tensor extensions listed above apply to the same scalar Eq.(3.137). If one suggests another scalar CE instead of Eq.(3.137) (which adequately accounts for the mechanical response of a given material under uniaxial loading) one will encounter the same problem. The available set of experimental data is sparse and it does not provide a reliable basis for a theory that is capable of describing the behaviour of a material under complex stresses.

The present status of the problem of tensor extension of CEs for superplastics and related difficulties are examined in Chap. 5. Here, it is noted that the following assumption is always made in the literature on superplasticity while using the scalar CEs of superplastic flow to analyse non-uniaxial and non-uniform stress–strain states. The scalar properties of the constitutive equations for superplastics are obtained from the corresponding scalar (uniaxial) CE, by introducing in this CE the equivalent stress, strain and strain rate (see, for example Eqs. (3.137) and (3.138)). *In other words, the tensor properties of the constitutive equations for superplastics are the same as those of an isotropic viscous liquid. That is, the deviators of stress and strain rate are proportional.*

$$\frac{S_{ij}}{\sqrt{S_{mn}S_{mn}}} = \frac{\xi_{ij}}{\sqrt{\xi_{pq}\,\xi_{pq}}}$$

or in the most common form

$$S_{ij} = 2\mu\xi_{ij} \tag{3.141}$$

In superplasticity, this equation characterizes the scalar properties of the material and is given by the relation

$$\mu = \frac{\sigma_e}{3\xi_e}$$

Comparing Eqs. (3.141) and (3.139), one can conclude that Eq. (3.139b) can always be used. It should be noted that replacement of

$$\xi_{ij} \;\; by \; \xi_{ij}^{p} = \xi_{ij} - \xi_{ij}^{e}$$

(ξ_{ij}^{e} is the elastic strain rate) gives a more accurate description.

3.5.2 Some Tensor Constitutive Equations

The hypothesis that leads to Eq. (3.141) is already in common use. First approximate experiments [378] have revealed that this hypothesis fails near the break in the deformation trajectory. For that case, the hypothesis should be expanded in an appropriate way using past experience. As mentioned earlier, substitution of ξ^{p} in place of ξ makes the CE more realistic. Still it does not allow the correct distribution of superplastic properties in case of complex loading (e.g., in the vicinity of the break in the deformation trajectory) and makes the solution of the correpronding boundary value problems more complicated. If a complication of the boundary value problem using more adequate CEs is permitted, then there will be at least two ways of generalizing Eq. (3.141).

1. One of the variants of the flow theory to cover the case of large deformation may be used. In addition, the scalar properties in the chosen variant of the flow theory must be formulated in a manner that will take into account the high strain-rate sensitivity of superplastics. Some examples of a treatment of this kind can be seen in special publications.

2. It may employ a simplified variant of the general relationship given in Iljushin's theory of elastic-plastic processes (Chap.2.4). In this theory, the coplanarity hypothesis for the vectors $\vec{\sigma}$, $d\vec{\sigma}$, $d\,\vec{\ni}$ is used [379–381]. That allows the general relations to be reduced to

$$\dot{\vec{\sigma}} = N\,\vec{p}_1 - M\vec{\sigma}^0,\ \ \vec{\sigma}^0 \equiv \frac{\vec{\sigma}}{\sigma},\ \ \vec{p}_1 = \frac{d\,\vec{y}}{ds} \tag{3.142}$$

where the dot denotes a derivative with respect to s (Chap.2.). Equation (3.142) is known as the three-term formula of Iljushin. The coefficients N and M represent the functionals of the deformation process; the following expressions for M are valid.

$$M = N\cos\vartheta - \dot{\sigma},\ \cos\vartheta \equiv \cos\vartheta_1 = \vec{\sigma}_0\vec{p}_1 \tag{3.143}$$

Therefore, one can rewrite Eq. (3.142) as

$$\dot{\vec{\sigma}} = N\ \vec{p}_1 - (N\cos\vartheta - \dot{\sigma})\vec{\sigma}_0 \tag{3.144}$$

The three term Eqs.(3.144) are well investigated for small deformations. Various simplifications that allow the construction of specific variants for M and N [220, 248] are available. There are also generalisations of the three term equation for large deformations for the class of tensor measures of rotational type of stress–strain state. Therefore, it is reasonable to use them in mechanics of superplasticity. It is necessary to note that the Prandtl–Reuss equation (Chap. 2) can be considered as a particular case of the three-term formula (3.144).

3.6 Material Constants from Technological Tests

Mathematical modelling of metalworking processes includes the formulation of the boundary value problems (BPs). Choice of suitable constitutive equations represents the central problem in formulating the BP. Evidently, an adequate model for any metalworking process can not be developed if one uses an inadequate CE; in this case even the use of powerful computers will not help. Even when one uses an adequate CE, one has to develop reliable methods to determine the material constants. This problem may prove to be rather difficult to solve even when one can realize experimentally a uniform stress–strain state (e.g., on account of the large number of material constants to be determined). If a non-uniform stress–strain state is present during an experiment, the problem of determining the material constants becomes significantly more difficult. There are only a few examples where one can estimate the stress–strain state analytically or by using effective numerical procedures. In practice, often, a number of additional hypotheses are assumed. Superplasticity is characterised by finite strain behaviour. So it is not possible to attain in practice a uniform stress–strain state. Similarly, one can not achieve experimentally the loading paths that allow the 'decoding' of the data recorded (except for the case of fixed-end torsion). Therefore, the following scheme to determine experimentally the material constants from technological experiments may be considered.

Technological experiments are to be performed so as to simulate the experimental conditions found in the technological process to be modelled. The similarity could be based on the material, geometry, loading conditions, etc. A simplified mathematical model of the technological process is to be then developed. This model should relate the experimental data recorded during the technological experiments to the values of the material constants to be determined. Evidently, one has to simplify the formulation of the BP with the aim of obtaining an analytical solution. In particular, one can simplify the boundary conditions and/or make some assumptions concerning the velocity field present in the material being formed. The following advantages of this procedure can be noted.

1. Deformation trajectories in Iljushin space realised during technological experiments are close to those found in technological processes. In this case there is no need to establish the field of applicability of the CE used.
2. Errors introduced into the model of the technological process due to a simplification of the formulation of the BP can be somewhat compensated by the estimated values of the material constants. Evidently, the values of these material constants may not be the same as the ones determined from uniaxial tension tests. Such a discrepancy is to be attributed to the inadequacy of the simplified model for the technological process.
3. The initial structural state of the specimen used in technological experiments will often be closer to that seen during the technological process than would be the case with uniaxial testing.

4. The method of heating the specimen as well as the influence of the surrounding atmosphere are likely to be closer in this case to industrial practice than during tesile testing. In particular, one can take into account preliminary heating of performs, heating needed to achieve phase equilibrium, etc.
5. Technological experiments may be used as an alternative method of determining material properties when it is not possible to do so using tensile testing. (For example, when it is difficult to prepare standard specimens for tensile testing and/or it is difficult to prevent oxidation.)
6. Finally, technological experiments could form a part of a new process being developed.

3.6.1 Inverse Problems

From a mathematical point of view, determination of the material constants of a CE from experimental data can be considered as a typical example of the so-called 'inverse problem', which is well-known in mathematical physics. It is necessary to develop special methods to decode the experimental data in order to determine the material constants.

Experiments in which a uniform stress–strain state is present are the simplest to decode. In this case the mathematical formulation of the problem can be as follows: one knows directly from the experimental data the strain history, $\varepsilon_{ij} = \varepsilon_{ij}^{\exp}(t)$, and the stress history $\sigma_{ij} = \sigma_{ij}^{\exp}(t)$. The concrete functional form of the CE is known. However, the values of the material constants are unknown. The problem is to fit the experimental data to obtain an appropriate set of values for the material constants. In other words, one has to select the values of the material constants so that when the known functions $\varepsilon_{ij} = \varepsilon_{ij}^{\exp}(t)$ are substituted into the CE, the material response $\sigma_{ij}^{calc}(t)$ should be very close to the experimentally determined functions

$$\sigma_{ij} = \sigma_{ij}^{\exp}(t); \qquad \left|\sigma_{ij}^{calc}(t) - \sigma_{ij}^{\exp}(t)\right| \le \delta\sigma^*$$

where $0 \le \delta << 1$, σ^* is some characteristic value of the stress.

During superplastic flow it is not possible to achieve a uniform stress–strain state. In spite of considerable elongation, the stress and strain in the gauge length of deformed specimens is not uniform since, in most cases, short gauge length specimens are used. Compression tests at high temperatures are accompanied by considerable non-uniformity due to the action of contact friction forces. A classic method of studying complex loading problems in the theory of plasticity involving infinitesimally small strains is to use standard thin-walled tube specimens. However, superplastic deformation of such specimens will invariably be accompanied by shape distortions. Therefore, one will have to use thick-walled or solid cylindrical specimens for investigating the mechanical response of superplastics under multiaxial loading conditions. In this case one encounters another kind of inverse problem which is of interest to mechanics of superplasticity. This problem includes both an inverse boundary value problem and the problem of determining the material constants of the CE. In general, this is an ill-proposed problem in mathematical physics. Such problems can be solved only if some additional assumptions are made. Presently this problem is stated in detail.

It is possible to determine the stress–strain state throughout a specimen if the following information is available.

1. Integral loads on the grip ends: axial force $P(t)$ and torque $\overline{M}(t)$.
2. Total elongation $\Delta L(t)$ and the angle of rotation $\varphi(t)$.
3. Lateral surface of specimen is not subjected to surface loading.

It is emphasized that the values of the material constants of the CE are not known and so they should also be determined.

One can conclude from the above considerations that the problem stated is incorrect for the following two reasons: (i) Boundary conditions are given using an integral form of presentation, besides their being overdetermined; (ii) The CE is not fully determined (the values of the material constants are not available). Such a situation is typical of a wide range of problems in mechanics of solids (and consequently, mechanics of superplastic flow) where the values of material constants are to be determined from experimental results pertaining to a non-uniform stress–strain state.

As stated earlier, ill-defined inverse problems can not be solved successfully without adopting additional hypotheses. As for the above problem an additional kinematic hypothesis may be adopted, e.g., plane sections remain planar. In the next sections it is shown that the simplified mathematical model of superplastic forming of a thin circular membrane, as well as that of a long rectangular membrane can be used to evaluate the material constants.

In general, all metalworking processes are characterised by a non-uniform stress –strain state. As mentioned earlier, extending the scalar CE to include non-uniform stress–strain states is non-trivial as it does not lead to a unique solution. Therefore, multiaxial loading to study the tensor properties of the CEs is necessary. The use of standard thin walled tubes for testing gets ruled out because it is not possible to avoid shape distortions under large deformation. Therefore, it will be necessary to use thick-walled or solid cylindrical specimens in multiaxial loading experiments. In this case, the measured values of the axial force P and the torque $\overline{M}$ represent the overall response of the superplastic. This creates a problem while decoding the experimental time dependence of the axial force and the torque into the time dependence of the stress tensor components. This is an example of a complex and typical inverse problem in mechanics of superplasticity.

Since difficulties exist in the identification of CEs for the non-uniform stress–strain state and also different problems arise with respect to the correct formulation of the boundary value problem (while using a given CE) for specific technological processes, it is desirable to determine the material constants with the same equipment that is used in the technological process. In this case one has to develop methods for decoding the results to estimate numerically the material constants that enter the CE for superplastics. From a mathematical point of view, this is also an inverse problem. And its solution is possible if a simplified model for the technological process is constructed and a full constitutive equation is vailable. Hence, the following general approach can be suggested to solve such inverse problems:

1. Development of a simplified model for the technological process;
2. Development of methods to determine the material constants directly from the technological experiment, these methods being based on the above mentioned simplified model.
3. Calculation of material properties and technological parameters.

In the following Sections the standard power law $\sigma = K\xi^{m}$ is considered along with the model of Padmanabhan and Schlipf. It is shown that the simplified mathematical model of superplastic forming of a thin circular membrane, as well as that of a long rectangular membrane, can be used to find out the corresponding material constants.

3.6.2 Constant Pressure Forming of a Rectangular Membrane

As superplastic deformation is carried out at high homologous temperatures the specimen is thermally equilibrated for about 15 minutes before the commencement of a mechanical test. Further, high temperature exposure will affect the percentage elongation. If a specimen is from a thin sheet, of say 1.0 mm thickness, the properties can be severely altered by oxidation. Application of a protective coating, on the other hand, can alter the force–time diagram. Use of a protective atmosphere is incompatible with most testing machines in view of the extreme tensile elongation.

As steep temperature gradients are to be avoided throughout a test, extremely short specimens have been used to report elongations in excess of 4500%. The analysis of these tests has been faulty: the uniaxial stress state assumed would not be present; without significant grip flow such elongations can not be achieved (i.e., volume constancy assumed in the analysis is not valid). In contrast, in a test using a technological equipment, a protective environment as well as thermo-mechanical processing schedules very close to industrial practice can be introduced.

The procedure of determining the superplastic properties of a thin sheet material using a constant pressure test on a long rectangular membrane is described below. The analysis of results is based on the following assumptions:

1. The material is uniform, isotropic and incompressible.
2. Elastic strains are negligible.
3. Strain rates are sufficiently small so that equilibrium equations can be considered instead of equations of movement in a continuum.
4. The initial thickness of the sheet is sufficiently small so that bending and shearing effects are negligible and membrane theory may be used.
5. The temperature is uniform.

Let the long rectangular membrane form into a die of semi-width W and depth D (Fig. 3.37). This problem has already been considered, see, e.g., [184, 282, 382, 383]. Further attention will be focussed on the constants K and m in Eqs. (3.4).

Let the length of the membrane L be sufficiently large ($W/L << 1$). Assuming that the initial thickness is small ($h_0 << W$), it can be concluded that plane strain condition will prevail during forming at any cross-section of the membrane. The two components of the strain rate tensor are the circumferential ξ_c and the normal ξ_n. The tangential strain rate $\xi_t = 0$. From the incompressibility condition $\xi_n + \xi_c = 0$. Therefore,

$$\xi_t = -\xi_n = -\frac{\dot{s}}{s} \tag{3.145}$$

When the balance of forces on a small element of the membrane (Fig. 3.38) is considered, projections of the equilibrium condition on the X and Y axes lead to

$$d(\sigma_c s) = 0, \sigma_c = \frac{p\rho}{s} \tag{3.146}$$

where σ_c is the circumferential stress and ρ is the radius of curvature. The following relations are then valid: $\sigma_t\, s = p\rho = C$, where C is constant along the shell profile. Thus for the case of uniform pressure ($p \neq p(\varphi)$), the radius of curvature of the median plane is constant along the shell profile. Consequently, the median plane represents a part of a cylindrical surface with a half arc angle α (Fig. 3.37). For this case, if the thickness of the membrane was uniform before bulging, then it would be uniform after bulging also. This coupled with the incompressibility condition ($s_0 W = sR\alpha$) gives

$$s = \frac{s_0 W}{R\alpha} = s_0 \left(\frac{\sin\alpha}{\alpha} \right) \tag{3.147}$$

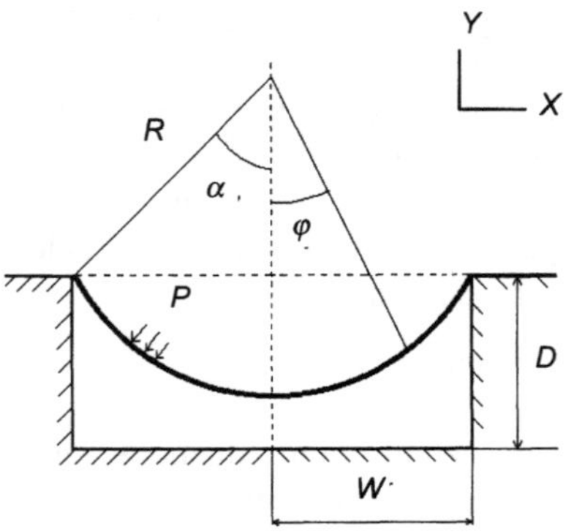

Fig. 3.37. Schematic of a long rectangular box section

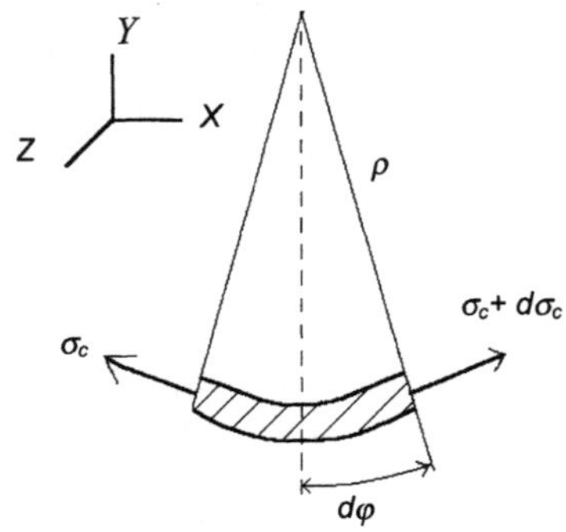

Fig. 3.38. Equilibrium of a membrane element

The effective strain rate then is

$$\xi_e = \sqrt{2\xi_{ij}\xi_{ij}/3} = 2\xi_t/\sqrt{3} = \frac{2}{\sqrt{3}}\left(\frac{1}{\alpha} - \cot\alpha\right)\dot{\alpha} \tag{3.148}$$

The value of the axial stress can be found from the proportionality between the stress and the strain rate deviator.

$$\xi_1 = S_1 \equiv \sigma_1 - \frac{1}{3}(\sigma_1 + \sigma_c), \xi_1 = 0 \Rightarrow \sigma_1 = \frac{1}{2}\sigma_c$$

Then, the effective stress σ_e is given by

$$\sigma_e = \sqrt{\frac{3}{2}S_{ij}S_{ij}} = \frac{\sqrt{3}}{2}\sigma_t = \frac{\sqrt{3}}{2}\cdot\frac{pR}{s} \tag{3.149}$$

Using Eq.(3.4) as the CE, one obtains

$$\sigma_e = K\xi_e^m = K\cdot\left[\frac{2}{\sqrt{3}}\cdot\left(\frac{1}{\alpha} - \cot\alpha\right)\dot{\alpha}\right]^m = \frac{\sqrt{3}}{2}\cdot\frac{pW}{s_0}\cdot\frac{\alpha}{\sin^2\alpha} \tag{3.150}$$

This equation is an ordinary differential equation of the first order as to the unknown function $\alpha(t)$. Its solution can be found by numerical methods. In particular, for the case of constant gas pressure forming $p = p_0 =$ constant

$$t\frac{\sqrt{3}}{2}\cdot\left(\frac{\sqrt{3}}{2}\cdot\frac{pW}{s_0K}\right)^{1/m} = J_m(\alpha) \overset{def}{=} \int_0^\alpha \left[\frac{\sin^2 x}{x}\right]^{1/m}\left(\frac{1}{x} - \cot x\right)dx \tag{3.151a}$$

Eq. (3.151a) gives the dependence of $\alpha(t)$ (in an implicit way). Let the reference value of stress be σ_q and that of strain rate be ξ_q. That is,

$$\sigma_q = \frac{\sqrt{3}p_0W}{(2s_0)} = K(\xi_q)^m, \xi_q = \left(\frac{\sigma_q}{K}\right)^{\frac{1}{m}} \tag{3.152}$$

Then one can rewrite Eq. (3.151a) as

$$t^* = \frac{\sqrt{3}\xi_q t}{2} = J_m(\alpha) \tag{3.151b}$$

where t* is normalised time. Time dependence of the dome height H is defined by the expression

$$H = W\ tan\left(\frac{\alpha}{2}\right) \tag{3.153}$$

where $\alpha(t)$ is from Eq. (3.151a). The duration of the process is defined by the condition $H(t_k) = D$. It is easy to show that $\alpha_k = 2\arctan^{-1}(H/W)$. Combining Eqs. (3.4), (3.151) and (3.152) the following relations for the strain rate are obtained.

$$\frac{\sigma_e}{\sigma_q} = \left(\frac{\xi_e}{\xi_q}\right)^m = \frac{\alpha}{sin^2\alpha} = 2\,tan^{-1}\,\overline{H}\cdot\left[\frac{1+\overline{H}^2}{2\overline{H}}\right]^2 \tag{3.154}$$

where $\overline{H} = (H/W)$ is the relative height. Theoretical dependence of $H(t)$ calculated according to Eqs. (3.151)–(3.153) is presented in Fig. 3.39 in normalised co-ordinates. Theoretical dependence of the relative strain rate on the relative dome height is presented in Fig. 3.40. One can conclude from this figure that the strain rate varies significantly during superplastic forming. This was pointed out by Jovane [384] for the case of constant gas pressure forming of a circular diaphragm.

The next step is to evaluate *K* and *m* in Eq. (3.4) using the above simplified model. From Eq. (3.151) the duration of constant pressure forming to the same relative height is defined by the expression pt^m = constant and so the value of m can be determined as

$$m = \ln\left(\frac{p_2}{p_1}\right)\Big/\ln\left(\frac{t_2}{t_1}\right) \tag{3.155}$$

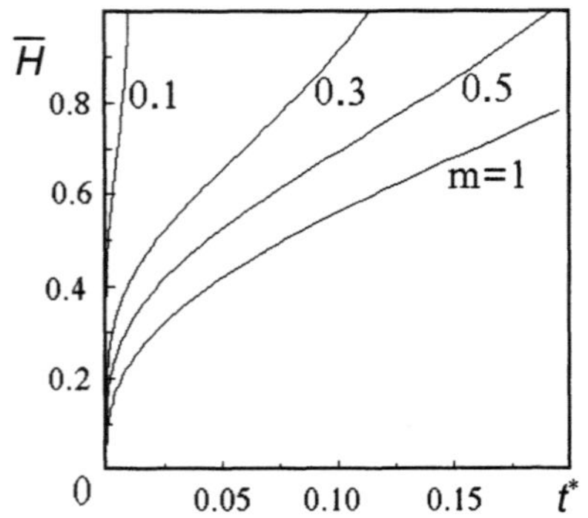

Fig. 3.39. The dependence of the relative height on normalised time calculated according to Eqs (3.151) and (3.153) for various *m* values (numbers near the curves)

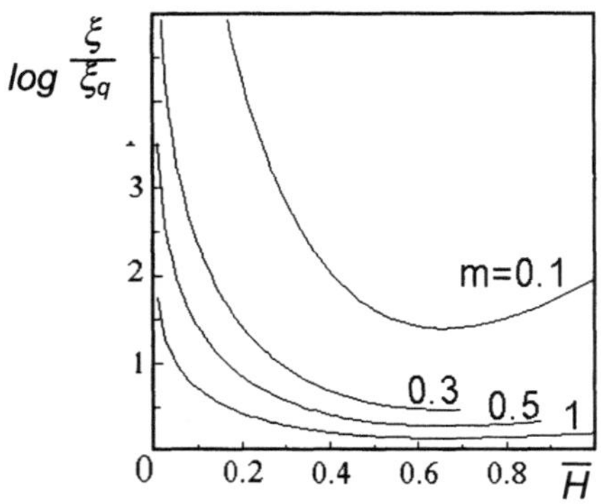

Fig. 3.40. The dependence of the relative strain rate on the relative height calculated according to Eq. (3.154) for various m values (numbers near the curves)

Table 3.13 Material constants K and m for Ti–6Al–4V alloy sheet of 0.8 mm thickness. Constant gas pressure tests using rectangular membranes

i–j	1–2	1–3	2–3	Average
m	0.452	0.454	0.456	0.454
K (MPa.s^m)	623.1	623.9	623.9	623

t_1 = 1380 s.; t_2 = 658 s.; t_3 = 300 s.

where t_1 and t_2 are the durations of superplastic forming to the same depth D under constant pressures $p = p_1$ and $p = p_2$ respectively. That is, $H_1(t_1) = H_2(t_2) = D$.

The value of K can be found from Eq. (3.151) if m is known. But, it is more reliable to find K from

$$K = \frac{K_1 + K_2}{2} \tag{3.156}$$

where

$$K = \frac{\sqrt{3}}{2} \cdot \frac{p_f W}{s_0 K} \cdot \left[\frac{\sqrt{3}}{2} \cdot \frac{t_f}{J_m(\alpha_k)}\right]^m, i = 1,2 \tag{3.157}$$

If the durations of constant gas pressure forming to different relative heights are known, the value of m can be calculated from the relation

$$m = \frac{\ln(p_2 / p_1)}{\ln(t_1 J_m(\alpha_2) / t_2 J_m(\alpha_1))} \tag{3.158}$$

The value of K is

$$K_i = \frac{\sqrt{3}}{2} \cdot \frac{p_1 W}{s_0 K} \cdot \left[\frac{\sqrt{3}}{2} \cdot \frac{t_i}{J_m(\alpha_i)}\right]^m \quad (i = 1, 2) \tag{3.159}$$

Evidently, an iterative procedure is involved. When $H_1 = H_2$ ($\alpha_1 = \alpha_2$), Eqs. (3.158) and (3.159) will coincide with Eqs. (3.155) and. (3.157). Vasin et al. [93] (see also [57]) have analysed their results using Eqs.(3.155), (3.156) and the findings are reported in Table 3.13

It was found that the theoretical duration calculated with $m = 0.454$, $K = 623$ MPa.s^m in Eqs. (3.151) and (3.153) are 1380, 658 and 300 s for gas pressures 5, 7 and 10 atm respectively. Thus, the theoretical estimates practically coincided with the experimentally measured values. One can also evaluate numerically the strain rate intervals of constant pressure forming. According to Eq. (3.150)

$$\log \xi_e = \frac{1}{m} \cdot \log\left[\frac{\sqrt{3}}{2} \cdot \frac{p_0 W}{K s_0} \cdot \frac{\alpha}{\sin^2 \alpha}\right] \tag{3.160}$$

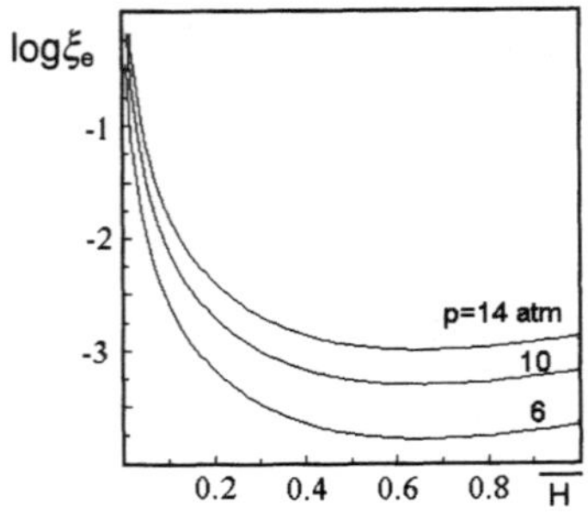

Fig. 3.41. Dependence of equivalent strain rate ξ_e on relative height calculated using Eqs. 3.148 –154. $m = 0.47$, $K = 645$ MPa·s^m (Different constant pressures p_o used in the calculations are indicated near the curves.)

In Fig. 3.41 the dependence of the equivalent strain rate on the relative height is presented. From the figure, it is clear that ξ_e varies from ~10^{-4} s^{-1} to about 10^{-1} s^{-1}. It is approximately constant in the range $1 > \overline{H} > 0.4$

3.6.3 Constant Pressure Forming of a Circular Membrane

In [57] the material constants *K* and *m* have been estimated from results pertaining to constant gas pressure forming of a circular membrane (Table 3.14). Using the above values of *m* and *K*, behaviour in an approximately constant strain rate forming test can be predicted and compared with the experimental results (Table 3.15). One may consider the constant strain rate test as a crucial experiment while the constant gas pressure test may be regarded as the basic one.

Table 3.14 *K* and *m* values for constant gas pressure forming of circular sheets [57]

i–j	1–2	1–3	2–3	Average
m	0.429	0.431	0.432	0.431
K (MPa.s^m)	410.7	410.9	410.8	411

$t_1 = 1500$ s.; $t_2 = 685$ s.; $t_3 = 300$ s.

3.6.4 Model of Padmanabhan and Schlipf

Repeating the above analysis with CE Eq.(3.115), one obtains

$$\xi_p = 2\dot{\alpha}(1/\alpha - \cot\alpha) = a\sigma^2 + b\sigma + c =$$

$$a\left(\frac{p_0 R_0}{2s_0}\right)^2 \cdot \left(\frac{\alpha^2}{\sin^3\alpha}\right) + b\frac{p_0 R_0}{2s_0} \cdot \frac{\alpha^2}{\sin^3\alpha} + c \qquad (3.161)$$

Table 3.15 Duration of constant ξ_q forming of a circular diaphragm [57]

Strain rate ξ_q (s^{-1})	Stress σ_q (MPa)	Duration (calc.) (Min–s)	Duration (exp.) (Min–s)	Error %
$4 \cdot 10^{-4}$	14	37–36	39–56	6
$8 \cdot 10^{-4}$	19	18–48	19–50	5
$1.2 \cdot 10^{-3}$	23	12–30	14–00	11

where ξ_p is the equivalent strain rate at the pole. Equation (3.161) is an ordinary differential equation as to the unknown function $\alpha(t)$. Solution of this equation satisfying the initial condition $\alpha(0) = 0$ results in

$$t = \int_0^{\alpha} \frac{2\left(\frac{1}{x} - \cot x\right)}{a\left(\frac{p_0 R_0}{2s_0}\right)^2 \frac{x^4}{\sin^6 x} + b\frac{p_0 R_0}{2s_0}\frac{x^2}{\sin^3 x} + c} dx \tag{3.162}$$

Time dependence of dome height is defined by the relation

$$h_p(t) = \tan\left[\frac{\alpha(t)}{2}\right] \tag{3.163}$$

where $\alpha(t)$ is from Eq.(3.162). One can see that it is rather difficult to determine the values of a, b and c using Eq. (3.162). Therefore, it is reasonable to use the constant gas pressure test as a crucial experiment to verify the Padmanabhan–Schlipf model.

4 Boundary Value Problems in Theory of Superplastic Metalworking

Boundary Value Problems (BPs) in mechanics of superplasticity are not well developed. Inadequacies in the development of the constitutive equations for superplastics are examined below using simple examples.

The topic of boundary value problems was discussed qualitatively in Chap. 2. BPs are solved, as a rule, using numerical methods. As already mentioned, the choice of a constitutive equation to be included in the formulation of a BP is the central problem in mechanics of solids. It is noted correctly in [385], "if the governing law is established, the appropriate mathematical description of the phenomenon can be always obtained, and the true nature comes into view in the processing and arrangement of the digits; this is the central doctrine of computational mechanics". Most reports in which numerical methods are used to solve the BPs of metalworking in the superplastic state deal with non-linear viscous liquids. Some examples are considered in this chapter and some model tasks are completed.

4.1 General Formulation of the Boundary Value Problem for Metalworking Processes

4.1.1 Basic Concepts and Principal Equations

Euler's method of description of motion in a continuum is used. All equations are written with respect to a fixed Cartesian system of co-ordinates $x = (x_1, x_2, x_3)$. The velocity field $\overline{V}$, with components V_1, V_2, V_3 is assumed further as the main kinematic characteristic of the process. In view of the incompressibility condition, one can write the laws of conservation of mass and momentum as

$$div\overline{V} = \frac{\partial V_i}{\partial x_i} \equiv V_{i,i} \equiv \xi_v = 0 \qquad (4.1)$$

$$\frac{\rho dV_i}{dt} \equiv \rho\left[\frac{\partial V_i}{\partial t} + \frac{v_j \partial V_i}{\partial x_j}\right] = -\frac{\partial p_H}{\partial x_i} + \frac{\partial S_{ij}}{\partial x_j} + \rho F_i \qquad (4.2a)$$

where S_{ij} are the deviatoric components of the stress tensor, p_H is the hydrostatic pressure, F_i are the body forces and ξ_v is the volumetric strain rate. Summation rule with respect to repeated indices is assumed.

In many problems, body and inertia forces are small compared with other forces. Therefore, it is often possible to neglect body and inertia forces in Eq. (4.2a) and to consider the following equilibrium equations.

$$\frac{\partial S_{ij}}{\partial x_j} - \frac{\partial p_H}{\partial x_i} = 0 \tag{4.2b}$$

Tensor form of the constitutive equations is given by the condition of proportionality between the stress and the strain rate deviators

$$S_{ij} = \left(\frac{2\sigma_e}{3\xi_e}\right)\xi_{ij} \tag{4.3}$$

The scalar form of the constitutive equation is

$$\sigma_e = \phi(\xi_e, s, T) \tag{4.4}$$

where T is temperature. And

$$\begin{gathered} \xi_{ij} = \left(\frac{1}{2}\right)(V_{i,j} + V_{j,i}) - \left(\frac{1}{3}\right)\xi_v \delta_{ij} \\ \xi_e^2 = \left(\frac{2}{3}\right)\xi_{ij}\xi_{ij}, \sigma_e^2 = \left(\frac{3}{2}\right)S_{ij}S_{ij} \end{gathered} \tag{4.5}$$

$$s = \int_0^t \xi_e dt' \tag{4.6}$$

where the integration is effected along the path of material particle motion. According to the second law of thermodynamics, the rate of increase of entropy is defined by

$$\frac{\rho T d\eta}{dt} = -\frac{\partial q_i}{\partial x_i} + w^* \tag{4.7}$$

where η is the entropy, $\vec{q}$ is the vector of the heat flux, w^* is the so-called uncompensated heat of dissipation (a part of the work done by the internal forces which transforms into heat). The following commonly recognised hypotheses are also assumed in case of metalworking processes.

1. Fourier's law of heat transfer

$$\vec{q} = -\lambda\, grad\ T \tag{4.8}$$

2. Certain portion k of the work done by the internal shear stress is dissipated ($0 < k \leq 1$). Taking into consideration that $S_{ij}V_{ij} = (2\sigma_e/3\xi_e)\ V_{ij}\ V_{ij} = \sigma_e\xi_e$, one obtains

$$w^* = k\sigma_e\xi_e - Q_c \tag{4.9}$$

where Q_c is the heat release (or absorption) due to microstructural changes /transformations. If $Q_c = 0$, then Eq. (4.9) becomes

$$w^* = k\sigma_e\xi_e \tag{4.10}$$

Entropy depends on temperature only. (This is a consequence of the supposition that the internal energy depends on temperature.) Therefore,

$$\frac{Td\eta}{dt} = T\left(\frac{d\eta}{dT}\right)\left(\frac{dT}{dt}\right) = c_0\frac{dT}{dt} \tag{4.11}$$

where $c_0 \equiv T\ (\mathrm{d}\eta/\mathrm{d}T)$ is the coefficient of specific heat which, in general, is temperature dependent. Using equations (4.8–4.10) one can rewrite Eq. (4.7) as

$$\rho c_0\frac{dT}{dt} = \frac{\partial}{\partial x_i}\left(\lambda\frac{\partial T}{\partial x_i}\right) + k\sigma_e\xi_e \tag{4.12}$$

If c_0 and λ are assumed to be constants, the following standard equation of heat transfer is obtained.

$$\frac{dT}{dt} \equiv \frac{\partial T}{\partial t} + V_i\frac{\partial T}{\partial x_i} = a^2\Delta T + \zeta\sigma_e\xi_e \tag{4.13}$$

where $a^2 = \lambda/\rho c_0$, $\zeta = k/\rho c_0$, Δ is standard Laplace operator.

4.1.2 Initial and Boundary Conditions

Above equations determine the behaviour of the deforming body, if the initial and boundary conditions are also specified. Before the start of the deformation process, let the body be unstressed and undeformed (natural initial conditions).

$$t = 0 : \vec{V} = 0, \quad \sigma_{ij} = 0, \quad s = 0 \tag{4.14}$$

Let the initial temperature distributions in the material $T(x)$ and at the instrument $T^{\text{in}}(x)$ be known as well:

$$t = 0: \ T = T_0 \Psi(\vec{x}) \ \ and \ \ T^{in} = T_0^{in} \Psi^{in}(\vec{x}) \tag{4.15}$$

where T_0 and T_0^{in} are the characteristic temperatures of the deforming body and the instrument, respectively. $\Psi(x)$ and $\Psi^{\text{in}}(x)$ are given functions. Now the boundary conditions (BC) are specified.

Temperature BC: Let Γ be the total surface of the deforming body. The temperature BC may be of the following 4 types:

$$T\big|_{\Gamma} = T_{env} \tag{4.16a}$$

$$\lambda \frac{\partial T}{\partial n}\bigg|_{\Gamma} = -q^0(\vec{x}_b, t), \ \vec{x}_b \subset \Gamma \tag{4.16b}$$

$$\lambda \frac{\partial T}{\partial n}\bigg|_{\Gamma} = -\alpha\left(T(\vec{x}_b, t) - T_{env}\right), \ \vec{x}_b \subset \Gamma \tag{4.16c}$$

$$\lambda \frac{\partial T}{\partial n}\bigg|_{\Gamma} = \lambda^{in} \frac{\partial T^{in}}{\partial n}\bigg|_{\Gamma} \tag{4.16d}$$

where T_{env} is the environment temperature; λ, λ^{in} are heat conductivity coefficients for the deforming body and the instrument, respectively; $\vec{q}^{\,0}$ is a given vector of heat flux supplied to the deforming body; α is the heat transfer coefficient; $\partial/\partial n$ means a derivative with respect to the normal to the surface Γ. The temperature BC (4.16a–d) are referred to as the condition of II, III, IV and I respectively. The physical meanings of these conditions are as follows: the temperature on the contact surface (I) or the heat flux (II) is given on Γ. The temperature BC of type I can be used while modelling isothermal forming, while type II is pertinent to inductive heating (e.g., die-less wire-drawing). If cooling takes place on Γ, the BC of type III can be used (e.g., air cooling of sheet metal after rolling, or water cooling of wire during die-less wire drawing). The BC of type IV corresponds to contact between two solids (e.g., instrument and deforming body). If it is necessary to take into account heat release due to friction at the contact surface between the instrument and the body, one can modify Eq. (4.16d) as

$$\lambda \frac{\partial T}{\partial n}\bigg|_{\Gamma} = \lambda^{in} \frac{\partial T^{in}}{\partial n}\bigg|_{\Gamma} + q_{fr} \tag{4.17}$$

$$q_{fr} = \frac{\tau_m |\Delta \vec{w}|}{J} \tag{4.18}$$

where τ_m is the shear stress of the material just below the contact surface; $|\Delta \vec{w}|$ is the absolute value of the vector of relative velocity of sliding of the material with respect to the instrument, J is the mechanical equivalent of heat. It should be mentioned that one could consider a mixed temperature BC as different combinations of the above-described conditions (4.16a–d). For example, for the compression of a disc, type IV can be adopted at the contact 'instrument–disc' while a condition of type II can be assumed on the free surface of the disc. One can also assume conditions of different types to be present in different parts of the surface Γ. It is emphasised that in this case as well as in all the other above mentioned cases, we deal with certain simplifications. The real conditions of heat interaction at the contact surfaces are more complicated. For example, sometimes non-ideal contact is present when the temperature BC depends on the stress state at the contact surface (on the normal contact force for instance). Then the problem of determining the stress–strain state in a deforming body can become related to the temperature problem.

Consider now the mechanical BCs. As mentioned in Chap. 2, two kinds of mechanical BCs can be used: kinematic BCs (displacements or velocities are given at the boundary) and dynamic BCs (distributed forces are given on the boundary). Sometimes some intermediate situations are possible. In classical theory of elasticity, a mixed mechanical BC is given on two different parts of a deforming body. Also, a mixed mechanical BC can be given as a superposition of the kinematic and the dynamic BCs on the same surface. In metalworking problems different situations are possible.

Let the surface of the deforming body be $\Gamma = \Gamma_1 + \Gamma_2$, where Γ_1 is the free surface while Γ_2 is the contact surface 'instrument–material'. Then we have on Γ_1:

$$\Gamma_1 : \sigma_{ij} n_j = 0 \tag{4.19}$$

(n_j are the components of the unit vector, which is orthogonal to Γ_1). As far as the contact surface Γ_2 is concerned, some mixed BC can be considered. In practice the following two limiting cases are used: (i) the condition of full sticking of the material to the instrument; (ii) the condition of free sliding (without friction) of the material along the instrument.

For the first case, the velocities of the particles of the deforming material are equal those of the instrument.

$$\Gamma_2 : \overline{V} = \vec{f}(\vec{x}_b, t) \tag{4.20}$$

where $\vec{f}$ is a given function. For example, in the compression of a superplastic ceramic $YBa_2Cu_3O_7$, at the contact zone Γ_2 the condition of full adherence is often assumed. In the second case, on Γ_2 a mixed kinematic-dynamic BC is given: the condition of non-penetration of the material inside the instrument through Γ_2, and the condition that the shear force on the contact surface is equal to zero. If the instrument surface does not move and can be described by the equation $F(x_1, x_2, x_3) = 0$, then the kinematic condition on particle velocity is

$$\Gamma_2 : VgradF = 0 \;\; or \;\; V_n = \vec{V} \cdot \vec{n} = 0, \;\; \vec{n} = grad\left(\frac{F}{|grad\ F|} \right) \tag{4.21a}$$

If the boundary can be displaced during deformation and can be described by the equation $F_R(x_1, x_2, x_3) = 0$, and a material particle belonging to Γ_2 at moment t has the co-ordinates $\vec{x}$, that is, $F(\vec{x}, t) = 0$, at time $t + dt$ this particle will have the co-ordinates $\vec{x} + \vec{V}dt$. Taking into account the condition of non-penetration, one has $F_R\left(\vec{x} + \vec{V}dt, t + dt\right) = 0$. Consequently

$$\Gamma_2 : \; \frac{\partial F_R}{\partial t} + \vec{v} gradF_R = 0 \tag{4.21b}$$

The conditions of ideal smoothness (shear stresses are negligible) are written as follows.

$$\Gamma_2 : \; \sigma_{ij} n_j = 0 \qquad or \qquad S_{ij} n_j = p_H n_i \tag{4.22}$$

The compression of a disc between parallel platens using a good lubricant (to avoid barrelling) can be considered an example of such a BC. It should be mentioned that the limiting BC Eqs.(4.20)–(4.22) are often used while modelling superplastic forming of sheet materials.

The more realistic mechanical BC is the condition of contact friction sliding. The friction factor is an important concept in metalworking which has a profound influence on the contact conditions, the quality of the part and its surfaces, the power, force parameters and hence the technical/economic characteristics of the metalworking process. The friction factor, often, determines to a considerable extent the reliability and the capability of the technological equipment.

Specific friction force is the projection of the friction force and vector directed tangentially at the contact surface. Attempt is made to determine the value of the specific friction force both experimentally and theoretically. The theoretical approach is based on the physics of friction surfaces. It should be noted that these investigations are very complicated. Use of the results of the theoretical investigations is limited by unknown details of friction conditions for a technological process. From a mathematical point of view determination of the force due to friction is a very complex non-linear inverse problem. Known approaches for solving this problem consist of considering a series of simple problems for chosen modes of friction force and comparing the results of the calculations with the experimental data. Therefore, a number of empirical models of *friction* interaction are available. The role of friction in metalworking is widely discussed, see, e.g., [184, 239, 313]. Some phenomenological descriptions of contact *friction* forces are considered below.

Let the deforming material slide along the instrument. Mathematically this means that the material velocity $\vec{V}$ differs from the instrument velocity $\vec{w}$ by the value $\Delta \vec{w} = \vec{V} - \vec{w}$. The vector $\vec{\tau}_m$ of the contact friction force is given as

$$\frac{\vec{\tau}_m}{|\vec{\tau}_m|} = -\frac{\Delta\vec{w}}{|\Delta\vec{w}|}, \; |\vec{\tau}_m| \equiv \tau_m = \varphi\left(|\vec{\sigma}_N|, |\Delta\vec{w}|, T, \tau_s, \chi_k\right) \tag{4.23}$$

where $\vec{\sigma}_N$ is the vector of the normal stress at the contact surface, τ_S is the shear yield stress, χ_k are the characteristics of the material surface and its near-contact layers.The simplest forms of the function φ are presently considered.

1. The value of the shear stress at the interface is assumed to be directly proportional to the stress normal to the interface (Amonton–Coulomb law or Coulomb friction).

$$\tau_m = k\sigma_N, \quad \sigma_N = |\vec{\sigma}_N| \tag{4.24}$$

   where k is a constant, which is to be found experimentally.
2. The value of the shear stress at the interface is assumed to be directly proportional to the yield stress in shear τ_S (Siebel's law):

$$\tau_m = \psi\tau_S \tag{4.25}$$

   where ψ is an empirical coefficient. This condition is often used in the mathematical modelling of metalworking processes at elevated temperatures.

There are some combinations of these laws, e.g., the Unksow equation [313, 386]:

$$\tau_m = \begin{cases} \mu\sigma_N, & if \quad \sigma_N \le \dfrac{\tau_k}{\mu} \\ \tau_k, & if \quad \sigma_N > \dfrac{\tau_k}{\mu} \end{cases} \tag{4.26}$$

where μ and τ_k are empirical constants. Levanov has suggested the following friction law [313, 386, 387]:

$$\tau_m = k_{max}\left[1 - exp\left(-1.25\frac{\sigma_N}{\sigma_S}\right)\right]\frac{\tau_s}{\sqrt{3}} \tag{4.27}$$

where $k_{\max}$ is an empirical constant. Eq. (4.27) transforms to Eq. (4.24) at large σ_N, while at low σ_N it reduces to Eq. (4.25).

Sometimes, one observes the strong dependence of the friction force on the sliding velocity, which reveals the *hydrodynamic* character of friction. As a first approximation, in analogy with Newton's law applied to the boundary layer, one can write

$$\tau = \frac{\eta V_s}{h} \tag{4.28}$$

where η is the dynamic viscosity of the lubricant, h is the thickness of the lubricant layer and V_s is the relative velocity of sliding at the contact surface. The next step is to use a non-linear expression on the right hand side of Eq. (4.28). Such approaches are known in the literature. But, they are yet to be used widely.

At this stage, some special features of deformation of viscoplastic materials are considered. A special characteristic is the presence of 'frozen' domains in the deforming material (dead metal zones). Additional conditions of sticking or internal sliding can be imposed at the boundaries of these domains, see, e.g., [387, 388]. Physically, this would mean that the viscoplastic material has a rigid spatial structure, which can counter the external influence until some limiting stress is reached. Above this stress level, the rigidity is lost and the material behaves like a viscous liquid under the action of a stress. The effective stress would then be the difference between the applied stress and the above mentioned limiting stress. When the effective stress decreases to 0, solid-like response is restored.

If there is a temperature gradient in the deforming material, another type of boundary may be present: the boundary between a melted and an unmelted (solid) material. As a rule, the exact location of such a boundary is not known in advance. In this case, the following BC is adopted: the heat flux through such a boundary is abrupt, the difference being equal to the specific heat of melting q_{melt}

$$\lambda\left.\frac{\partial T}{\partial n}\right|_1 = \lambda\left.\frac{\partial T}{\partial n}\right|_2 + q_{melt} \tag{4.29}$$

In the above equation, 1 and 2, respectively, denote the unmelted and the melted regions across the interface.

In the theory of metalworking a number of boundary value problems are solved within the framework of the theory of ideal plasticity. It is interesting to note that different solutions for the same problem satisfying all the initial and boundary conditions can be obtained in this case depending on the choice of the surfaces of internal division into domains. Then, additional conditions like the principle of minimum expenditure of power by the external forces are brought in. Often, simplifications are also achieved through approximations about the stress state present. Sometimes, it is possible to obtain the solution for ideal plasticity as a limiting case of the solution for viscoplastic behaviour. For example, the constitutive equation for Bingham plastic behaviour is $\sigma = A + B\xi$, where A and B are material constants. One can find the solution for ideal plastic behaviour from the solution of the viscoplasticity problem as $B \to 0$.

4.1.3 Damage Accumulation

In the recent reviews of Oh [391] and Kolmogorov [392], some aspects of fracture in metal forming are considered. Oh has not examined Russian publications, while Kolmogorov has considered mainly Russian works.

Gubkin [393] has suggested the following condition for deformation without fracture.

$$\varepsilon_i = a\varepsilon_{\max} < \varepsilon_p \tag{4.30}$$

where ε_i is the characteristic strain calculated for the process under study, $\varepsilon_{\max}$ is the maximum principal strain, a is a coefficient which depends on the type of stress state and ε_p is a quantity determined from experiments. ε_p depends on the material, temperature and stress state. (Following Bridgman, Gubkin has pointed out the influence of hydrostatic pressure on plasticity.) Kolmogorov [392] has noted that Eq. (4.30) would be more accurate if instead of ε_i the Odqvist parameter is used. Later Smirnov-Alyaev and Rosenberg [394] also suggested the use of the Odqvist parameter as a measure of plasticity. Presnyakov [395] has considered, in contrast, empirical issues with regard to metal formability.

Presently it is recognised that plastic deformation is often accompanied by the continual generation and development of microcracks and pores. The growth of micro-discontinuities (micro-damage) ends with the formation of a macro-crack, i.e., with the irreversible loss of properties in the product. In Russia, Kachanov [396] and Rabotnov [397] were the first to describe this process in a simple (scalar) version. Damage ω may develop with time according to the kinetic equation $d\omega/dt = f(\omega \ldots)$, where the function f, determined by experiments based on theoretical considerations, depends on ω and other variables important for the process under study. Kachanov and Rabotnov have suggested the normalisation of ω setting $\omega=0$ in the initial state and $\omega=1$ at the instant of macrofracture. This idea was later developed by Kolmogorov and his colleagues [398–400]. This theory has become one of the widely used phenomenological models of damage accumulation during large plastic deformation. The numerous experimental verifications clearly demonstrate its applicability in case of simple (proportional) loading. A brief outline of this model is presented below.

Let Λ_p be the accumulated shear strain just before fracture. Λ_p depends on the stress state: $\Lambda_p = \Lambda_p\ (I_1, I_2, I_3)$, where I_1, I_2, I_3 are the invariants of the stress tensor. This relation can be rewritten as

$$\Lambda_p = \Lambda_p(\sigma_m, I_{2D}, I_{3D}) \tag{4.31}$$

where σ_m is the mean stress, I_{2D}, I_{3D} are the invariants of the stress deviator. Taking into account that Λ_p is a dimensionless parameter one can conclude on the basis of π–theorem [401, 402] that Eq. (4.31) can be presented as a dependence of Λ_p on two independent parameters. Kolmogorov has suggested the following two parameters: the index of the stress state scheme k where $k = \sigma_{\text{m}}/\tau_e$ and the Lode coefficient

$$\mu_\sigma = \left[\frac{2\sigma_3 - \sigma_1 - \sigma_2}{\sigma_1 - \sigma_2}\right]$$

where σ_1, σ_2, and σ_3 are the principal stresses. The values of k for uniaxial tension, simple torsion and uniaxial compression tests are equal to $1/\sqrt{3}$, 0 and $-1/\sqrt{3}$, respectively. As the value of k is small, the scheme of the stress state is believed to be 'soft'. The value of the Lode coefficient is equal to −1 for simple tension, $\mu_\sigma =$ +1 for simple compression; for torsion, and pure shear, it is equal to zero. Thus, expression (4.31) becomes

$$\Lambda_p = \Lambda_p(k, \mu_\sigma) \tag{4.32}$$

A plot of $\Lambda_p = \Lambda_p\ (k, \mu_\sigma)$ is known as the plasticity diagram.

To estimate the capacity of a material to deform without macroscopic fracture, it is necessary to suggest a model (criterion). For a particle in a material deforming under constant values of k and μ_σ, one can write the condition for deformation without fracture as $\Lambda < \Lambda_p$, or

$$\Psi = \frac{\Lambda}{\Lambda_p} < 1 \tag{4.33}$$

The parameter $\Psi \equiv \Lambda/\Lambda_p$ is referred to as the *rate of exhaustion of the plasticity resource.* Equation (4.33) can be extended for the general case as

$$\Psi = \int_0^t \frac{H(\tau)d\tau}{\Lambda_p} \tag{4.34}$$

where H is the effective shear strain rate and the integration is effected for a given particle along the trajectory of movement. The criterion for fracture is obvious.

$$\Psi = 1 \tag{4.35}$$

The *plasticity resource* of a cold worked material can be recovered if ψ_0, which was reached in preliminary deformation, is not too large. For example, by annealing low carbon steel at 600°C the plasticity resource can be fully restored. When $\psi_0 > 0.6$, such annealing is not sufficient. Kolmogorov has suggested a method of determining experimentally the function Λ_p $(k,\ \mu_\sigma)$ from torsion experiments, which are to be performed under a high hydrostatic pressure. Under torsion at atmospheric pressure

$$\sigma_1 = \tau_s; \qquad \sigma_2 = 0; \qquad \sigma_3 = -\tau_s$$

A hydrostatic pressure, p_H changes the principal stresses as follows.

$$\sigma_1 = \tau_s - p_H; \qquad \sigma_2 = -p_H; \qquad \sigma_3 = -\tau_s - p_H$$

Let τ_s be constant for all p_H. Taking into account $\tau_e = \tau_s$, one obtains

$$k = \sigma / \tau_e = -p_H / \tau_e; \qquad \mu_\sigma = (2\sigma_3 - \sigma_1 - \sigma_2)/(\sigma_1 - \sigma_2)$$

The following function can be determined experimentally.

$$\Lambda_p = \Lambda_p(k)\Big|_{\mu_\sigma = \ const}$$

The relation $\Lambda_p = \Lambda_p(k)$ at μ_σ = constant is a decreasing function and can be approximated by an exponential. (It is connected with the fact that other things being equal an increase in compressive stresses enhances the plasticity resource.) Λ_p is a measure of plasticity. It depends on chemical composition, structure, temperature, strain rate, stress–strain state, deformation and stress history, specimen/workpiece geometry and environment.

The model suggested by Kolmogorov [313] has been extended by others. Bogatov et al. [403] have proposed a non-linear model of plastic collapse which is described by a power relationship $\kappa = b\Lambda^a$. The critical value of $\kappa = \kappa^*$ corresponding to macrofracture ($\Lambda = \Lambda_p$) will be given by $\kappa^* = b\ (\Lambda_p)^a$. It has also been shown that the increment $d\kappa$ of plastic collapse is proportional to the increment in plastic strain $d\kappa = \alpha d\Lambda$, where the modulus of plastic collapse is given by $\alpha = ba\Lambda^{a-1}$. The rate of exhaustion of the plasticity resource will be obtained from

$$\Psi = \int_0^t \frac{a\Lambda^{a-1}dt'}{\Lambda_p^a} \tag{4.36}$$

In general, the value of Λ_p in Eq. (4.36) depends on the thermomechanical parameters of the process of plastic deformation and therefore is to be determined experimentally. For the linear case ($a = 1$) one obtains Eq. (4.34) from Eq. (4.36). Kolmogorov's model has been validated for the case of simple loading. For complex (non-proportional) loading the criterion for fracture is to be generalised appropriately, see, e.g., [313, 380].

During high temperature plastic deformation crystal defects are eliminated by static recovery and recrystallisation or dynamic recovery and/or dynamic recrystallization. This is taken into account by introducing a special function $B(\tau)$ so that

$$\Psi = \int_0^t B(\tau)\frac{H(\tau)d\tau}{\Lambda_p} \tag{4.37}$$

where $0 < B(\tau) \leq 1$. In multistage metalworking processes (rolling, extrusion, forging, etc.) a particle in a material passes through the deformation zone in a very short time so that static recovery of the plasticity resource takes place mainly during the pauses between individual steps. Partial or even full recovery can take place during such pauses, especially at high temperatures. In this case, the fracture criterion Eq.(4.35) can be extended to non-proportional loading by dividing the deformation process into a number of stages, for each of which Eq. (4.33) is applicable. Then,

$$\Psi = \sum_{i=1}^{n} \Psi_i^{a_i} < 1$$

where $a_i \geq 1$. If the parameters k and μ_σ can be assumed to be approximately constant for every stage, one can rewrite the condition of fracture as

$$\Psi = \sum_{i=1}^{n} \left(\frac{\Delta\Lambda_i}{\Lambda_{p_i}}\right)^{a_i} \tag{4.38}$$

where n is the number of stages.

Considering low cycle fatigue involving torsion–reverse torsion in a symmetric cycle as an example, the amplitude of plastic deformation $\Delta\Lambda$ can be related to the number of cycles N as [313]

$$\Delta\Lambda\ N^{\gamma} = c \tag{4.39}$$

where γ and c are empirical constants. Eq. (4.39) is known as the Coffin–Manson relation [210]. Taking into account that each cycle includes two stages of plastic deformation, torsion and reverse torsion ($n = 2N$), condition (4.38) can be rewritten as

$$n\left(\frac{\Delta\Lambda}{\Lambda_p}\right)^a = 1 \ or \ (2N)^{1/a}\left(\frac{\Delta\Lambda}{\Lambda_p}\right) = 1 \ or \ \Delta\Lambda(N)^{1/a} = \Lambda_p 2^{1/a} \tag{4.40}$$

One can see that Eq. (4.40) coincides with the empirically established Eq. (4.39).

If we consider the two-stage deformation of a cylindrical rod in tension followed by torsion, the criterion of fracture can be written as

$$\Psi_1^{a_1} + \Psi_2^{a_2} = 1$$

For a common steel $a_1 = 2.1$ and $a_2 = 1.9$ [313].

It should be noted that the Kolmogorov–Bogatov theory does not represent the only way of describing damage accumulation. There are also other approaches in the Russian literature. However, when the behaviour of materials under different complex loading conditions are investigated, these theories are found to be inadequate. Iljushin [188, 380] has suggested a general approach that allows the description of damage in terms of special tensor characteristics. The tensor criteria of estimating damage accumulation were known before his work; but Iljushin appears to have been the first to formulate them in a sufficiently general form. Damage is tensorial in nature; it can be described by a scalar parameter only as a first approximation.

A solid may fail by any one of competing micromechanisms of fracture: cleavage, ductile fracture, rupture, intergranular creep fracture etc. (see, e.g., [404]). The overall behaviour of a material under a given stress state can be summarised in a fracture mechanisms map: a diagram with stress along one axis and temperature on the other, showing the field of dominance of each mechanism. Such maps are constructed by materials scientists. It is of interest to combine the efforts of specialists in different fields for developing fracture mechanics.

Rabotnov [195] has proposed coupled continuum damage constitutive equations, which can be written for uniaxial loading as

$$\xi^c = B_1[\sigma(1-\omega)]^n \ ; \quad \omega = B_2[\sigma(1-\omega)]^p \tag{4.41}$$

ξ^c is the creep strain rate, B_1 , B_2, n and p are constants and ω is a parameter that accounts for damage accumulation. These equations are widely used for describing creep and creep rupture behaviour of metals and alloys, see, e.g., [309, 405]. Creep fracture under both uniaxial tension and multiaxial stress states are considered in [406], where void growth is analysed and mechanistic models are presented to facilitate the extrapolation of creep data. Thus, the incorporation of the efforts of materials scientists into those of specialists in mechanics is of importance in studying damage accumulation as well as developing mechanics of superplastic flow. This comment is particularly relevant because many materials are known to cavitate during superplastic deformation.

4.2 Model Boundary Value Problems in Mechanics of Superplasticity

In most cases, the BPs in mechanics of superplasticity can not be solved analytically. Therefore, a number of numerical methods are used to solve them. As there is a multitude of BPs in mechanics of solids, no universal method of solution exists. The quest for a method of solution for a given BP is non-trivial. For example, if the mechanical response of a superplastic alloy exhibits a mechanical threshold, the mathematical type of equation of motion Eq.(4.2) becomes qualitatively different from when a mechanical threshold is absent. Thus, the methods of solving the BPs differ considerably for solids and liquids (no mechanical threshold). Descriptions of the many methods of solving the BP are available in literature but it is beyond the scope of this book. In this section some simple examples are considered to demonstrate clearly the ways of solving model BPs. A software intended to solve the BPs in metalworking may be regarded as a computer-aided more powerful version of these examples. That is, the software solves the same equations using numerical methods under more general assumptions. Here, three examples are considered: Couette flow of superplastics between coaxial rotating cylinders (Sect. 4.2.1); combined loading of a solid cylindrical specimen (Sect.4.2.2) and free forming of a spherical and a cylindrical shell (Sect.4.2.3). For clarity some details of calculation are given. The solutions obtained can also be used to test and validate any software used in the mathematical modelling of technological processes of metalworking.

4.2.1 Couette Flow of Superplastics

Movement of a superplastic between two infinite coaxial cylinders rotating about their common axis with angular velocities ω_a and ω_b is considered. The radii of the cylinders are a and b ($b > a$). Infinitesimally small deformations of an isotropic incompressible material will be examined. Such a problem for a newtonian viscous fluid is known as Couette flow [200]. Couette flow is one of the very few cases where one can obtain an analytical solution for the Navier–Stokes equations (differential equations of motion of a newtonian liquid). As noted in [200], Navier-Stokes equations can be solved exactly for a few cases only, e.g., a disc rotating in an infinite liquid; flow at a diffuser and a confuser; sunk flow. Such solutions are of interest, at least, from the viewpoint of methods/techniques. Couette flow is considered here within the framework of the general formulation of BP given in Sect. 4.1. (Damage accumulation is not considered.)

Let us choose the cylindrical co-ordinates r, φ, z with z-axis coinciding with the common axis of the cylinders. The equations of motion (4.2a) for such a system of co-ordinates are given by

$$\frac{\partial\sigma_{rr}}{\partial r}+\frac{1}{r}\frac{\partial\sigma_{r\varphi}}{\partial\varphi}+\frac{\partial\sigma_{rz}}{\partial z}+\frac{\sigma_{rr}-\sigma_{\varphi\varphi}}{r}+\rho F_r=\rho\, a_r \tag{4.42a}$$

$$\frac{\partial\sigma_{\varphi r}}{\partial r}+\frac{1}{r}\frac{\partial\sigma_{\varphi\varphi}}{\partial\varphi}+\frac{\partial\sigma_{\varphi z}}{\partial z}+\frac{2\sigma_{\varphi r}}{r}+\rho F_\varphi=\rho\, a_\varphi \tag{4.42b}$$

$$\frac{\partial\sigma_{zr}}{\partial r}+\frac{1}{r}\frac{\partial\sigma_{z\varphi}}{\partial\varphi}+\frac{\partial\sigma_{zz}}{\partial z}+\frac{\sigma_{zr}}{r}+\rho F_z=\rho\, a_z \tag{4.42c}$$

where σ_{rr}, $\sigma_{\varphi\varphi}$, σ_{zz}, $\sigma_{r\varphi}$, σ_{rz}, $\sigma_{\varphi z}$ are the stress tensor components; a_r, a_φ, a_z are the acceleration vector components; F_r, F_φ, F_z are the components of the vector of body forces and ρ is the density. The components of acceleration in a cylindrical system of co-ordinates are [402]

$$a_r=\frac{\partial u_r}{\partial t}+u_r\frac{\partial u_r}{\partial r}+u_\varphi\frac{1}{r}\frac{\partial u_r}{\partial\varphi}+u_z\frac{\partial u_r}{\partial z}-\frac{u_\varphi^2}{r} \tag{4.43a}$$

$$a_\varphi=\frac{\partial u_\varphi}{\partial t}+u_r\frac{\partial u_\varphi}{\partial r}+\frac{u_\varphi}{r}\frac{\partial u_\varphi}{\partial\varphi}+u_z\frac{\partial u_\varphi}{\partial z}+\frac{u_r u_\varphi}{r} \tag{4.43b}$$

$$a_z=\frac{\partial u_z}{\partial t}+u_r\frac{\partial u_z}{\partial r}+\frac{u_\varphi}{r}\frac{\partial u_z}{\partial\varphi}+u_z\frac{\partial u_z}{\partial z} \tag{4.43c}$$

where u_r, u_φ, u_z are the velocity components.

Further we shall use tensor constitutive equations (4.3); the scalar properties of the continuum are assumed to be described by the relations (4.4). The following three variants are considered for the case of uniaxial deformation in order to select a concrete functional to be used in Eq.(4.4).

A. Newtonian viscous liquid:

$$\sigma=3\mu\xi \quad or \quad \tau=\mu\dot{\gamma} \tag{4.44a}$$

where $\tau=\sigma/\sqrt{3}$ is the shear stress and $\dot{\gamma}=\xi\sqrt{3}$ is the shear strain rate; μ (= constant) is the viscosity coefficient.

B. Shvedov-Bingham plastic [313]:

$$\sigma = \sigma_0 + 3\mu\xi \quad or \quad \tau = \tau_0 + \mu\dot{\gamma} \tag{4.44b}$$

where $\sigma_0 = \tau_0\sqrt{3} =$ constant is the mechanical threshold.

C. Standard power relation used for superplastics:

$$\sigma = K\xi^m \quad or \ \tau = K'\dot{\gamma}^m \tag{4.44c}$$

m and $K = K'(\sqrt{3})^{m+1}$ are material constants. In particular, Eq. (4.44c) with $m = 1$ reduces to Eq. (4.44a).

The conditions of the problem reduce derivatives with respect to φ and z as equal to zero. If stationary regime alone is considered, then a derivative with respect to t also equals zero. In this situation the velocity field will be

$$u_r = u_z = 0,\ u_\varphi = u_\varphi(r) \tag{4.45}$$

Neglecting body forces and from Eqs. (4.42) and (4.43)

$$\frac{d\sigma_{rr}}{dr} + \frac{\sigma_{rr} - \sigma_{\varphi\varphi}}{r} = \rho\left(-\frac{u_\varphi^2}{r}\right) \tag{4.46a}$$

$$\frac{d\sigma_{r\varphi}}{dz_r} + \frac{2\sigma_{r\varphi}}{r} = 0 \tag{4.46b}$$

The strain rate tensor components in a cylindrical system of co-ordinates can be calculated in accordance with the following formulae [402].

$$\begin{aligned} \xi_{rr} &= \frac{\partial u_r}{\partial r} = 0; \quad \xi_{\varphi\varphi} = \left(\frac{1}{r}\right)\frac{\partial u_\varphi}{\partial \varphi} + \frac{u_r}{r} = 0; \\ \xi_{\varphi z} &= \left(\frac{1}{2}\right)\left[\frac{\partial u_\varphi}{\partial z} + \left(\frac{1}{r}\right)\frac{\partial u_z}{\partial \varphi}\right] = 0; \end{aligned} \tag{4.47a}$$

$$\begin{aligned} \xi_{zz} &= \frac{\partial u_z}{\partial z} = 0; \ \xi_{rz} = \left(\frac{1}{2}\right)\left[\frac{\partial u_r}{\partial z} + \frac{\partial u_z}{\partial r}\right] = 0; \\ \xi_{r\varphi} &= \left(\frac{1}{2}\right)\left[\left(\frac{1}{r}\right)\frac{\partial u_r}{\partial \varphi} + r\frac{\partial}{\partial r}\left(\frac{u_\varphi}{r}\right) = \left(\frac{r}{2}\right)\frac{d}{dr}\left(\frac{u_\varphi}{r}\right)\right] \end{aligned} \tag{4.47b}$$

Thus, the only non-zero component of the strain rate tensor will be

$$\xi_{r\varphi} \equiv \xi_{12} = \left(\frac{1}{2}\right)\left(\frac{du_\varphi}{dr} - \frac{u_\varphi}{r}\right) \tag{4.48}$$

According to this relation the effective shear strain rate $\dot{\gamma}_e$ is

$$\dot{\gamma}_e \equiv \sqrt{2\xi_{ij}\xi_{ij}} = r\frac{d}{dr}\left(\frac{u_\varphi}{r}\right) \tag{4.49}$$

From CEs Eq.(4.3) and Eq.(4.47) it follows that the stress deviator components are $S_{ij} = \sigma_{ij} + P_H\delta_{ij} = 0$, excepting $S_{12} \equiv \sigma_{r\varphi}$. Taking into account that $S_{11} = S_{22} = 0$, one has $\sigma_{rr} = \sigma_{\varphi\varphi} = -P_H$ and Eq. (4.46a) leads to

$$\frac{dp}{dr} = \rho\frac{u_\varphi^2}{r} \tag{4.50}$$

The effective shear stress τ_e will be

$$\tau_e \equiv \sqrt{\frac{1}{2}S_{ij}S_{ij}} = \sigma_{r\varphi} \tag{4.51}$$

Integrating Eq.(4.46b) gives

$$\sigma_{r\varphi} = \frac{C_1}{r^2} \tag{4.52}$$

where C_1 is a constant (which, in general, can be time dependent). Consider only the kinematic BCs, because such BCs are used in classical problems of Couette flow in a newtonian liquid.

$$u_\varphi(a) = a\omega_a; \quad u_\varphi(b) = b\omega_b \tag{4.53}$$

The scalar constitutive equation (4.4) in the form of Eqs. (4.44), recorded for the effective shear stress and the effective shear strain rate, leads to the following.

4.2.1.1 Newtonian viscous liquid

Substituting Eq.(4.49) and Eq.(4.51) into Eq.(4.44a) one can write

$$\tau_e = \mu\dot{\gamma}_e \Rightarrow \mu r\frac{d}{dr}\left(\frac{u_\varphi}{r}\right) = \frac{C_1}{r^2} \tag{4.54}$$

Integration of this ordinary differential equation gives

$$u_\varphi(r) = -\frac{C_1}{2\mu r} + C_2 r \tag{4.55}$$

The integration constants C_1 and C_2 are found by substituting Eq.(4.55) in Eq.(4.53) as

$$C_1 = 2\mu \frac{a^2 b^2}{b^2 - a^2}(\omega_b - \omega_a); \qquad C_2 = \frac{b^2\omega_b - a^2\omega_a}{b^2 - a^2} \tag{4.56}$$

Solutions for many particular cases are presented in [200].

4.2.1.2
Shvedov–Bingham Plastic

Substituting Eq.(4.49) and (4.51) into Eq.(4.44b) one can write

$$\tau_e = \tau_0 + \mu\dot{\gamma}_e \Rightarrow \tau_0 + \mu r \frac{d}{dr}\left(\frac{u_\varphi}{r}\right) = \frac{C_1}{r^2} \tag{4.57}$$

The solution of this ordinary differential equation is

$$u_\varphi(r) = -\frac{C_1}{2\mu r} + C_2 r - \left(\frac{r\tau_0}{\mu}\right)\ln r \tag{4.58}$$

The integration constants C_1 and C_2 are found by substituting Eq.(4.58) in Eq.(4.53) as

$$\begin{aligned} C_1 &= 2\mu \frac{a^2 b^2}{b^2 - a^2}(\omega_b - \omega_a) + 2\tau_0 \frac{a^2 b^2}{b^2 - a^2}\ln\frac{b}{a} \\ C_2 &= \frac{b^2\omega_b - a^2\omega_a}{b^2 - a^2} + \frac{\tau_0}{\mu}\frac{b^2 \ln b - a^2 \ln a}{b^2 - a^2} \end{aligned} \tag{4.59}$$

One can see that as $\tau_0 \to 0$, Eq. (4.58) reduces to Eq. (4.55).

4.2.1.3
Non-linear Viscous Material

Substituting Eq.(4.49) and Eq.(4.51) into Eq.(4.44c) the following ordinary differential equation for the unknown function $u_\phi(r)$ is obtained.

$$r\frac{d}{dr}\left(\frac{u_\varphi}{r}\right)=\left(\frac{C_1}{K'r^2}\right)^{\frac{1}{m}} \tag{4.60}$$

The solution of this equation is

$$u_\varphi(r)=C_1 r^{1-\frac{2}{m}}+C_2 r \tag{4.61}$$

where $C_1 = -(m/2)\,(C_1/K')^{1/m}$ and C_2 are integration constants, which can be found from the BC (4.53) as

$$\begin{aligned} C_1 &= \omega_a - \omega_b a^{\frac{2}{m}}\left[1-\left(\frac{a}{b}\right)^{-\frac{2}{m}}\right] \\ C_2 &= \omega_b - \omega_a a^{\frac{2}{m}}\left[1-\left(\frac{a}{b}\right)^{-\frac{2}{m}}\right] \end{aligned} \tag{4.62}$$

It is easy to see that when $m = 1$ the solution Eq.(4.61) with C_1 and C_2 from Eq. (4.62) is transformed into Eq.(4.55) with C_1 and C_2 given by Eq.(4.56). Thus, the well-known solution to the problem of Couette flow can be derived from (4.61) as a particular case ($m = 1$).

Substituting the solution obtained for a non-linear viscous material into Eq. (4.49) one gets

$$\dot{\gamma}_e=\left(\frac{2}{m}\right)\left((\omega_b-\omega_a)\left(\frac{a}{r}\right)^{\frac{2}{m}}\right)\left(1-\left(\frac{a}{b}\right)^{\frac{2}{m}}\right)^{-1} \tag{4.63}$$

The solution obtained is analysed in [94]. It is shown that strain localisation increases as m decreases and as $m\to 0$ the deformation is localised in the vicinity of the inner cylinder. A similar conclusion can be reached from Eq. (4.63). According to Eq.(4.63) $\dot{\gamma}_e \sim r^{-2/m}$. Hence, if for a newtonian viscous liquid $\dot{\gamma}_e \sim 1/r^2$ then for a material with m, say, equal to 0.1, $\dot{\gamma}_e \sim 1/r^{20}$, that is, a very sharp dependence is seen in $\dot{\gamma}_e(r)$.

4.2.2 Combined Loading of a Cylindrical Rod by Axial Force and Torque

A uniform cylindrical rod of initial length L_0 and radius r_0 is considered. Let this rod deform under an axial force P and a torque $\overline{M}$. For simplicity, infinitesimally small strains in an isotropic and incompressible material are examined. The loading conditions are assumed to be determined by constant linear ($V = V_0 =$ constant) and angular ($\omega = \omega_0$ = constant) relative velocities at the ends of the rods. The problem is to determine the stress–strain state in the rod during deformation as well as the values of the axial force P and torque $\overline{M}$ needed to achieve the given conditions.

Let r, φ, z be a cylindrical system of co-ordinates with the z-axis directed along the axis of the rod. Assuming $L_0 >> r_0$ we obtain that deformation is uniform along the specimen (i.e., it is not dependent on z). Assuming the material to be incompressible one can write the velocity field u_r, u_φ, u_z as

$$u_r = -\frac{rV}{2L}, \; u_\varphi = \frac{\omega z r}{L}, \; u_z = \frac{zV}{L} \tag{4.64}$$

where L is the current length of the rod; r is the current distance of the point under consideration from the axis of symmetry and V, ω are the relative axial and angular velocities at the ends of the deforming rod. Then, the strain rate tensor components can be calculated using Eqs. (4.47) as

$$\begin{aligned}
&\xi_{rr} = \frac{\partial u_r}{\partial r} = -\frac{V}{2L}; \quad \xi_{\varphi\varphi} = \left(\frac{1}{r}\right)\frac{\partial u_\varphi}{\partial \varphi} + \frac{u_r}{r} = -\frac{V}{2L}; \xi_{zz} = \frac{\partial u_z}{\partial z} = \frac{V}{L}; \\
&\xi_{r\varphi} = \left(\frac{1}{2}\right)\left[\left(\frac{1}{r}\right)\frac{\partial u_r}{\partial \varphi} + r\frac{\partial}{\partial r}\left(\frac{u_\varphi}{r}\right)\right] = 0; \; \xi_{rz} = \left(\frac{1}{2}\right)\left[\frac{\partial u_r}{\partial z} + \frac{\partial u_z}{\partial r}\right] = 0; \\
&\xi_{\varphi z} = \left(\frac{1}{2}\right)\left[\frac{\partial u_\varphi}{\partial z} + \left(\frac{1}{r}\right)\frac{\partial u_z}{\partial \varphi}\right] = \frac{r\omega}{2L}
\end{aligned} \tag{4.65}$$

The equivalent strain rate ξ_ε will be

$$\xi_e = \sqrt{\frac{2\xi_{ij}\xi_{ij}}{3}} = \frac{1}{L}\sqrt{\frac{V^2 + r^2\omega^2}{3}} \tag{4.66}$$

The Odqvist parameter s can be found from Eq. (4.6). Integrating with $V = V_0$ = constant, $\omega = \omega_0$ = constant, one obtains

$$s = \left| 2\sqrt{1+\frac{\theta^2}{3}} - 2\sqrt{1+\frac{\theta^2}{3(1+e)}} + \ln(1+e) + \ln\frac{1+\dfrac{\theta^2}{6(1+e)} + \sqrt{1+\dfrac{\theta^2}{3(1+e)}}}{1+\dfrac{\theta^2}{6} + \sqrt{1+\dfrac{\theta^2}{3}}} \right| \tag{4.67}$$

where $e = V_0 t/L_0$ and $\theta = r_0\omega_0/V_0$. The parameter e is engineering strain in a tensile test, while θ equals the ratio of the circumferential velocity to the axial velocity for

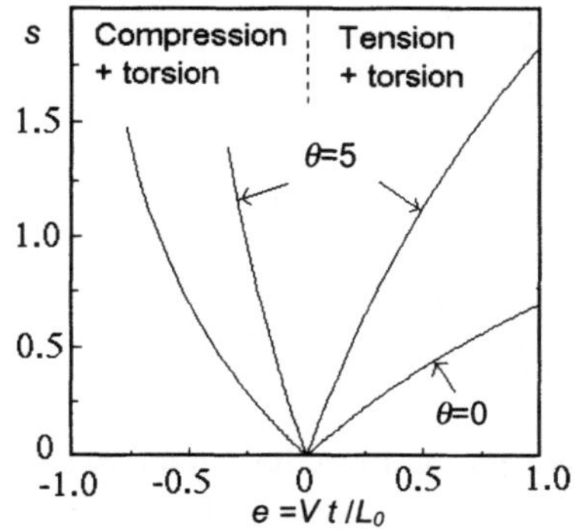

Fig. 4.1. The variation of the value of the Odqvist parameter s with the relative strain calculated in accordance with (4.67) with different θ (numbers near the curves)

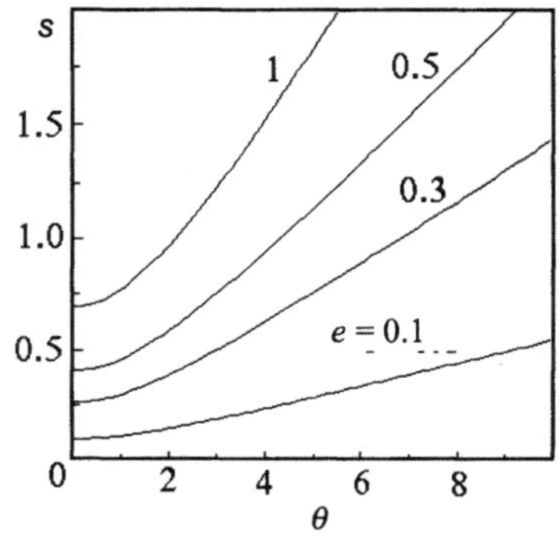

Fig. 4.2. The variation of the value of the Odqvist parameter s with θ calculated in accordance with (4.67) with different e (numbers near the curves)

the outside surface of the specimen at $t = 0$. θ is characteristic of the relative role of torsion: $\theta = 0$ corresponds to simple tension, while $\theta >> 1$ corresponds approximately to simple torsion.

It should be noted that r_0 can be treated as a lagrangian co-ordinate of the point under consideration (see Appendix where the validity of expressions Eq.(4.66) and Eq.(4.67) is proved for finite strains). Eq.(4.67) has been used in [170, 171] to estimate the strain state of a cylindrical specimen.

The dependence of the Odqvist parameter s on e and θ Eq.(4.67) are shown in Figs. 4.1 and 4.2. It is seen that s increases monotonically with $|e|$ and θ. The function $s(e)$ is not symmetric about the vertical axis. This asymmetry reflects the following well-known fact: for large deformations the tensile engineering strain $e_{tens} = (L-L_0)/L_0$ is not equivalent to the compressive engineering strain $|e_{comp}| = (L_0 - L)/L_0$; e.g., if for $e = e_{tens} = +0.5$ (tension) we have $s = \ln(1.5) \cong 0.4055$, but for $e = e_{comp} = -0.5$ (compression), $s = \ln(2) \cong 0.6931$. *In addition*, if one takes into consideration the influence of contact friction forces in real experiments, then this difference will be even more significant (at the central zone). Presently some applications of Eq. (4.67) are considered.

Simple tension: $V = V_0 =$ constant, $\omega = 0$ ($V_0 > 0$). In this case one can obtain the following limiting case as $\omega \to 0$.

$$\xi_e = \frac{V}{L}; s \to ln(1+e) = ln\left(\frac{L}{L_0}\right)$$

Compression without barrelling: $V = V_0 =$ constant, $\omega = 0$ ($V_0 < 0$). Let $V_0 = -u$, where $u > 0$. Then,

$$\xi_e = \frac{u}{L}; s = -\ln\left(\frac{1-ut}{L_0}\right) = \ln\left(\frac{L_0}{L}\right)$$

Fixed end torsion: $V = 0$, $\omega = \omega_0$ = constant; $L = L_0$ = constant.

$$\xi_e = \frac{r_0\omega_0}{L_0\sqrt{3}}, \qquad s = \frac{r_0\omega_0 t}{L_0\sqrt{3}} = \frac{r_0\varphi}{L_0\sqrt{3}}$$

where r_0 is the distance of the point under consideration from the symmetry axis at $t = 0$; $\varphi = \omega_0 t$ is the angle of twist. Thus, formula (4.67) covers the well-known expressions used in literature for calculating the true strain and the strain rate in simple tests like uniaxial tension, uniaxial compression without barrelling and fixed-end torsion. The problem of estimating the strain state for the case of large plastic deformation is discussed in [95, 407]. The power /force parameters are now considered. From the proportionality between the stress and the strain rate deviator Eq.(4.3) the following relations are derived.

$$S_{rr} = S_{\varphi\varphi}, \quad \frac{S_{\varphi z}}{S_{zz}} = \frac{\xi_{\varphi z}}{\xi_{zz}} = \frac{\left(\dfrac{r\omega}{2l}\right)}{\left(\dfrac{V}{l}\right)} = \frac{r\omega}{2V}, \quad S_{\varphi z} = \left(\frac{r\omega}{2V}\right) S_{zz} \tag{4.68}$$

Then the equivalent stress is according to equation (4.67) and the relation $S_{rr} + S_{yy} + S_{zz} = 0$ is given by

$$\sigma_e = \frac{3}{2}\left|S_{zz}\right|\sqrt{1 + \frac{r^2\omega^2}{3V^2}} \tag{4.69}$$

Equations of motion Eq.(4.2a) for the conditions of the problem are reduced to the following equilibrium equations.

$$\frac{\partial \sigma_{rr}}{\partial r} = 0, \Rightarrow \sigma_{rr} = \sigma_{rr}(z) \tag{4.70a}$$

$$\frac{\partial \sigma_{\varphi z}}{\partial z} = 0, \Rightarrow \sigma_{\varphi z} = \sigma_{\varphi z}(r) \tag{4.70b}$$

$$\frac{\partial \sigma_{zz}}{\partial z} = 0, \Rightarrow \sigma_{zz} = \sigma_{zz}(r) \tag{4.70c}$$

Boundary conditions: As the cylindrical surface is free from loading the following is satisfied.

$$\sigma_{rr}\big|_{r=R} = 0$$

where R is the instantaneous radius of the rod. Consequently $\sigma_{rr} = 0$ and $\sigma_{\varphi\varphi} = \sigma_{rr} = 0$. From $S_{zz} = (2\sigma_{zz} - \sigma_{rr} - \sigma_{\varphi\varphi})/3$ it follows that $\sigma_{zz} = (3/2)\, Szz$. From Eq. (4.69) one obtains

$$|\sigma_{zz}| = \frac{\sigma_e}{\sqrt{1 + \dfrac{r^2\omega^2}{3V^2}}} \tag{4.71}$$

Then the value of the axial force will be

$$P_z = \iint\limits_S \sigma_{zz}\, dS = \iint\limits_S \frac{\sigma_e}{\sqrt{1 + \dfrac{r^2\omega^2}{3V^2}}}\, dS\ sign\,(P_z)$$

where the region of integration is a circle of radius R. Substituting Eq.(4.66) one can obtain on integration

$$|P_z| \equiv |P_k| = K\left[\frac{|V|}{L_0(1+e)}\right]^m \frac{\pi r_0^2}{1+e} \frac{3(1+e)}{\theta^2} \frac{2}{m+1} \left\{\left[1 + \frac{\theta^2}{3(1+e)}\right]^{\frac{m+1}{2}} - 1\right\} \tag{4.72}$$

For simple tension, from Eq. (4.72) as $\omega \to 0$

$$P_0 = K\left[\frac{V}{L_0(1+e)}\right]^m \frac{\pi r_0^2}{1+e} \tag{4.73}$$

It is easy to see that the axial force P_0 equals the stress multiplied by the cross-sectional area. Let the normalised axial force $\overline{P}$ be introduced as

$$\overline{P} = \frac{P_k}{P_0} = \frac{3(1+e)}{\theta^2} \frac{2}{m+1} \left\{\left[1 + \frac{\theta^2}{3(1+e)}\right]^{\frac{m+1}{2}} - 1\right\} \tag{4.74}$$

Many applications of this formula can be considered. For example, in a newtonian viscous liquid ($\sigma_e = K\xi_e$, $m = 1$) torsion does not influence the magnitude of the axial force, so that $P_k = P_0$ for any θ as $V \neq 0$. Notice that $P_k/P_0 \to 0$ when $\theta \to \infty$, if $m < 1$. From the law of conservation of energy the equation for the balance of power is

$$N_{in} = \iiint_V \tau_e \dot{\gamma}_e dV = \iiint_V \sigma_e \xi_e dV = PV + \overline{M}\omega \tag{4.75}$$

where N_{in} is the power of internal forces; $\overline{M}$ is a torque and P is an axial force. Substituting Eq.(4.44) and Eq.(4.66) on the left-hand side of Eq.(4.75) and comparing the left and right hand sides of Eq.(4.75) $\overline{M}$ is estimated as

$$\overline{M} = \left[\frac{V}{L_0(1+e)}\right]^m \frac{\pi r_0^2}{1+e} \frac{3(1+e)}{\theta^3} \frac{4Kr_0}{(m+1)(m+3)} \times \left\{1 - \left[1 + \frac{\theta^2}{3(1+e)}\right]^{\frac{m+1}{2}} \left[1 - \frac{m+1}{2} \frac{\theta^2}{3(1+e)}\right]\right\} \tag{4.76}$$

For a newtonian liquid ($m = 1$) from this formula

$$\overline{M}_N = \frac{\pi r_0^2}{1+e} \frac{KV}{L_0} \frac{\theta\, r_0}{6(1+e)^2} \tag{4.77}$$

When $V = 0$ (simple torsion) Eq.(4.77) is reduced to

$$\overline{M}_{N_0} = \frac{\pi r_0^4}{6L_0} K\omega \tag{4.78}$$

For an independent validation of the results obtained, Eq. (4.78) is derived in another way. Let a cylindrical container be filled with a liquid and the parameter μ be the viscosity coefficient. Friction along the cylindrical walls of the container is negligible. Then, the twisting moment $\overline{M}^0$ needed to rotate the upper surface of the liquid with an angular velocity ω with respect to the bottom layer of the liquid column (when $\omega = 0$) will be given by

$$\overline{M}^0 = \iint_S \tau\, r\, dS \tag{4.79}$$

For a newtonian liquid $\tau = \mu\dot{\gamma}$. The shear strain rate $\dot{\gamma} = r\omega/L_0$. Then, the double

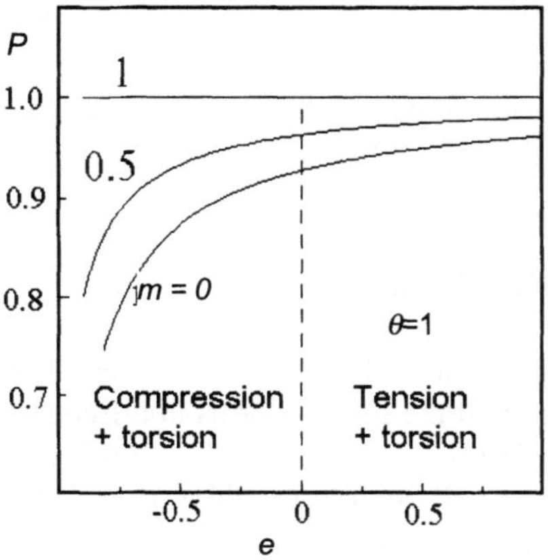

Fig. 4.3. The variation of the normalised axial force $\overline{P}$ with the relative strain $e = V_0t/L_0$, calculated in accordance with (4.74) with $\theta = 1$ and different *m values* (numbers near the curves)

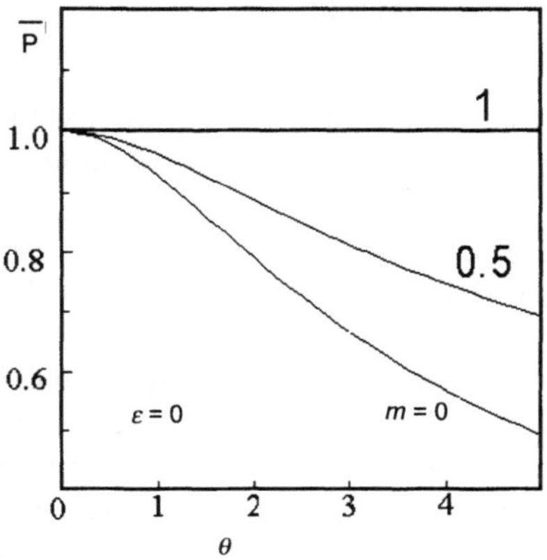

Fig. 4.4. The variation of the value of the normalised axial force $\overline{P}$ with the parameter $\theta = r_0\omega_0/V_0$, calculated in accordance with (4.74) with $e = 0$ and different m values (numbers near the curves)

integral in Eq.(4.79) is calculated in a polar system of co-ordinates (r, φ) as

$$\overline{M}^0 = \left(\frac{\mu\omega}{L_0}\right)\int_0^{2\pi} d\varphi \int_0^{r_0} r^3 dr = \pi r_0^4 \left(\frac{\mu\omega}{2L}\right)$$

As $\mu = K/3$, this formula coincides with Eq. (4.77). For a Shvedov–Bingham plastic Eq.(4.44b) the following expressions can be derived.

$$\overline{M} = \frac{\pi r_0^3}{1+e}\sigma_0\left\{\frac{4(1+e)}{\theta^3}\left[1-\sqrt{1+\frac{\theta^2}{3(1+e)}}\left(1-\frac{\theta^2}{6(1+e)}\right)\right]+\frac{AV}{\sigma_0 L_0}\frac{\theta}{6(1+e)^2}\right\}$$
$$P_k = \frac{\pi r_0^2}{1+e}\left\{\frac{AV}{L_0(1+e)}+\frac{3(1+e)}{\theta^2}2\sigma_0\left[\sqrt{1+\frac{\theta^3}{3(1+e)}}-1\right]\right\} \qquad (4.80)$$

where $A = 3\mu$. The theoretical dependences of the normalised axial force $\overline{P}$ on the relative strain e and the parameter θ calculated in accordance with Eq. (4.74) are presented in Figs. 4.3 and 4.4. The following conclusions emerge.

1. For compression with torsion ($e < 0$; $V < 0$), the normalised force $\overline{P}$ decreases with an increase in the absolute value of ε (Fig.4.3).

2. The influence of torsion on the magnitude of the axial force is greater when the value of m is less (Fig.4.4).

These results will be used to discuss the decrease in the axial force when torsion is superimposed in the combined stamping of discs (see Chap.5).

4.2.3 Free Bulging of Spherical and Cylindrical Shells

In this section, superplastic forming (SPF) of a spherical shell as well as an infinite cylindrical shell are considered. It is assumed that the shell deforms in a free state, i.e., without clamping and other restrictions. From the symmetry of the problem, it may be assumed that the thickness of the shell will be uniform along its profile at any instant of time. It is assumed further that the thickness h of the shell is small compared with the minimum radius R ($h/R << 1$), so that the principal equations of the membrane theory are valid for all particles of the shell. The notation used below and the basic assumptions are the same as those used in Sect. 3.6.2. Some information about the principal equations of the membrane theory is also given in Sect. 4.3.3.1 (for a shell which is a surface of revolution). Superplastic forming of a shell under constant effective strain rate as well as constant pressure are considered. Free forming leads to unambiguous exact solutions.

4.2.3.1 Free Forming of a Sphere

For a spherical shell the principal radii ρ_t and ρ_c are equal. $\rho_t = \rho_c = R$ and from the equation of the membrane theory

$$(\sigma_t/\rho_t)+(\sigma_c/\rho_c)=p/h \tag{4.81}$$

It follows that

$$\sigma_t=\sigma_c=\frac{pR}{2h} \tag{4.82}$$

As $\sigma_n = 0$,

$$\sigma_e=\sqrt{\frac{3}{2}S_{ij}S_{ij}}=\frac{pR}{2h} \tag{4.83}$$

It is easy to show that at any point the main strain rates are $\xi_t=\xi_c=\dot{R}/R$. The normal component of the strain rate tensor is $\xi_n=\dot{h}/h$. Using the incompressibility condition, $\xi_n= -\xi_t- \xi_c= -2\xi_t$. The equivalent strain rate can be found as

$$\xi_e = \sqrt{\frac{2}{3}\xi_{ij}\xi_{ij}} = \left(\frac{2}{R}\right)\dot{R} = -\frac{1}{h}\dot{h} \tag{4.84}$$

Assuming $\xi_e = \xi_0 =$ constant, where ξ_0 is the optimal strain rate, the following time dependences for the thickness and the radius of the shell can be found.

$$h(t) = h_0 \exp(-\xi_0 t) \;\; ; \;\; R(t) = R_0 \exp\left(\frac{\xi_0 t}{2}\right) \tag{4.85}$$

If the scalar properties of the material are determined by $\sigma_e = f(\xi_e)$, then the pressure–time cycle is defined by Eqs. (4.83) and (4.85).

$$p(t) = \left(\frac{2h_0}{R_0}\right)\sigma_0 \, exp\left[-\frac{3\xi_0 t}{2}\right] \tag{4.86}$$

Considering a spherical shell loaded by internal pressure (so that pressure at $t = 0$ is p_0) and assuming that there is neither additional pumping of gas into the shell during bulging nor a change in temperature, from Boyle's law it can be written that $pV =$ constant, where $V_e = 4\pi R^3/3$ is the volume of the sphere. From Eq. (4.83) and the incompressibility condition ($4\pi R^2 h =$ constant)

$$\sigma_c = \frac{pR}{2h} = const = \frac{p_0 R_0}{2h_0} \tag{4.87}$$

Thus, in this case the shell will deform *under constant stress*. If the stress is unambiguously related to the strain rate (i.e., there is neither strain hardening due to grain growth nor strain softening due to grain refinement), then the strain rate will also be constant. Therefore, the deformation of a spherical shell under the above regime of loading will be characterised by constant σ and ξ regardless of the form of the function f in the relationship $\sigma = f(\xi)$ or $\xi = f^{-1}(\sigma)$.

The steady state (stationary) values of the stress, σ_s, and the strain rate, ξ_s, are determined by the relations

$$\sigma_s = \left(\frac{R_0}{2h_0}\right)p_0; \;\; \xi_s = f^{-1}(\sigma_s) \tag{4.88}$$

In particular, for the power law of superplastics $\xi_s = C(\sigma_s)^n$, where $C = 1/K^n$, $n = 1/m$. The initial gas pressure is so chosen as to satisfy the condition $\sigma_s = \sigma_0$, $\xi_s = \xi_0$, where σ_0, ξ_0 correspond to the point of inflection in the sigmoidal $\log\sigma$–$\log\xi$ curve. From Eq. (4.88)

$$p_0 = \frac{2h_0}{R_0}\sigma_0 \tag{4.89}$$

4.2.3.2 Free Forming of an Infinite Cylindrical Shell

From Eq. (4.81) when $\rho_t = \infty$, $\rho_c = R$, the following relationship is obtained.

$$\sigma_c = \frac{pR}{h} \tag{4.90}$$

Since deformation along the shell is absent, $\xi_t = 0$. Therefore, from the condition of incompressibility $\xi_c + \xi_n = 0$ and the relation $\xi_n = \dot{h}/h$, the equivalent strain rate is obtained as

$$\xi_e = \sqrt{\frac{2}{3}\xi_{ij}\xi_{ij}} = \sqrt{\frac{2}{3}2\xi_c^2} = \frac{2}{\sqrt{3}}\xi_c \tag{4.91}$$

The condition of proportionality between stress and strain rate deviators Eq.(4.3) gives $S_{tt} \equiv S_t = 0, S_c = -S_n$. As $|\sigma_n| \leq p$, it is assumed that $\sigma_n = 0$ and consequently

$$\begin{gathered}\sigma_m \equiv \frac{1}{3}\sigma_{ii} = \frac{1}{3}(\sigma_c + \sigma_t);\ S_t = 0 \Rightarrow \sigma_t - \sigma_m = 0 \Rightarrow \sigma_t = \frac{1}{2}\sigma_c, \\ S_c = \frac{1}{2}\sigma_c,\ \sigma_e = \sqrt{\frac{3}{2}S_{ij}S_{ij}} = \frac{\sqrt{3}}{2}\sigma_c\end{gathered} \tag{4.92}$$

Substituting Eq.(4.90) in Eq.(4.92), it is obtained that

$$\sigma_e = \sqrt{3}pR/2h \tag{4.93}$$

Consider now the condition $\xi_e = \xi_0 =$ constant for which case the following expression results.

$$h(t) = h_0\, exp\left(-\frac{\sqrt{3}\xi_0 t}{2}\right) \tag{4.94a}$$

With the assumptions that are discussed above, the law of evolution of R will be

$$R(t) = R_0 \, exp\left(\frac{\sqrt{3}\xi_0 t}{2} \right) \tag{4.94b}$$

The pressure–time cycle can be found from Eqs. (4.93) and (4.94) as

$$p(t) = \left(\frac{2h_0}{R_0\sqrt{3}} \right) \sigma_0 \, exp\left[-\sqrt{3}\xi_0 t \right] \tag{4.95}$$

It is easy to show that such a pressure–time cycle can be realised in practice by free bulging at a constant temperature. Let $p\ (0) = p_0$ and no additional gas supply is there at $t > 0$. For this scheme, the stress is constant with time.

$$\sigma_e = \frac{\sqrt{3}pR}{2h} = \frac{\sqrt{3}p_0 R_0}{2h_0} = \text{constant} \tag{4.96}$$

p_0 is chosen from the condition

$$p_0 = \left(2h_0 / \sqrt{3}R_0 \right) \sigma_0 \tag{4.97}$$

which ensures the deformation of the shell at $\sigma_e = \sigma_0 =$ constant and $\xi_e = \xi_0 =$ constant, where σ_0, ξ_0 correspond, as before, to the point of inflection in the sigmoidal curve.

Next, constant pressure forming of a shell under constant pressure $p = p_f =$ constant is considered. Here additional pumping of gas inside the forming chamber or heating of the forming gas is necessary. The standard power law $\sigma_e = K\xi_e^m$ is assumed. From Eq. (4.82) and the incompressibility condition $4\pi R^2 h =$ constant, after integration the following time dependence for the shell thickness is found.

$$h(t) = h_0 \left[1 - \frac{3t}{2m} \left(\frac{p_f R_0}{2h_0 K} \right)^{\frac{1}{m}} \right]^{\frac{2m}{3}} = h_0 \left[1 - \frac{3\xi_{in} t}{2m} \right]^{\frac{2m}{3}} \tag{4.98}$$

where $\xi_{in} = (\sigma_{in}/K)^n$; $n = 1/m$; $\sigma_{in} = pR_0/h_0$ and σ_{in} and ξ_{in} are the initial values of stress and strain rate, i.e., $\sigma_{in} = \sigma(0)$; $\xi_{in} = \xi(0)$. From Eq. (4.97), $h \to 0$ as $t \to t_{cr} = 2m/3\xi_{in}$. Thus, the critical time is maximum for a newtonian material ($m = 1$) and tends to 0 as $m \to 0$. It should be noted that the condition $\xi_e =$ constant $= \xi_0$ which was analysed above, has only mathematical interest.

4.3 Numerical Solving of Boundary Value Problems in Superplasticity

4.3.1 Features of Boundary Value Problems in Mechanics of Superplasticity

It is possible to suggest many formulations of the BPs in mechanics of superplasticity. A feature of the phenomenon is that the strain rates are rather small, usually about 10^{-5}–$10^{-1}s^{-1}$. (This statement does not apply to high strain rate superplasticity where strain rates in excess of 1.0 s^{-1} are reported.) Therefore, the BPs in superplasticity can be considered *quasi-static*, rather than dynamic. Thus, the equations of motion (4.2a) can normally be reduced to Eqs. (4.2b). Of course, there are exceptions, e.g., impulse loading of superplastics during magnetic impulse stamping or application of a shock wave impact to achieve a fine grained microstructure. These are problems to be solved dynamically by taking into account the inertia forces in Eq. (4.2a). So far as 'traditional' structural superplasticity is concerned, neglecting the inertia forces is a reasonable simplification of the BPs.

Another typical feature is that the optimal temperature range of superplastic flow is rather narrow. Therefore, one can consider isothermal BPs so that the equation of heat transfer reduces to an identity. This allows considerable simplification of the BPs, which become unbounded. Moreover, there is no need to solve separately the temperature problem.

A characteristic feature of the mechanical behaviour of superplastics is a strong dependence of the flow stress on the strain rate; the strain dependence in contrast is often weak. Therefore, one may consider the BPs of viscoplastic materials for understanding superplasticity. In this case, the stress depends on temperature and strain rate only. Routine analysis of this BP, including group analysis, has been completed [408]. One can develop a general method to evaluate all material functions and constants for viscoplastic flow. There are also many special methods of solving the BPs. In particular, the variation methods are well developed [409], which include the methods of variation inequalities [410, 411].

The next feature of the phenomenon of superplasticity is the extreme deformation to fracture. Therefore, it is not correct to consider infinitesimally small strains while solving the BP. Thus, the problem of choosing the measures of stress and strain becomes important. As a rule, elastic strains can be neglected and the condition of incompressibility can be assumed to be a good approximation. However, neglecting the elastic strains prevents a consideration of the problem of residual stresses. This problem may be of importance in some cases [412]. But there are reports that there are no residual stresses in superplastically formed components.

Thus, a typical BP in mechanics of superplasticity includes the equilibrium equation (4.2b) and the isothermal problem for an incompressible material. Therefore, the BP will be distinguished by the choice of constitutive equations, including the way the finite strain behaviour of superplastics is taken into account. The choice of boundary conditions and possibly, the accounting of the effects of anisotropy could be different There are many methods of solving the BPs, see, e.g., [413]. In this connection, it is to be noted that the most popular method is the finite element method (FEM). This method can be applied to any arbitrary problem. Examples are considered in Sect. 4.3.2. There are also other methods. The choice of a method is decided by the class of the BPs and the mathematical type of equations, e.g., for magnetic impulse stamping the appropriate method is the method suggested by Wilkins [414]. Some simplifications can be adopted while solving certain classes of BPs. For example, in superplastic sheet metal forming membrane theory is often used. There are a limited number of BPs for which it is possible to find exact analytical solutions; some such cases are discussed in Sect. 4.3.3.2.

Many practical problems that exploit superplastic behaviour have been solved using simplified engineering methods. Many publications are available in this regard. Therefore, the next chapter is devoted to a discussion of such simplified models. Presently, without going into the details and the peculiarities of numerical methods, some typical formulations of the BPs available in the literature are discussed. In quite a few cases the formulation is not complete. Hence understanding becomes difficult.

4.3.2 Finite Element Modelling of Superplastic Metalworking Processes

Biba et al. [415] have developed a software FORM–2D for analysing superplastic metalworking. The material is considered to be an incompressible non-newtonian liquid; elastic strains are neglected. The authors believe that their approach is similar to those of Kobayashi et al. [416, 417], Zienkiewicz and Godbole [418] and Gun [419, 420]. Finite element discretization is developed in [416, 418, 420]. In [415] the general equations (4.1)–(4.5) become

$$\sigma_{ij,j} = 0;\ \xi_{ij} = \frac{1}{2}\left(V_{i,j} + V_{j,i}\right);\ S_{ij} = \left(\frac{2\sigma_e}{3\varepsilon_e}\right)\xi_{ij};\ V_{ii=0} \tag{4.99}$$

The scalar properties of the material are assumed to be described by Eq. (4.4); but in practice s is excluded. Eq. (4.4) is reduced to the standard form

$$\sigma_e = B(\xi_e)^m \tag{4.100}$$

The equation of heat transfer is written in form Eq.(4.12) with Coulomb's law Eq. (4.24) being assumed as the mechanical BC. The thermal BCs are as follows: for a free surface the BC of types II and III (convection and radiation) are assumed; for the contact surface between the instrument and the material a BC of type IV is assumed. Finite element method is used to solve the above BP. The following two cases are considered [421]: elongation of a cylindrical specimen and its compression.

While solving the first problem two types of materials were considered: a low carbon steel and Ti–6Al–4V. It was shown that a localised neck formed in the steel specimen while the titanium sample deformed uniformly. Compression of an axisymmetric part in a nickel based alloy, which may have a complex configuration, was the second example considered using the FORM–2D software. The material properties were determined experimentally. Based on the results of the analysis, suggestions for modifying the equipment could be made. From these, it was suggested that the software FORM–2D could be used for modelling superplastic forming processes.

The change of shape of the deforming specimen was well approximated by the model. As elastic strains were neglected, residual stresses could not be estimated. Although the strains are estimated by integrating the velocity field, the model considers only the strain rate. None of the above are serious limitations but the constitutive equation used has a very restricted range of applicability and this limits the usefulness of the analysis.

Chumachenko et al. [422] have formulated the BP of superplastic metalworking based on the constitutive equation suggested by Smirnov [2, 339] (see Chap 3 also). The vector properties of the material are assumed to be the same as for an isotropic viscous liquid, i.e., similar to Eq. (4.3). Then, one can write that $S_{ij} = 2\mu\xi_{ij}$. Correspondingly, the scalar properties are described by the relation $\sigma_e = 3\mu\xi_e$. The expression for 'viscosity' μ is

$$\mu = \frac{1}{3}\frac{\sigma_s}{\xi_e}\left(\frac{\sigma_r + k\xi_e^m}{\sigma_s + k\xi_e^m}\right) = \Psi(\xi_e) \tag{4.101}$$

where σ_s, σ_r, k and m are material constants. To avoid difficulties in numerical calculations, two assumptions are made.

1. Equation (4.101) is valid for $\xi_e > \delta_0 > 0$ only, where δ_0 is small. If $\xi_e < \delta_0$, then it is assumed that $\mu = \mu_0 =$ constant.
2. The material is compressible so that the volumetric parts of the stress and the strain rate tensors are related linearly through Hooke's law: $\sigma_m = K\Theta$, where σ_m is the mean stress, Θ the relative change in volume (volumetric strain) and K is the bulk modulus.

Assumptions 1 and 2 facilitate the reduction of the integro-differential equations (which are derived in a common scheme of finite element discretization) to differential equations of the same order. Thus, the common formulation of the BP is

$$\sigma_{ij,j} = 0; \quad \sigma_{ij} = S_{ij} + \sigma_m\delta_{ij}; \quad S_{ij} = 2\mu\left(\dot{\varepsilon}_{ij} - \frac{\delta_{ij}\dot{\Theta}}{3}\right); \dot{\Theta} = V_{i,i} \tag{4.102a}$$

$$\sigma_m = K_v \Theta = K \int_0^t \dot{\Theta} dt'; \ \dot{\varepsilon}_{ij} = \frac{1}{2}\left(V_{i,j} + V_{j,i}\right); \ \xi_e = \sqrt{\frac{2}{3}\dot{\varepsilon}_{ij}\dot{\varepsilon}_{ij}} \qquad (4.102b)$$

Hydrostatic pressure is accumulated from step to step and this is taken into account as follows:

$$\sigma_m = K_v \int_0^t \dot{\Theta} dt' = K_v \left[\int_0^{t_1} \dot{\Theta} dt + \int_{t_1}^{t_2} \dot{\Theta} dt + \cdots \int_{t_{n-1}}^{t_n} \dot{\Theta} dt \right] = K_v \sum_{i=1}^{n-1} \Delta t_i \dot{\Theta}_i + K \Delta t_n \dot{\Theta}_n$$

$$= \sigma^* + K_v \Delta t \dot{\Theta} \qquad (4.103)$$

Depending on the technological process, the kinematic, dynamic and mixed boundary conditions can be given at the boundary of the deformed body. Combining Eq. (4.102) and Eq. (4.103) for any step of loading

$$\sigma_{ij} = \mu\left(V_{i,j} + V_{j,i}\right) + \left[\left(K\Delta t - \frac{2\mu}{3}\right)V_{k,k} + \sigma^*\right]\delta_{ij} \qquad (4.104)$$

where Δt is the time interval corresponding to the current step and σ^* is the accumulated value of the mean stress and μ from Eq. (4.101). Based on the above BP, special software SPLEN was developed [422]. The boundary condition used for shear stress τ^L was

$$\tau^L = 0.58\, k_{fr} \sigma_e^k \left[1 - exp\left(-1.25 \cdot \frac{|\vec{p}|}{\sigma_e^k}\right)\right] \qquad (4.105)$$

where σ_e^k is the equivalent stress at the contact layer; $\vec{p}$ is the normal loading vector, k_{fr} is a general characteristic of the contact surface. The same equation has also been used later in [423]. The latter Eq.(4.102) is valid only for incompressible materials and this is not consistent with assumption 2 above, which assumes the material to be compressible. Further, the mathematical method used with Eq. (4.101) leads to: $\xi_e > \delta_0$, $\sigma_e = 3\mu(\xi_e)\xi_e$ and $\sigma_e = 3\mu\xi_e$ for $\xi_e < \delta_0$. This implies that $\sigma(0) = 0$. In other words, the constitutive equation for a superplastic solid used by Smirnov is reduced to a liquid.

Romanjuk [424] has considered deformation using absolutely rigid dies of arbitrary shape. In general, the contact surface is unknown. The formulation of the BP [424] is as follows.

$$\sigma_{ij,j}(x,t)=0;\ \ \sigma_{ij}=\sigma_m\delta_{ij}+S_{ij};\ \ S_{ij}(x,t)=2\mu\left(\xi_e^2(x,t)\right)\xi_{ij}(x,t);$$

$$\sigma_m(x,t)=K\int_0^t div\bar{V}(x,s)ds;\ \ \dot{\varepsilon}_{ij}=\frac{1}{2}\left(V_{i,j}+V_{j,i}\right);\ \ \dot{\varepsilon}_{ij}=\xi_{ij}+\left(\frac{1}{3}\right)V_{i,i} \tag{4.106}$$

$$\xi_e^2=C_3\xi_{ij}\xi_{ij}$$

Here, $x(x^1,x^2,x^3)$ is the orthonormalised Eulerian system of co-ordinates. Let the boundary of deforming domain be $\Gamma=\Gamma_\sigma\cup\Gamma_V\cup\Gamma_K$. Surface loading forces P_i and velocities u_i are given on the surfaces Γ_σ and Γ_v, respectively. Then the BCs are

$$\begin{aligned}&T_i(x,t)=\sigma_{ij}(x,t)n_j(x,t)=P_i(x,t), at\ x\in\Gamma_k;\\&V_i(x,t)=u_i(x,t), at\ x\in\Gamma_k\end{aligned} \tag{4.107}$$

The following decompositions are used.

$$\vec{V}=\vec{V}_n+\vec{V}_\tau\ ;\ \vec{V}_n=\left(\vec{V}\cdot\vec{n}\right)\vec{n}=V_n\vec{n}\ ;\vec{V}_\tau=\left(\vec{n}\times\vec{V}\right)\times\vec{n}=V_\tau\vec{\tau}$$

$$\vec{T}=\vec{T}_n+\vec{T}_\tau\ ;\ \ \vec{T}_n=(\vec{T}\cdot\vec{n})\,\vec{n}=T_n\vec{n};\ \ \ \vec{T}_\tau=(\vec{n}\times\vec{T})\vec{n}=T_\tau\vec{\tau}_T$$

The boundary conditions on the unknown contact surface Γ_K are as follows.

$$\begin{aligned}&V_{reln}(x,t)\le 0;\ T_n(x,t)\le 0;\ V_{reln}(x,t)T_n(x,t)=0 \text{ at } x\in\Gamma_k\\&\vec{V}(x,t)=\vec{V}_{die}(x,t)+\vec{V}_{rel}(x,t)\end{aligned} \tag{4.108}$$

where $\vec{V}_{\text{die}}(\vec{x},t)$ is the velocity of the die (as mentioned earlier, the die is assumed to be absolutely rigid). The following limits are imposed on the tangential stress at the contact surface T_τ.

$$|T_\tau(x,t)|\le F,\ \ F\ge 0\,,\ x\in\Gamma_K$$

where F is a friction force determined according to some friction law. From the law of conservation of energy [306]

$$\vec{T}_\tau\cdot\vec{V}_{\text{rel}\tau}\le 0$$

Assuming $\vec{\tau}_{fr}=-\vec{\tau}_v$,

$$T_\tau(x,t)V_{rel\ \tau}(x,t)\le 0;\ \left(|T_\tau(x,t)|-F\right)V_{rel\ \tau}=0\ x\in\Gamma_k \tag{4.109}$$

The details of the calculations are given in [424].

Recently Efimov et al. [388] have presented a formulation of the BP for materials of high strain rate sensitivity. The mathematical correctness of this formulation has been tested and the convergence of the numerical solutions to the exact solution investigated. The inverse problem of mechanics of solids has been formulated and this allows the determination of some parameters of the friction law from the results of technological tests. Some numerical calculations have also been presented.

Argyris and Doltsinis [29] and later Doltsinis [425] developed a computational technique within the framework of finite element discretization. The following constitutive equations (written for the uniaxial case) were used [29, 425].

$$\sigma = f(\xi, d), \; \dot{d} = g(\xi, t) \tag{4.110}$$

The following simple functional forms for f and g were used.

$$f = kd^n \xi^m \; ; \; g = d_0 \left(\frac{N}{t_0} \right) \left(\frac{t}{t_0} \right)^N \tag{4.111}$$

These scalar constitutive equations were extended to multiaxial loading using equations of (Saint-Venant) type (4.3). To use standard computational procedures in the treatment of incompressible processes, the so-called penalty approach was used. In this approach the incompressibility condition div $\vec{V} = 0$ was replaced by the following relation

$$\sigma_H = 3\overline{K}\dot{\varepsilon}_m; \; \dot{\varepsilon}_m = \frac{1}{3}\dot{\varepsilon}_{ii}; \; \overline{K} = \frac{2\mu}{3} \cdot \frac{1+\nu}{1-2\nu} \tag{4.112}$$

where ν is Poisson's ratio, $\boldsymbol{\sigma_H}$ is the hydrostatic part of the Cauchy stress tensor. For an incompressible material $\nu \to 0.5$ and correspondingly the penalty coefficient $\overline{K} \to \infty$. The viscosity μ of a superplastic can be determined from uniaxial experiments by introducing the equivalent Mises stress as follows.

$$\sigma_e = 3\mu\xi_e \equiv f(\xi_e, d)$$

The following relations are assumed in [29] as boundary conditions

$$\vec{F}_n = -k_n \vec{V}_n, \quad \vec{F}_t = -k_t \vec{V}_t \tag{4.113}$$

where $\vec{F}_t$ and $\vec{F}_n$ are the tangential and the normal contact forces, respectively; $k_n \to \infty$ is a penalty parameter; the value of k_t is restricted by the condition

$$k_t = \frac{|\vec{F}_t|}{|\vec{v}_t|} \le k_{max}$$

A number of superplastic forming processes were investigated and compared with experimental results where possible [29]. The bulging of a circular diaphragm from a sheet of alloy Ti–6Al–4V of 1.8mm thickness and average grain size 8 μm is considered in [425].

There are a few other reports where information on the formulation of the BP is absent, but still one can outline a common character. Some papers, which also contain information about the CEs used, will be discussed below.

Chandra et al. [426–434] have used the finite element method to describe superplastic sheet metal processes in terms of the two parameter scalar equation $\sigma = K\xi^m$ and the equations of Saint-Venant type (4.3). A number of technological processes have been modelled. A similar approach based on the same constitutive equation has been adopted in [435–437].

Chennot and Bellet [438–440] have assumed that the total strain rate is equal to the sum of the elastic and the viscoplastic strain rates, i.e.,

$$\dot{\varepsilon}_{ij} = \dot{\varepsilon}_{ij}^{e} + \dot{\varepsilon}_{ij}^{vp} \tag{4.114}$$

It is assumed that Hooke's law is valid for the elastic strain rate and there is a viscoplastic potential Φ

$$\dot{\varepsilon}_{ij}^{vp} = \frac{\partial \Phi}{\partial \sigma_{ij}} \tag{4.115}$$

where Φ is a function of the equivalent stress and equivalent viscoplastic strain rate. Accumulated strain is obtained by integration along the deformation path of a particle, viz.,

$$\varepsilon_{ij}^{vp} = \int_0^t \dot{\varepsilon}_{ij}^{vp} dt' \tag{4.116}$$

The problems of choice of stress and strain measures for finite deformation (which is not trivial), a justification and domain of adequacy of the CEs used (in particular, the validity or otherwise of the same equations for describing both active and passive loading), and the thermodynamics aspects have not been discussed. Similar remarks may also be made about the other papers that are discussed here. Moreover, in this paper and some others the integration step (or differentiation with respect to a variable being a functional of the process) has not been clarified (see Sect. 3.3.2).

Bonet and Wood [441–444] have suggested the following approach assuming the material properties to be described by a variant of the model of Hamilton [360, 445]. The extension of uniaxial relations to multiaxial loading is achieved by rewriting them using the equivalent Mises stress σ_e and strain rate ξ_e and also introducing an appropriate viscoplastic potential, which corresponds to the relation among $\dot{\varepsilon} \sim \sigma \sim d$ in the model of Hamilton.

Superplastic forming of a sheet of Ti–6Al–4V into a mould is considered as an example. The second example considers the superplastic forming of the same alloy sheet into a combined configuration of two parts of different depths. Sadeghi and Pursell [446, 447] have used the same constitutive equation given by Hamilton [360, 445]; but a variant of the flow theory is considered in [446] instead of a viscoplastic potential. The yield surface is determined by the function $f(\sigma_{ij}, \varepsilon_{kl}, k) = 0$ where k is the strain hardening parameter. When the material is considered to be incompressible, the stress and strain deviators are assumed to be proportional, but the 'viscosity coefficient' depends on strain and strain rate.

Recently Wood and Bonet [448] have presented the following variants of the CEs.

$$\xi = \Phi(\sigma, d, \alpha_1, \alpha_2, ..., \alpha_n, T) \tag{4.117a}$$

or

$$\sigma = \Phi^{-1}(\xi, d, \alpha_1, \alpha_2, ..., \alpha_n, T) \tag{4.117b}$$

where d is the average grain size, α_1, α_2,.....,α_n are the internal variables. Evolution equations are to be given for d and α_1, α_2...α_n. Viscoplastic potential Ψ is given for stresses so that

$$\sigma = \frac{\partial \Psi(\xi_e, d, \alpha_1, \alpha_2, ..., \alpha_n, T)}{\partial D} \tag{4.118}$$

where ξ_e is the equivalent strain rate and D is the strain rate tensor. The authors use in the above equations the generalised form of writing of a relation between the stress (σ) and the strain rate (D) tensors.

There are many other reports in the literature dealing with finite element modelling of superplastic metalworking processes. A full review of all these formulations is beyond the scope of the present book. It is pertinent to note that recently a special conference was held [449] in which the various approaches to the numerical modelling of superplastic flow was considered. Further details can also be found in [416–421, 450–455].

4.3.3
Numerical Models of Superplastic Sheet Forming Processes

The best known applications of the phenomenon of superplasticity deal with superplastic forming processes (SPF) for sheet materials. Complex articles have been produced in single or two stage forming. SPF and superplastic forming concurrent with diffusion bonding (SPF/DB) are two such processes. They offer substantial cost savings by reducing the number of detail parts and mechanical fasteners, by achieving more efficient load carrying structures, and by increasing fly-to-buy ratio. SPF and SPF/DB processes are accepted as advanced technologies for forming different aerospace components. Therefore, the problem of modelling such processes adequately is of great practical importance (see next Chapter also). The models are of two types: (i) those based on finite element considerations, (ii) those based on thin shell theory. The mathematical models based on finite element consideration of an appropriate BP were discussed above (see Sect. 4.3.2). The models based on thin shell theory will be examined in detail in Chap.5. The main equations of thin shell theory are briefly outlined below and some examples of their numerical solution are also given.

4.3.3.1 Principal Equations of Membrane Theory

The thickness of alloy sheets is usually small compared with typical dimensions of the die. Therefore, bending and shearing effects are negligible. Then one can use the membrane theory in order to describe the stress state in the deformed sheet. The main principles of the theory are stated below.

Consider the equilibrium of a shell which is a surface of revolution: (Fig. 4.5). Here, the principal radii of curvature are

$$\rho_c = \frac{r}{\sin\varphi}, \quad \rho_t = \frac{1}{\cos\varphi}\cdot\frac{dr}{d\varphi} \tag{4.119}$$

where the principal radii ρ_t and ρ_c are not independent. Eliminating φ

$$\frac{r}{\rho_c}\cdot\frac{d\rho_c}{dr} = 1 - \frac{\rho_c}{\rho_t} \tag{4.120}$$

The stress state of deformed shell is described by two non-zero principal stresses σ_t and σ_c; σ_t lies in the plane of Fig. 4.5 while σ_c is directed normal to this plane (along the circumferential or hoop direction). It is assumed that these stresses are uniformly distributed along the shell thickness. Thus, one can write an expression for the equivalent stress as

$$\sigma_e = \sqrt{\frac{3S_{ij}S_{ij}}{2}} = \sqrt{\sigma_t^2 + \sigma_c^2 - \sigma_t\sigma_c} \tag{4.121}$$

Kinematics of the deformed state is determined by the tangential ξ_t, circumferential ξ_c and normal ξ_n components of the strain rate tensor $\dot{\varepsilon}_{ij}$. Using the incompressibility condition the equivalent strain rate is obtained as

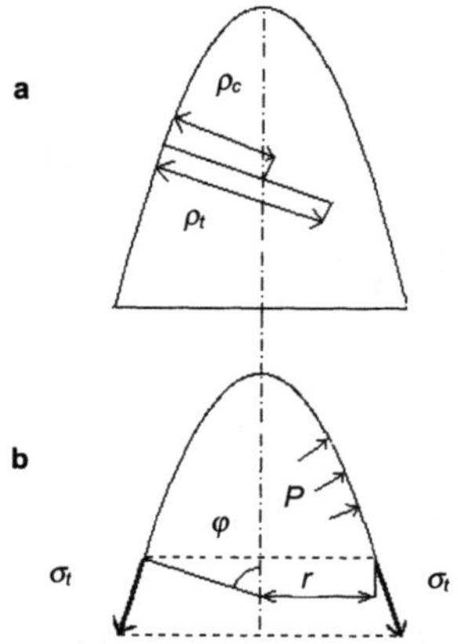

Fig. 4.5. Schematic of an axi-symmetric shell loaded by internal pressure

$$\xi_e = \sqrt{\frac{2}{3}\xi_{ij}\xi_{ij}} = \frac{2}{\sqrt{3}}\sqrt{\xi_t^2 + \xi_c^2 + \xi_t\xi_c} \tag{4.122}$$

The equilibrium equations can be written now as [184]

$$\frac{\sigma_t}{\rho_t} + \frac{\sigma_c}{\rho_c} = \frac{p}{h}; \quad (\sigma_t h) 2\pi z \sin\phi = Q_z \equiv p\pi r^2 \qquad (4.123a,b)$$

where h is the thickness of the shell. The first of Eqs. (4.123) is often referred to as Laplace equation. From Eqs.(4.123a,b)

$$\sigma_t = \frac{p\rho_c}{2h} \qquad (4.124a)$$

$$\sigma_c = \frac{p\rho_c}{2h}\left(2 - \frac{\rho_c}{\rho_t}\right) \qquad (4.124b)$$

the geometry, the thickness of the shell and the value of the pressure are known, then one can determine σ_t and σ_c using these equations. For a spherical shell,

$$\rho_c = \rho_t = R \ , \quad \sigma_c = \sigma_t = \frac{pR}{2h}$$

If $\rho_c > \rho_t$, then $\sigma_c < \sigma_t$ and vice versa. Taking account of Eq.(4.124) and denoting ρ_c / ρ_t by $\overline{\rho}$ one obtains

$$\{S_{ij}\} = \frac{p\rho_c}{2h}\begin{bmatrix} \frac{1}{3}\bar{\rho} & 0 & 0 \\ 0 & 1-\frac{2}{3}\bar{\rho} & 0 \\ 0 & 0 & -1+\frac{1}{3}\bar{\rho} \end{bmatrix} \tag{4.125}$$

Then the equivalent stress Eq.(4.121) is

$$\sigma_e = \frac{p\rho_c}{2h}\sqrt{3-3\bar{\rho}+(\bar{\rho})^2} \tag{4.126}$$

In order to determine the strain rate state and the equivalent strain rate it is necessary to use a hypothesis concerning the vector properties of the material. If the hypothesis of proportionality of the stress and the strain rate deviator Eq.(4.3) is assumed, from expressions Eqs. (4.122) and (4.125)

$$\xi_e = \frac{2\xi_t}{\bar{\rho}}\sqrt{3-3\bar{\rho}+(\bar{\rho})^2} = \frac{2\xi_c}{3-2\bar{\rho}}\sqrt{3-3\bar{\rho}+(\bar{\rho})^2} \tag{4.127}$$

In particular, for a spherical shell $\rho_c = \rho_t = R$ and $\sigma_e = \sigma_c = \sigma_t = pR/2h$; $\xi_e = 2\xi_c = 2\xi_t$. For a cylindrical shell $\rho_c = R$, $\rho_t \to \infty$ and so $\sigma_t = pR/2h$; $\sigma_c = pR/h$; $\sigma_e = p\sqrt{3}\,R/2h$; $\xi_e = 2\xi_c/\sqrt{3}$; $\xi_t \to 0$.

4.3.3.2
Numerical Solutions of the Principal Equations of Membrane Theory

Storakers [456] has analysed the plastic deformation of a circular membrane subjected to one-sided hydrostatic pressure. This analysis is based on Hencky–Mises stress–strain equations.

$$\varepsilon_{ij} = \frac{3}{2}\frac{\varepsilon_e}{\sigma_e}S_{ij} \tag{4.128}$$

where ε_e is the equivalent strain (infinitesimally small strains are considered). The basic assumptions are: a) elastic strains are omitted; b) final strain is independent of the rate of straining; c) the material is incompressible and isotropic. The material is assumed to have a parabolic scalar stress–strain relationship

$$\sigma_e = \sigma_0\varepsilon_e^n \tag{4.129}$$

Equations (4.128) and (4.129) were employed along with Eq. (4.124) to derive the system of two differential equations, which could then be solved numerically using the Runge–Kutta–Gill method.

The analysis presented in [456] deals with the plastic deformation of a strain rate insensitive material. But, superplastic flow is characterised by strong strain rate sensitivity and limited strain hardening. Nevertheless, this work is of interest as an example of a correct approach to the problem of bulging of a circular membrane. This example will be used later.

Brandon, Lucent and Octane [457] have examined the creep bulging of clamped circular thin shells subjected to one-sided hydrostatic pressure. The following assumptions are involved: (i) the material is isotropic and incompressible; (ii) the dome is a surface of revolution; (iii) the membrane theory holds and gives sufficient accuracy for the stress distribution all over the dome except near the clamped boundary.

The system of equations (4.124) is numerically solved [457] for the secondary creep law which, for uniaxial loading, can be written as $\xi = C\sigma^n$ and extended to the three dimensional case as

$$\xi_e = C\sigma_e^n \ ; \quad \xi_{ij} = \frac{3}{2} C\sigma_e^{n-1} S_{ij} \tag{4.130}$$

where σ_e and ξ_e are found from Eqs (4.121) and (4.122), respectively; no distinction is made between the strain rate tensor and its deviator because of assumption (i). In this case, Eq. (4.130) can be presented as

$$\xi_t = \xi^* \bar{\rho}; \ \xi_c = \xi^* \cdot (3 - 2\bar{\rho}); \ \xi_n = -\xi^* \cdot (3 - \bar{\rho}) \tag{4.131}$$

where

$$\xi^* = \frac{C}{2} \cdot \left(\frac{p\rho_c}{2h} \right)^n \cdot \left[3 - 3\bar{\rho} + (\bar{\rho})^2\right]^{\frac{n-1}{2}} \tag{4.132}$$

The system of equations Eqs.(4.131) and (4.132) was solved in [457] using numerical procedures. Experimental verification was not done. The following generalisation can be suggested. It is easy to see that constitutive equations (4.128) do not apply to superplastic flow. However, the following relationships are obtained immediately from Eq. (4.127).

$$\begin{aligned} \xi_t &= \frac{1}{2} \xi_e \bar{\rho} \left[3 - 3\bar{\rho} + (\bar{\rho})^2\right]^{-1/2} \\ \xi_c &= \frac{1}{2} \xi_e (3 - 2\bar{\rho}) \left[3 - 3\bar{\rho} + (\bar{\rho})^2\right]^{-1/2} \end{aligned} \tag{4.133a,b}$$

where the equivalent strain rate ξ_e is an arbitrary function of the flow stress. If one assumes $\xi_e = C\sigma_e^n$, one can obtain the expressions for ξ_t, ξ_c and $\xi_n = -\xi_t - \xi_c$ coinciding with Eqs. (4.131) and (4.132). One can also use any other functional form for $\xi_e = f(\sigma_e)$ in Eq.(4.133).

Malinin [184] has considered the same problem, using the following power relationship.

$$\sigma = A\varepsilon^n \xi^m \tag{4.134}$$

It is evident that this relation can be considered as a generalisation of the above-mentioned approaches of Storakers [456] and Brandon et al. [457]. Assuming that the total strain rate is equal to the sum of the elastic and creep strain rates, this equation can be written for the uniaxial case as

$$\xi = \xi^e + \xi^c = \frac{\dot{\sigma}}{E} + \xi^c \tag{4.135}$$

where E is Young's modulus. Then one can extend Eq. (4.135) for the multiaxial case as

$$\xi_{ij}^c = \frac{3\xi_e^c}{2\sigma_e} S_{ij} \ ; \sigma_e = A\left(\xi_e^c\right)^m \left(s^c\right)^n, \ P \equiv s^c = \int_0^t \xi_e^c \, dt' \tag{4.136}$$

and Hooke's law is used for ξ_{ij}^e. For secondary creep elastic and plastic strains are neglected in [184] so that Eqs. (4.136) deal with the total strain rates and the total strains. Then the superscript c can be omitted. A number of examples are considered in [184]. In general, the solution is found by numerical procedures. For the deformation of long rectangular membrane, however, an analytical solution can also be found.

It is necessary to note that the constitutive Eq. (4.134) has only a restricted range of applicability (Chap. 3). Nevertheless, the results obtained can be used for analysing superplastic metalworking processes. But, the method of generalising equation Eq.(4.134) for the case of multiaxial loading used in [184] is not unique. The equations of thin shell theory Eq.(4.124) are often solved by using some approximate methods that are based on additional hypotheses concerning the stress or the strain state in the deformed shell. Such approaches will be considered in detail in Chap.5. Here, only one example of such an approach, suggested by Ghosh and Hamilton [288], is considered.

Superplastic bulging of a thin circular diaphragm is considered. It is assumed that the shape of the deformed membrane at any instant is a part of a sphere. This assumption has often been used while modelling superplastic bulge forming of a circular membrane and it has been confirmed experimentally in superplastics having a sufficiently high strain rate sensitivity index, m (see Chap.5 also). The following considerations are involved [282]. For a spherical cap the membrane equilibrium equation at the apex is

$$\sigma_t^a = \frac{pR}{2h_a} \tag{4.137}$$

where σ_t^a, h_a are respectively the tangential stress and the thickness of the shell at the dome apex, R is the radius of the dome. The equilibrium equation for vertical forces at the clamped edge is

$$\sigma_t^b = \frac{pR}{2h_b} \tag{4.138}$$

where σ_t^b, h_b are respectively the tangential stress and the thickness of the shell near the clamped edges.

The vector properties of the superplastic material are assumed to be described by the equations of Saint-Venant type (4.3), the term 'associated flow rule' being used in [282]. Then, taking into account that at the dome apex $\xi_t^a = \xi_c^a$ while at the clamped edges the hoop strain rate $\xi_c^b = 0$ one can find, in accordance with Eq.(4.3), that $\sigma_t^a = \sigma_c^a$ at the dome apex and $\sigma_t^b = 2\sigma_c^b$ at the clamped edges. Then, according to Eqs. (4.121) and (4.122), at the dome apex

$$\sigma_e^a = \sigma_t^a = \sigma_c^a \;\; ; \;\; \xi_e^a = 2\xi_t^a = 2\xi_c^a \tag{4.139}$$

and at the clamped edges

$$\sigma_e^b = \frac{\sqrt{3}}{2}\sigma_t^b = \sqrt{3}\sigma_c^b \;\; ; \;\; \xi_e^b = \frac{2}{\sqrt{3}}\xi_t^b \tag{4.140}$$

From Eqs. (4.139) and (4.140)

$$\begin{aligned} h_a &= h_0 \exp\left(\varepsilon_n^a\right) = h_0 \exp\left(-2\varepsilon_t^a\right) = h_0 \exp\left(-\varepsilon_e^a\right) \\ h_b &= h_0 \exp\left(\varepsilon_n^b\right) = h_0 \exp\left(-\varepsilon_t^b\right) = h_0 \exp\left(-\frac{\sqrt{3}\varepsilon_e^b}{2}\right) \end{aligned} \tag{4.141}$$

According to Eqs. (4.137) and (4.138) $\sigma_t^a h_a = \sigma_t^b h_b = pR/2$ and

$$\frac{2}{\sqrt{3}}\sigma_e^b \exp\left(-\varepsilon_t^b\right) = \sigma_e^a \exp\left(-2\varepsilon_t^a\right) = \frac{pR}{2h_0} \tag{4.142}$$

(from Eqs. (4.128) and (4.129)). The following constitutive equation is used in [288]

$$\sigma_e = \sigma_0 + B\varepsilon_e \tag{4.143a}$$

where

$$\log\sigma_0 = A_0 + A_1 \cdot \log\xi_e + A_2 \cdot (\log\xi_e)^2 + A_3 \cdot (\log\xi_e)^3 + A_4 \cdot (\log\xi_e)^4 \tag{4.143b}$$

The material constants A_0, A_1, A_2, A_3, A_4, and B are obtained experimentally. For a desired value of strain rate at the dome apex $\xi_e^a = \xi_{\rm opt}$ = constant, where $\xi_{\rm opt}$ corresponds, e.g., to the point of inflection on the sigmoidal $\log\sigma_e - \log\xi_e$ curve, one can calculate $\xi_e^a = \xi_{\rm opt}$. Then σ_e^a is determined from Eq. (4.143a). Further it is necessary to obtain an appropriate solution at the edge which can be computed from Eqs. (4.140) and (4.143) so that Eq. (4.142) is satisfied.

According to Eq. (4.142), the time dependence of the radius of the dome $R = R(t)$ is to be found in order to calculate the pressure–time cycle. For this, the following empirical equation is suggested in [282].

$$R\, sin^{-1}\left(\frac{R_0}{R}\right) = \int_0^t (exp\,\varepsilon_t)dx = \frac{R_0}{3}\left(2\,exp\,\varepsilon_t^a + exp\,\varepsilon_t^b\right) \tag{4.144}$$

It is important to point out that Ghosh and Hamilton appear to have been the first to realise that the strain state changes from the balanced biaxial at the centre to plane strain near the clamped edge. This helps to overcome a serious limitation in the analyses of Jovane [384] and many later workers which predicted uniform dome thickness during bulging. (Also, as the die is rigid, the hoop strain should be zero but it is significant in Jovane's analysis.) Also, the constitutive equation (4.143) used in [282] takes into account both strain and the strain rate hardening, which allows the calculation of the pressure–time cycles at relatively low strain rates where the grain growth and flow (strain) hardening can be strong. An alternative constitutive equation is also shown to be applicable for the approach developed in [282]. Superplastic forming of a long rectangular membrane is also considered in [282]. The thinning process during blow forming has been divided in that case into three main stages.

Apart from some minor errors, it is easy to show that the second principal equation (4.124b) is not obeyed at the clamped edge within the framework of the Ghosh–Hamilton model [282]. This is a serious limitation. This problem is discussed further in Chap.5.

It is necessary to consider the full formulation of the BP while developing a mathematical model for any technological process. In particular, one should check thoroughly the validity of the equations of motion; the temperature and mechanical boundary conditions are to be analysed and verified; appropriate constitutive equations are to be chosen; conservation of energy and the applicability of the basic concepts for the given problem are to be examined etc. If some contradictions are present, they should be removed in a refinement of the model. In the present example, e.g., one can relax the condition imposed on the shape of the deformed shell. Alternatively, one may reject the simplified Eq.(4.124) and solve the exact system of Eq. (4.2).

Akkus et al. [458] have considered the bulging of a tube into a spherical shape. The principal equations of thin shell theory (4.124) have been solved numerically assuming the standard power law $\sigma = K\xi^m$. Internal pressure bulge tests were also conducted on pipes of a titanium alloy which were formed into a spherical die with a view to experimentally validate the analysis.

Throughout this chapter the material has been assumed to be isotropic. But sheet materials usually produced by rolling have anisotropic properties. Let the parameter Y be the mathematical measure of such anisotropy. For example, Y is the ratio of the yield stresses measured along and transverse to the rolling direction. ($Y = 1$ for an isotropic material.)

Wang and Shammany [459] have applied Hill's incremental theory for anisotropic materials to solve the problem of axisymmetric stress state:

$$\frac{\xi_t}{\sigma_t - \frac{Y}{1+Y}\sigma_c} = \frac{\xi_c}{\sigma_c - \frac{Y}{1+Y}\sigma_t} = \frac{\xi_e}{\sigma_e} \tag{4.145}$$

where

$$\sigma_e = \sqrt{\sigma_t^2 + \sigma_c^2 - \frac{2Y}{1+Y}\sigma_t\sigma_c} \tag{4.146a}$$

$$\xi_e = \frac{1+Y}{\sqrt{1+2Y}}\sqrt{\xi_t^2 + \xi_c^2 + \frac{2Y}{1+Y}\xi_t\xi_c} \tag{4.146b}$$

It is easy to show that Eqs. (4.146a) and (4.146b) with $Y = 1$ reduce to Eqs. (4.140) and (4.139), respectively. Also, Eqs. (4.65) with $Y = 1$ are equivalent to Eq. (4.146b). Later, Eqs. (4.145) and (4.146) were solved numerically [460] assuming the standard power law $\sigma = K\xi^m$.

5 Mathematical Modelling of Superplastic Metalworking Processes

Superplastic metal working processes have been modelled mostly based on semi-empirical approaches and using simplifying assumptions [461–466]. In this chapter, such models known in the literature are reviewed. Attention is focussed on presenting the ideas from the viewpoint of mechanics of solids. The merits and limitations of these analyses are discussed with a view to improving them in the future. A comprehensive review of the industrial applications of superplasticity is beyond the scope of this book.

5.1 Modelling of Superplastic Bulk Forming Processes

5.1.1 General Comments

Extreme elongation without localised necking due to the presence of tensile forces and very low working stresses are two important properties of superplastic alloys, see, e.g., [1–8]. It is important to exploit these properties in metalworking. In particular, conventional compressive working processes, e.g., forging, extrusion, can benefit from the low flow stresses associated with superplastic deformation. Since most of the existing commercial metal forming processes make use of compressive forces, it becomes necessary to study the mechanical response of superplastics in compression. The first experiments in compression were made [465] on a superplastic Al–$CuAl_2$ eutectic alloy that included both uniaxial testing and closed-die forging. Smirnov [2] believes that exploitation of superplasticity in industry has been mostly through bulk forming processes. But, according to Sherby and Wadsworth [14] press forging of bulk materials in the superplastic state is still in its infancy. Nevertheless, it is commonly recognised that superplastic bulk forming entails a number of advantages in comparison with conventional methods of metal forming. In particular, the following benefits have been identified.

1. The capacity of materials to deform plastically without failure is one to two orders of magnitude greater in the superplastic state than in the conventional state.
2. High stress relaxation capability of superplastics leads to the elimination of residual stresses in parts and ensures high dimensional accuracy.

3. The relatively low flow stress in the range of 5 to 30 MPa has been used to form difficult-to-work superalloys (the Gatorizing process) and other strong materials or in embossing fine details in other applications [210, 466].
4. Superplastic deformation randomises the microstructure of the material, which allows parts with uniform isotropic properties to be obtained. It is especially attractive for producing large size parts, e.g., discs of gas turbine engines for the aerospace industry.
5. As already shown by Tresca and Bridgman, application of high hydrostatic pressures leads to large ductilities even in brittle materials, e.g., marble. Thus, the use of a hydrostatic pressure with superplastic alloys can further enhance the scope of application of superplastics in industry. The enormous ductility helps reduce the number of steps in a forming process. Multi-stage operations can be converted into single step processes. Assembly and labour costs can be considerably reduced.
6. The low flow stresses allow the use of cheap, novel die materials and when expensive die materials are used their life will be prolonged.
7. The post-forming properties are far more uniform than in conventional processes because superplastic forming (SPF) is done at relatively slow strain rates and nearly steady state conditions prevail during manufacture.
8. When the formed component is used at low temperatures, the fine grain size ensures that superior properties are available in service (Hall–Petch effect etc.).
9. The low flow stresses, on the other hand, ensure that inexpensive equipment can be used for SPF.

Figure 5.1 is the picture of a stamping shop in Ufa, Russia devoted to the manufacture of bulk formed parts in the superplastic state.

However, there are also some disadvantages. The strain rates are relatively low so that the common power drop hammers or gravity drop hammers are not acceptable (as the available strain rates on those machines are too high [465]). Therefore, mechanical or hydraulic presses capable of maintaining low velocities of cross head/ram movement to provide optimal rates of forming have to be specially constructed. Also, the cross head will have to be held (incorporating a non-return check valve in the hydraulic circuit) for 1–3 hours in the same position to complete forming. Most importantly, the relatively low strain rate reduces the rate of production and so, SPF can be used only for niche markets involving small and medium batch production. As the temperature interval of superplastic forming is usually narrow, special equipment to maintain constant temperature is needed. (Open-die forging as well as closed-die forging can be used to produce parts.) Preprocessing to convert the material into the superplastic state as well as the special nature of the alloys make the price of starting material higher.

Also, when the service conditions include high temperature, post-SPF treatments are needed to destroy the fine grain structure if creep during service is to be avoided. This step also will add to the cost of manufacture. Notwithstanding these drawbacks, in many applications superplasticity has been exploited, e.g., super-

Fig. 5.1. Stamping shop for producing parts in a superplastic state(Courtesy: A.G. Ermatchenko)

plastic bulk forming of alloys of aluminium, titanium, iron, nickel, etc. [2]. Even bevel gears have been warm forged using fine-grained ultra-high carbon steel.

Another possibility is the press forging of bulk materials into complex dies. It has been shown [465] that die profiles that include sharp corners and considerable depth can be reproduced accurately using very low loads. Thin webs, which would be difficult to forge in the conventional state without micro- or macro-crack formation, can be easily formed. (The SPF route cuts the cost of manufacture by about 50% [3], the reduced production rate can be offset by increasing the number of installations still ensuring overall favourable economics [465].) It has also been shown that turbine discs, blades, etc. can be produced in a single operation using fine grained nickel-based alloys or heat-resistant titanium alloys [6, 468–470]. Considerable experience has been accumulated in the last three decades in producing aircraft engine compressor discs and blades (see Fig. 5.2) [471–473].

As emphasised in [6], quantitative studies that evaluate the role of variables in superplastic forming are sparse. Finite element methods used to model bulk forming were considered in Chap. 4. In this chapter, some simplified mathematical models of superplastic bulk forming known in the literature are examined.

5.1.2 Compression of a Disc using Platens

Mathematical modelling of the process of compression of a disc between flat platens is a traditional problem in metalworking. Many publications deal with the

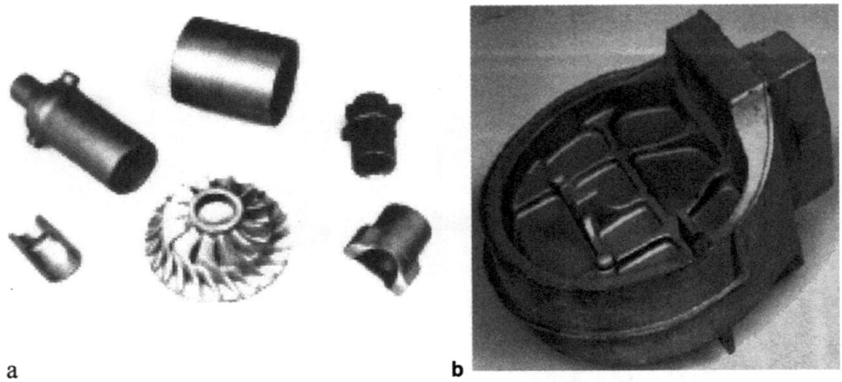

a b

Fig. 5.2. Products of Superplastic metal forming technologies : **a** tumbler, frame, bush sleeve, discs made in Russian alloys VT9, VT5-1, VT3-1; **b** frame, VT6 (Ti-6Al–4V) alloy (Courtesy: Dr. A.G. Ermatchenko)

various aspects of this process as applied to non-superplastic (conventional) materials. Tang [474] appears to have been the first to examine the compression of a disc of a superplastic material between two flat platens. The main features of this elementary analysis are as follows.

Let the velocity of the moving die $v > 0$ be directed along the z-axis. Barrelling is assumed to be absent. Then the velocity field can be written in a cylindrical system of co-ordinates r, φ, z with the z-axis coinciding with the disc axis as

$$u_r = \frac{rv}{2h}, \quad u_\varphi = 0\ , \quad u_z = -\frac{zv}{h} \tag{5.1}$$

where h is the current disc height. The components of the strain rate tensor are

$$\xi_r = \xi_\varphi = -\frac{1}{2}\xi_x = \frac{v}{2h};\ \xi_{r\varphi} = \xi_{z\varphi} = \xi_{rz} = 0 \tag{5.2}$$

Effective stain rate ξ_e can be found as $\xi_e = -\xi_z = v/h$. Thus, the kinematics of the process is known. Steady state creep equations are used in [474–476] as

$$\xi_{ij} = \frac{3}{2}C\sigma_e^{n-1}S_{ij} \tag{5.3}$$

Then,

$$S_r = S_\varphi;\sigma_\varphi \cdot \sigma_z = -p,\ \sigma_{r\varphi} = \sigma_{z\varphi} = \sigma_{rz} = 0$$

(from Eqs. (5.2) and (5.3)), where p is the hydrostatic pressure. Consequently, the equivalent stress σ_e is

$$\sigma_e = \sqrt{\frac{3S_{ij}S_{ij}}{2}} = \sigma_r - \sigma_z = \sigma_r + p_H \tag{5.4}$$

The system of equations written in a cylindrical system of co-ordinates as in Eqs. (4.42), leads to identities with respect to Eqs.(4.42b) and (4.42c) and the condition

$$\frac{\partial \sigma_r}{\partial r} = 0 \tag{5.5}$$

with regard to Eq. (4.42a). The mechanical boundary condition is

$$\sigma_r = 0 \ at \ r = r_0 \tag{5.6}$$

where r_0 is the current radius of the disc. It follows that $\sigma_r \equiv 0$ at $0 \leq r \leq r_0$. Consequently, (from Eq.5.4) $p = \sigma_e$ which is independent of r. The axial force is

$$P_0 = p_H \pi r_0^2 = \pi r_0^2 \sigma_e = \pi r_0^2 K \xi_e^2 = \pi r_0^2 K \left(\frac{v}{h} \right)^2 \tag{5.7}$$

Tang [474] has assumed that there is friction at the contact surface 'die–disc' and this stress is proportional to the average value of $p = p_{avr}$ or directly to p. This hypothesis leads immediately to the conclusion that $\sigma_{rz} \neq 0$. Using this assumption Tang obtained the formulae to calculate the value of the axial force over the contact surface. Then, the above mentioned elementary analysis was used. However, the original hypothesis of Tang that barrelling was absent is not consistent with the latter assumption of non-zero σ_{zr}. When $\sigma_{rz} \neq 0$, $\xi_{rz} \neq 0$ since

$$\xi_{rz} = \frac{3}{2} C \sigma_e^{n-1} \sigma_{rz} \tag{5.8}$$

As $\xi_{rz}/\xi_z = \sigma_{rz}/\sigma_z$, $\xi_{rz} = f\xi_z$ (close to the contact surface 'die–disc'), it is evident that at mid cross-section of the plane of symmetry $\xi_{rz} = 0$. From this, it follows that $\sigma_{rz} \neq 0$. This will inevitably lead to barrelling.

5.1.3 Forging of a Disc by Rotating Dies

5.1.3.1 Formulation of the Simplified Boundary Value Problem

Forging large components, e.g., discs used in gas turbine engines, from difficult to form materials like Ni-based superalloys and Ti-alloys is an important application of superplasticity. The force requirement for such a job when the material is in the non-superplastic state would exceed the capacity of most presses. Also, as a rough rule, the cost of a press increases as the square of its increase in capacity [465]. On both counts, it is highly desirable to reduce the forming (axial) force. When the required press capacity is not available, coupling torsion with compression should be explored.

One way of reducing the axial force is to form superplastically at a constant temperature. The axial force can also be reduced significantly, if the upper die is rotated with respect to the lower one about the common axis which is perpendicular to the contact surface 'die–workpiece' (rotary forging) see, e.g. [313, 386]. The theoretical analysis in this regard [313, 386, 477, 478] deals with conventional strain hardening materials. In this section, die forging with torsion of a superplastic material of high strain rate sensitivity is analysed.

The main reason for the decrease in the axial force is suggested to be as follows [313, 386, 477, 478]. During compression considerable friction forces arise at the contact surface 'die–disc'. The vector of contact friction forces $\vec{f}$ is directed radially. If the angular movement of the die is added to this, the vector $\vec{f}$ acquires an additional tangential component. But the friction forces arising at the contact surface between the die and the disc obeys the condition

$$\left|\vec{f}\right| \le f_{\max} \tag{5.9}$$

where $f_{\max}$ is an empirical constant (see Chap. 4 also). In compression without torsion, the vector $\vec{f}$ has only one non-zero component f_r. In compression with torsion it has two non-zero components: the radial f_r' and the tangential f_t' components. From condition (5.9) one can conclude that $f_r' < f_r$. From this it follows that the net resistance to the flow of material along the contact surface will be less for the case of compression with torsion. It is believed that during compression with torsion the axial force will be the lowest if the die surface is rough and no lubricant is used [386].

It is difficult to solve this problem by directly integrating the equilibrium equations, as one has to solve in this case a 3D-boundary value problem. The following assumptions are made in [386]: (i) barrelling is negligible; (ii) the law of friction is $\tau_{fr} = \psi\tau_s$, where ψ is the friction coefficient, τ_s is the shear stress; the vector of friction forces changes direction in accordance with the rate of relative sliding 'a' of the die with respect to the disc (a equals the ratio of the angular velocity of the die to that of the contact surface of the disc $0 \le a \le 1$). Thus, one can see that the kinematics of the process is not strictly determined, but it depends on the value of the coefficient a, which is yet to be determined.

The main idea used in [386] to solve the simplified boundary value problem is to apply the so-called *Upper Bound Technique, which* is well-known in the theory of metal working [210, 479, 480]. According to this method, the actual velocity field (which is an exact solution of the corresponding boundary value problem) corresponds to that for which the total power of plastic deformation is a minimum.

For the simplified boundary value problem under consideration, the total power of plastic deformation is minimised in [386] with respect to the parameter a; such a velocity field is termed a kinematically admissible velocity field [210, 479, 480].

A similar analysis is presented below for a superplastic material. The scalar constitutive equation is taken to be

$$\tau_s = \tau_0 + \mu\dot{\gamma} \tag{5.10}$$

where τ_s is the shear yield stress, $\dot{\gamma}$ is the shear strain rate τ_0 and μ are material constants. Let a cylinder of diameter $D_0 = 2r_0$ and height $2h_0$ be subjected to compression with torsion along the z-axis, which is perpendicular to the contact surfaces. Let $v > 0$ and ω respectively be the axial and the angular velocities of the moving die. Let $a = \omega_t/\omega$ the parameter of sliding; ω_t the angular velocity of the disc contact surface rotation, $h = h_0 - vt$ the current semi-height of the disc, t time, $\varepsilon = vt/h_0$ the engineering strain, $\theta = r_0\omega/v$ is the relative velocity of die rotation. Here, the boundary value problem will be considered in a way similar to that given in [386] but using constitutive equation (5.10) instead of the assumption that the material exhibits linear deformation hardening.

The velocity field satisfying the incompressibility condition and the kinematic boundary conditions can be written in a cylindrical system of co-ordinates as

$$u_r = \frac{rv}{2L};\ u_\theta = \frac{a\omega rz}{L};\ u_z = -\frac{zv}{L}, L = 2h \tag{5.11}$$

where u_r, u_φ and u_z are the velocity vector components. Then, the equivalent shear strain rate $\dot{\gamma}_e$ can be found as (see also Chap. 4 and Appendix)

$$\dot{\gamma}_e = \frac{v}{h}\sqrt{3 + \frac{a^2r^2\omega^2}{v^2}} \tag{5.12}$$

It is to be noted that the expression used in [386] for effective shear strain rate is different from that used in this book (Eq. (5.12)). But the velocity fields used here and in [386] are the same. Such a difference is related to the difference in the shear strain rate components used in [386] ($2\xi_{ij}$ instead of ξ_{ij}).

5.1.3.2
Solving the Simplified Boundary Value Problem

Total power needed for plastic deformation is

$$N = N_{in} + N_{fr} \tag{5.13}$$

where N_{in} and N_{fr} are the power due to internal forces and friction forces, respectively. It should be mentioned that neglecting barrelling is acceptable in this case for the following reasons. The method of solution is based on the upper bound technique. Thus, an upper estimate of the axial force is obtained. As far as an exact solution of the boundary value problem is concerned, it will correspond to a lower axial force and the presence of barrelling.

The power due to internal forces can be found as

$$N_{in} = \iiint_V \sigma_e \xi_e dV = \iiint_V \tau_e \dot{\gamma}_e dV = \int_0^{2\pi} d\varphi \int_0^h dz \int_0^{r_k} (\tau_0 + \mu \dot{\gamma}_e)\dot{\gamma}_e r dr \tag{5.14}$$

where σ_e , ξ_e are, respectively, the equivalent stress and strain rate, τ_e is the equivalent shear stress, $r_k = r_0 / \sqrt{1+e}$ is the current disc radius. Integrating Eq.(5.14) with $h = h_0 - vt = h_0(1+e)$, one obtains that

$$N_{in} = \tau_0 \pi r_0^2 v\sqrt{3}\left\{\frac{2}{a^2\theta^2}\left[\left(1+\frac{a^2\theta^2}{3(1+e)}\right)^{3/2} - 1\right] + \frac{\zeta\sqrt{3}}{1+e}\left[1+\frac{a^2\theta^2}{6(1+e)}\right]\right\} \tag{5.15}$$

where $\theta = r_0\omega/v$; $\zeta = \mu v/\tau_0 h_0$. The power of friction forces is

$$N_{fr} = \iint_S \tau_{fr} u'_s dS \tag{5.16}$$

where the integration is to be done over the contact surface (circle of radius r_k), while u'_s is the relative velocity of a particle of the disc with respect to the die.

$$u'_s = \sqrt{(u'_r)^2 + (u'_\varphi)^2} = \sqrt{(rv/2h)^2 + [(1-a)\omega r]^2} \tag{5.17}$$

Substituting Eq. (5.17) in Eq. (5.16)

$$N_{fr} = \tau_0 \pi r_0^2 \frac{v\sqrt{3}}{1+e} \Gamma \psi \left[\frac{w}{3\sqrt{3}(1+e)^{3/2}} + \frac{\zeta I}{1+e}\right] \tag{5.18}$$

where $\Gamma = \sqrt{1+4(1+e)^2(1-a)^2\theta^2/w^2}$, $w = r_0/h_0$ and I is the following integral

$$I = \frac{1}{r_0^2 h_0} \int_0^{\frac{r_0}{\sqrt{1+e}}} \sqrt{1+\frac{a^2\omega^2 r^2}{3v^2}} r^2 dr$$

Calculating I using standard procedures one obtains

$$I = \frac{w}{8}\left(\frac{3}{a^2\theta^2}\right)^{3/2}\left\{\left[1+\frac{2a^2\theta^2}{3(1+e)}\right]\frac{a\theta}{\sqrt{3}\sqrt{1+e}}\sqrt{1+\frac{a^2\theta^2}{3(1+e)}} - \right.$$
$$\left. -\frac{1}{2}\ln\left|1+\frac{2a^2\theta^2}{3(1+e)}+\frac{2a\theta}{\sqrt{3}\sqrt{1+e}}\sqrt{1+\frac{a^2\theta^2}{3(1+e)}}\right|\right\}$$

a is to be found from the condition

$$N = N_{in} + N_{fr} \rightarrow \min \tag{5.19}$$

Since the analytical expressions for N_{in} and N_{fr} are rather complicated, it is reasonable to find their values using numerical procedures.

The value of twisting moment $\overline{M}$ can be estimated on the basis of the following vector equation

$$\vec{M} = \iint_S \vec{\tau}_{fr} \times \vec{r} dS \tag{5.20}$$

Integrating Eq. (5.20) and keeping in mind that the vector of the friction forces is opposite to the vector of relative motion of the particles of the disc with respect to the die, one obtains that

$$\overline{M} = 2\pi r_0^2 h_0 \tau_0 \sqrt{3}\psi \frac{\sqrt{\Gamma^2 - 1}}{\Gamma}\left\{\frac{w}{3\sqrt{3}(1+e)^{3/2}} + \frac{\zeta I}{1+e}\right\} \tag{5.21}$$

In order to find the value of the axial force, the power balance (the law of conservation of energy) is written.

$$P_k V + M\omega = N = N_{in} + N_{fr} \tag{5.22}$$

where P_k is the axial force during compression with torsion. Combining Eqs. (5.15), (5.18), (5.21) and (5.22)

$$P_k = \frac{\sqrt{3}\tau_0 \pi r_0^2}{1+e}\left\{\frac{2(1+e)}{a^2\theta^2}\left[\left(1+\frac{a^2\theta^2}{3(1+e)}\right)^{3/2} - 1\right] + \frac{\zeta\sqrt{3}}{1+e}\left[1+\frac{a^2\theta^2}{6(1+e)}\right] + \right.$$
$$\left. \left[\Gamma\psi - (1+e)\psi\theta\frac{2\sqrt{\Gamma^2-1}}{\Gamma w}\right]\left[\frac{w}{3\sqrt{3}(1+e)^{3/2}} + \frac{\zeta I}{1+e}\right]\right\} \tag{5.23}$$

For the case of compression without torsion in Eq. (5.23) $\theta \rightarrow 0$ and one obtains

$$P_0 = \lim_{\theta \to 0} P_k = \frac{\sqrt{3}\tau_0 \pi r_0^2}{1+e}\left[1+\frac{\zeta\sqrt{3}}{1+e}\right]\left[1+\frac{\psi w}{3\sqrt{3}(1+e)^{3/2}}\right] \tag{5.24}$$

5.1.3.3
Analysis of the Solution Obtained

As an example, the compression of a disc of diameter $D_0 = 200$ mm and height $2h_0 = 320$ mm made of a superplastic titanium alloy (Ti–6.6%Al–3.5%Mo–1.7%Zr–0.27%Si) is considered. The values of the material constants entering Eq. (5.10) were calculated using standard regression procedures. $\tau_0 = 23.1$ MPa, $\mu = 133$ MPa·s, for a forming temperature of 950°C. The values of the friction coefficient ψ, the axial velocity v and the angular velocity n (rev/min) of the moving disc were varied within the intervals 0.05–0.8; 0.48–48 mm min^{-1} and 0–50 rev.min^{-1}, respectively.

The results of the calculations show that compression with torsion leads to a significant reduction in the axial force, the value of the normalised force $\overline{P} = P_k/P_0$ decreasing with increasing n, ψ, $|e|$ (Fig. 5.3). However, with large values of n, ψ, $|e|$ the value of the twisting moment $\overline{M}$ needed to provide the torsion of the die increases sharply. Also, at large strains the strain rate hardening of a superplastic material decreases the efficiency of the process, The following interesting feature can be seen in Fig. 5.3b. Starting from approximately $n \approx 0.05$ rev/min, a further increase in the value of the angular velocity becomes ineffective in decreasing the axial load, independent of the value of the friction coefficient. This can be explained by stating that the sliding parameter decreases with increasing n.

Thus, it is clear that a decrease in the axial force can be achieved by increasing the friction coefficient, the angular velocity of the moving die and the absolute value of the engineering strain. The situation is similar to what has been earlier

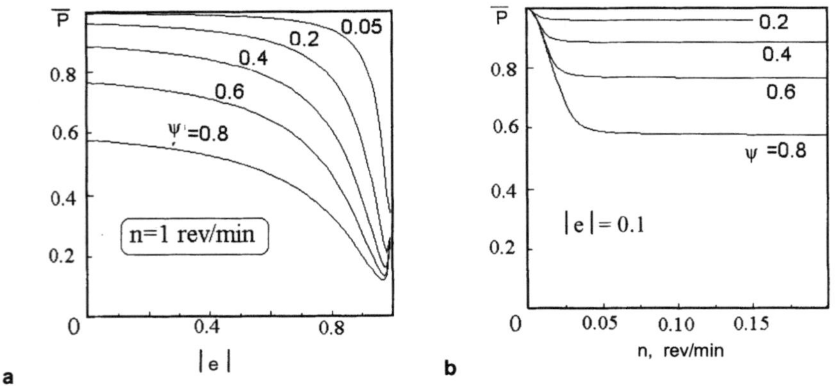

Fig. 5.3. The dependence of the normalised axial force $\overline{P} = P_k/P_0$ on the engineering strain e **a** and the angular velocity of the die n **b** calculated according to Eqs. (5.23) – (5.24) with $D_0 = 200$ mm; $2h_0 = 320$ mm; $v = 0.48$ mm/min and various values of the friction coefficient ψ

suggested for linearly strain hardening materials. Evidently, this result can be used to explain superplastic bulk forming of difficult-to-work materials also.

One can see from Fig. 5.3 that when $e < 0$ (compression with torsion), the normalised axial force $\overline{P}$ decreases as the absolute value of e is increased. This decrease can not be attributed to a change in the vector of contact friction forces, since the influence of friction is eliminated. Thus, the value of the axial force during compression with torsion decreases not only due to changes in the vector of contact friction forces, but also on account of multiaxial loading.

Another conclusion follows: the influence of torsion on the axial force is greater when the strain rate sensitivity index, m, is smaller. Actually, for a newtonian liquid ($m = 1$) it is insensitive to torsion. For a material with $m = 0.5$ (typical of many superplastics), the effect of torsion is noticeable. For a perfectly plastic material ($m = 0$), the influence of torsion is maximum. From this, it follows that the value of m can be estimated during multiaxial loading from the effect of torsion on the magnitude of the axial force.

5.1.4 Extrusion

Extrusion is a process in which a block of metal is reduced in cross-section by forcing it through a die orifice under high pressure [210]. The two basic types of extrusion are direct extrusion and indirect extrusion (also called inverted or back extrusion). Both of them are often used to refine the grain structure of polycrystalline materials, which is necessary to convert them into the superplastic state.

There are many reports in the literature, see, e.g., [2, 3, 475, 476, 481–486], dealing with the extrusion of superplastics. Tang [475] suggested a simplified mathematical model of extrusion of superplastics starting from the constitutive equations (5.3). This analysis was later reproduced in [2]. It was assumed that the rate of superplastic deformation did not depend on the hydrostatic pressure. The equilibrium equations for a small part of the deforming material is then derived [475]. (The same equations will follow if the equilibrium equations were written in a spherical system of co-ordinates.) An independent validation of the approach suggested in [475] is yet to be done. The equations for direct extrusion derived in [475] are then used to analyse wiredrawing. These equations are not examined here since the formulation of the boundary value problem is not correct (in particular, the boundary conditions are written in terms of stresses).

Also, as constitutive equations (5.3) are assumed, the solution corresponds to the superplastic flow of a liquid. In this connection, it is pertinent to consider the experimental results of Smirnov [2] obtained during the isothermal extrusion of titanium alloy VT3–1(Ti–5.5Al–2Mo–2Cr–1.0Fe). Two strain rate intervals were considered: one corresponding to the range of optimal superplastic flow and another where the strain rates were about three orders of magnitude greater than optimal. It was shown that the axial velocity profile at the die channel had a distinctive 'core', which corresponded to the movement of undeformed material. The size of this core increased with strain rate. From this, it was clear that the superplastic titanium alloy behaved as a viscoplastic material rather than as a non-newtonian liquid.

Kellett et al. [486] have investigated the extrudability of yttria stabilised tetragonal zirconia polycrystalline material. A slab analysis [487] was used to derive a piston load–velocity relation

$$\frac{v_p}{d_p} = K\sigma_p^n K_{geom} \tag{5.25}$$

where v_p is the piston velocity, d_p is the piston diameter, K, $n = 1/m$ are material constants in the constitutive equation $\sigma = K\xi^m$. K_{geom} has been suggested to be a factor dependent on die geometry, friction behaviour and stress exponent. The stress exponent estimated in extrusion [486] was consistent with that obtained in uniaxial compression and tension.

5.1.5 Die-less Drawing

Die-less wire drawing [2, 3, 488–493] is a metal working process that exploits superplasticity. Smirnov [2] has experimentally investigated this process using some titanium alloys. The deformation state has been investigated following standard procedures of co-ordinate mesh formation in order to determine the principal logarithmic strains and strain rates. The dependence of the drawing force on temperature and reduction in cross-section has been discussed, along with some possible applications. Patents have also been obtained for this process of manufacture [492, 493].

The mathematical model of die-less wire drawing [2] is based on approximating the streamline (the trajectory of a given particle) by

$$y = \frac{\rho}{2\sqrt{\mu}}\left[\left(\sqrt{\mu}-1\right)\cos\pi\frac{z}{L_{dz}} + \left(\sqrt{\mu}+1\right)\right] \tag{5.26}$$

where γ is the distance from the axis of symmetry (z-axis). μ the coefficient of reduction in cross-section; ρ the distance from the axis of symmetry of a given streamline just before the deformation zone and L_{dz} is the length of the deformation zone. Eq. (5.26) is used in [2] to calculate the three main logarithmic strains and strain rates. The stress state was estimated on the basis of the known velocity field, the temperature and the strain rate dependences of the yield stress were obtained from uniaxial tensile tests.

An alternative model of die-less wire drawing has been suggested in [283]. A common hypothesis, which is often used in the theory of plasticity, is that a plane cross-section of the wire perpendicular to the axis of symmetry remains a plane perpendicular to the same axis of symmetry after deformation. Experimental support for this assumption exists (Fig.29 of [2]). In addition, the following assumptions have been made. (i) Stable flow is present; (ii) Material is isotropic and incompressible; the flow stress is a single valued function of temperature, strain and strain rate; (iii) Thermal strains are negligible in comparison with plastic strains; (iv) Temperature is uniform in a cross-section but varies along the z-axis, coincident with the symmetry axis. Thus, the kinematics is determined by a single-valued function R (z) corresponding to the steady state profile of the wire in the deformation zone.

From symmetry one can conclude (and this is strictly proved in [283] for the case of finite strain behaviour) that the principal positive strain rates ξ_1 and ξ_2 are equal, so that

$$\xi_1 = \xi_2 = -\frac{\xi_z}{2} = \frac{\dot{R}}{R} = \left(\frac{v}{R}\right)\frac{dR}{dz} \tag{5.27}$$

where v is the velocity of the wire in the z-direction. From the incompressibility condition $vR^2 = v_0 (R_0)^2 =$ constant

$$\xi_e = -\left[\frac{2v_0(R_0)^2}{R^3}\right]\frac{dR}{dz} \tag{5.28}$$

Then, the accumulated strain is

$$s = 2\ln\left(\frac{R_0}{R}\right) = \ln\left(\frac{v}{v_0}\right) \tag{5.29}$$

when deformed in the superplastic regime (at $\xi_e = \xi_{opt} =$ constant), the stationary profile will be given by

$$R(z) = R_0\left[1 + \xi_{opt}\frac{z}{v_0}\right]^{-1/2} \tag{5.30}$$

As the coefficient of reduction of cross-section μ equals $(R/R_0)^2$, the following expressions for strain and strain rate are valid:

$$\xi_e = \frac{(\mu - 1)v_0}{L_{dz}}, s(z) = \ln\left(1 + \frac{\xi_{opt} z}{v}\right) = \ln\left[1 + \frac{(\mu - 1)z}{L_{dz}}\right] \tag{5.31}$$

A typical feature of the boundary value problem of the die-less wire drawing process is that it is a bounded boundary value problem (see Chaps. 2 and 4) as one should solve the equation of heat transfer along with the equations of motion of a continuum. The temperature boundary conditions of type II are used in [283] for the heated part of the wire, while type III boundary conditions are used for the cooler part.

Extrusion of a superplastic material is considered in [494]. In this work, plane cross-sections are assumed to remain planar after extrusion in spite of the action of friction forces. Evidently, when friction is present at the contact surface, plane surfaces will not remain planar after deformation. So, the analysis presented in [494] is more relevant to die-less drawing than to extrusion. Nevertheless, the analysis may be regarded as an example of using the upper bound technique for modelling superplastic metalworking. (Another example was given in Sect. 5.1.3.)

Notwithstanding the many publications, e.g., [488–491], the commercial viability of the die-less drawing process is yet to be established, mainly because of difficulties associated with achieving reproducible stable flow conditions.

5.1.6 Roll Forming Processes

As mentioned above, while producing large components like discs for gas turbine engines, it is highly desirable to reduce drastically the forming forces and power consumption. Apart from press forging and compression using rotating dies, axisymmetric products in difficult-to-form materials can be manufactured by hot roll forming under superplastic conditions. This process combines many versions of localised/incremental deformation and has many advantages. In particular, material utilisation is extremely high, the forces on the dies and equipment are low and this near-net-shape forming technology significantly reduces machining and assembly (labour) costs. Highly flexible and automated equipment ensures relatively low investment. Thus, the cost of manufacture can be significantly reduced compared with production based on conventional techniques.

A general scheme of the technological process of local deformation of axisymmetric workpieces requires knowledge of the forces and moments on the rolls and the workpiece, as also the power required for rotating the workpiece and the rolls. The task is to evolve a rationale (in terms of productivity, power expended, quality of product, etc.) for the combination of geometric dimensions (shapes) of the workpiece and tooling, kinematic parameters of the process (velocity of roll feed, velocities of workpiece and roll rotation) and the workpiece, roll temperature in the face of constraints placed by the force and power requirements (force on rolls, torque). It is also necessary to determine the stress–strain state present in the material and the available plasticity.

In localised deformation, shear between two counter-acting rolls or between a roll and a shaped die is present. The local character of load application leads to a 3D stress–strain state in the deformation zone. The deformation process has a non-monotonic character; the material is subjected to loading and subsequent unloading during each revolution of the deforming disc. Consequently, the loading conditions have a quasi-periodic character. Due to these features, the 3D non-steady state boundary value problem should be considered under non-monotonic load application and very complex boundary conditions for developing the mathematical model. It should be pointed out that the rheological behaviour of a superplastic

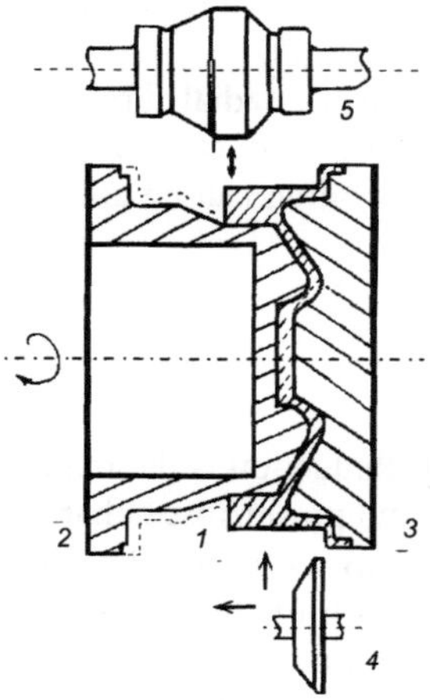

Fig. 5.4. Schematic of hot roll forming of a disc for use as an automobile wheel: *1* – preform; *2*, *3* – rotating dies; *4* – work roll; *5* – calibrating roll. Dashed line corresponds to the final geometry of the workpiece

under such non-monotonic loading is not yet investigated. The boundary conditions at the contact zone between the instrument (the roll) and the deforming disc are very complicated; all three components of the force on the roll are non-zero during deformation. Thus, the available theoretical and experimental investigations are not sufficient to decide if this problem can be analysed by a procedure similar to that used for making turbine discs out of Ti- and Ni- based heat-resistant alloys.

The problem is too complex to be solved by standard numerical procedures. Lack of reliable and adequate experimental data and poor understanding of the rheological behaviour of the material further complicate the analysis. Thus, simplified engineering methods, which lead to a solution of the boundary value problem as a first approximation is of interest. Subsequently, more rigorous numerical/analytical solutions can be attempted. Such an engineering approach is illustrated below using hot roll forming of a disc as an example.

The principal scheme of the process is presented in Fig. 5.4. Preform *1* is placed between two rotating dies *2* and *3*. During hot roll forming, roll *4* moves in the radial and axial directions; as a result, the profile of the detail shown by the dashed line can be obtained. A simplified approach is outlined in [283, 495]. The roll of maximum radius r_m can move independently in the axial as well as the radial directions with feed velocities v_z and v_r respectively.

The roll is assumed to be absolutely rigid and the contact 'roll-disc' is considered geometrically (as an intersection of the conical surface of the roll with the corresponding surface of the disc). For simplicity quasi-static isothermal condition is assumed, i.e., the temperature of the roll and the workpiece is the same and is constant with time. As for the rheological properties of the deforming material, a perfectly plastic solid is assumed; the flow stress is a single-valued function of temperature only. (Strain rate sensitivity is ignored.) If the stress does not exceed a limiting value (mechanical threshold), σ_s (i.e., if $\sigma < \sigma_s$), then deformation is absent; if $\sigma = \sigma_s$, deformation develops infinitely under $\sigma_s = \sigma_s(T)$.

To consider the embedding of the roll in the disc in the radial direction, the following method (equally applicable to embedding in the z-direction also) based on a comparison between this process and conventional rolling can be used. Thus, the length of the deformation zone is

$$L_{dz} = r_m \sin\alpha = \sqrt{2r_m \Delta h} \tag{5.32}$$

where Δh is the reduction; α is the angle of bite [210]. Reduction is the main technological parameter, which determines the kinematic and power/force characteristics of the process. The roll feed can be expressed in two ways: (i) Δh is the displacement of the roll in the radial direction per revolution of the disc (mm/rev); (ii) v_r is the velocity of the roll in the same direction (mm/min). These parameters are related as follows: $\Delta h = v_z \cdot t_{rev} = 2\pi v_z/\omega_z$, where ω_z and $t_{rev} = 2\pi/\omega_z$ are, respectively, the angular velocity and the time period of rotation of the disc about the z-axis. The area of contact as a first approximation is $S = b \cdot L_{dz}$, where L_{dz} is from Eq. (5.32). Effective roll width, b, is determined by its geometry and the current value of the depth of embedding. b increases with the degree of embedding in the disc. (This factor should be considered in numerical calculations.) Contact between the roll and the disc is of considerable practical significance. Kinematics would require that the angular velocity of the roll should match that of the disc at the contact surface. Strain per elementary step in deformation is calculated as $\varepsilon = \Delta h/h$ or $\varepsilon = ln\ (1+\Delta h/h)$, where h is the effective thickness of the material under the roll. The strain rate can be estimated in two ways. The instantaneous strain rate ξ_m is $\xi_m = \varepsilon/\Delta t$, where Δt is the time interval during which a particle of the disc has passed through the deformation zone; $\Delta t = L_{dz}/R_a\omega_z$ (where R_a is the distance from the disc axis). Apart from the instantaneous strain rate one can consider the average strain rate ξ_{av} defined as $\xi_0 = \varepsilon/t_{rev}$. Thus, one obtains the following relations for the kinematic parameters:

$$\varepsilon = \ln\left(1 + \frac{\Delta h}{h}\right), \quad \xi_m = \frac{\omega_z R_a \varepsilon}{L_{dz}}, \quad \xi_{av} = \frac{\varepsilon}{t_{rev}} = \frac{\varepsilon\omega_z}{2\pi} \tag{5.33}$$

In practice, the superplastic regime is defined on the basis of ξ_{av} rather than ξ_m. Thus, the angular velocity of the disc $\omega = 2\pi\xi_{opt}/\varepsilon$, where ε is from Eq. (5.33). It is to be noted that the strain rate can not be chosen independently of Δh. It is easy to show that $\xi_{av} = (1/t_{rev}) \ln (1 + v_z t_{rev}/h)$ and for small strains $\xi_{av} = v_z/h$.

The value of the force on the roll needed to provide the embedding into the disc is estimated as

$$P_{rol} = n_\sigma \sigma_s S \tag{5.34}$$

where n_σ is the so-called constraint factor[1]; S is the contact area and σ_s is the yield stress. n_σ is chosen empirically based on experience. σ_s is measured experimentally in uniaxial tension tests. The power on the roll will then be $N_{rol} = P_{rol}v_r$, where P_{rol} is from Eq. (5.34); v_r is given or can be calculated as $v_r = \Delta h/t_{rev} = \Delta h\omega_z/2\pi$. To estimate the steady state value of the twisting moment on the disc the following equilibrium equation is considered.

$$M_r = P_{rol}\psi l - \frac{M_z r_m}{R_a} \tag{5.35}$$

where M_r, M_z are the values of the twisting moments on the roll and the disc, respectively; ψ is the coefficient of the moment (lever) arm ($0 < \psi < 1$)[2].
The following three regimes of hot roll forming are possible:

1. Hot roll forming by rotating the disc ($M_r = 0$, $M_z > 0$);
2. Hot roll forming by rotating the roll ($M_r > 0$, $M_z = 0$);
3. Hot roll forming by rotating both the disc and the roll ($M_r > 0$, $M_z > 0$).

The force on the roll is not equal to zero for these regimes. The following comparison between the above regimes of hot roll forming and conventional metalworking is possible. The first variant corresponds to extrusion into a rotating die; the second is a variant of conventional rolling while the third corresponds to rolling with support. The last variant is used when it is necessary to decrease the torque to be applied to a disc in order to rotate it.

Power expended for rotating the roll (N_r) and the disc (N_z) are as follows: $N_r = M_r\omega_r$, $N_z = M_z\omega_z$, where ω_r and ω_z are the angular velocities of the roll and the disc, respectively. Power needed to overcome the friction in the bearings can be neglected as a first order approximation. But the power required to overcome friction between the roll and the disc, N_{fr}, is significant. N_{fr} depends on the contact area and the relative velocity at the interacting surfaces v_{rel}: $N_{fr} = \tau_{fr}\cdot S\cdot v_{rel}$. $S = bl$, $v_{rel} = |\omega_z R_a - \omega_r r_m|$ and assuming Coulomb friction $\tau_{fr} = \mu s$ (μ is the friction coefficient). Then, one has $N_{fr} = \mu\sigma bl\cdot|\omega_z R_a - \omega_r r_m|$. Thus, the total power for hot roll forming is equal to $N = N_{rol} + N_z + N_r + N_{fr}$ or

$$N = \frac{P_{rol}\Delta h\omega_z}{2\pi} + M_z\omega_z + \left(\frac{P_{rol}\psi l - M_z r_m}{R_a}\right)\omega_r + \mu\sigma bl|\omega_z R - \omega_r r_m| \tag{5.36}$$

From Eq. (5.36) it follows that the most favourable regime of loading ($N\to\min$) corresponds to the condition of the kinematic junction between the roll and the

[1] A similar approach is used in calculating the forging load in closed-die forging [210].
[2] It is usual in theory of rolling to consider the ratio of the moment arm to the projected length of the arc of contact; a typical value of ψ is 0.5 for hot rolling and 0.45 for cold rolling [11].

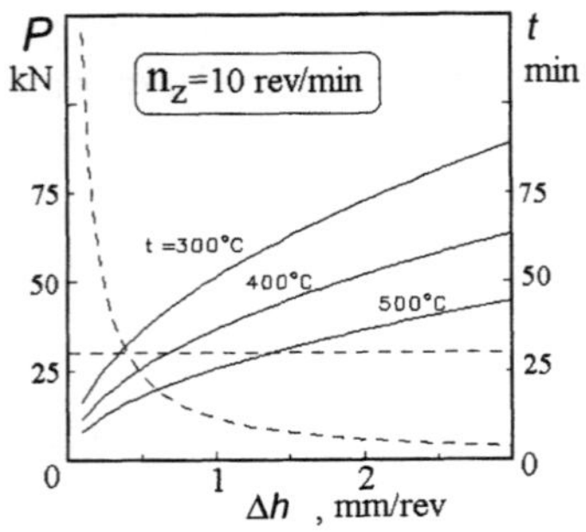

Fig. 5.5. Dependence of the force of deformation P on the roll feed Δh, calculated in accordance with Eqs. (5.32)–(5.34)

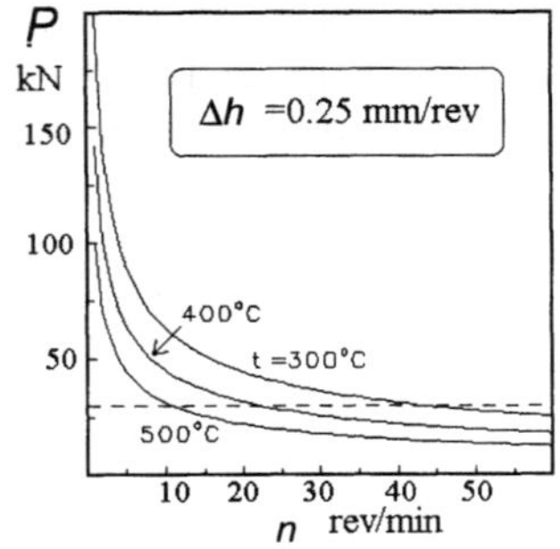

Fig. 5.6. Theoretical dependence of the force of deformation P on the angular velocity of the disc. Dashed line shows the upper limit of the available force on the equipment in this example

disc, which can be written mathematically as $\omega_z R = \omega_r r_m$. For this case one can find from Eq. (5.36) that

$$N = \frac{P_{rol}\Delta h \omega_z}{2\pi} + \frac{P_{rol}\psi R \omega_z}{r_m} \tag{5.37}$$

This equation does not include the values of M_z and M_r. The force on the roll P_{rol} according to Eq. (5.34) is determined by the material properties (σ_s), geometry of the roll (r_m and b) and the given value of the rolling reduction Δh. As ω_z is calculated from the strain rate ξ_{opt} one can conclude that the twisting moments M_z and M_r do not affect the total power requirement. This conclusion allows the choice of M_z and M_r based on the capabilities of the available gears.

As an example, the embedding of a roll made of a heat resistant Ni-based alloy is considered. The geometry of the roll is as follows: $r_m = 75$ mm, $b = 20$ mm. The material to be deformed is an aluminium alloy. The yield stress was chosen from literature [496] as $\sigma_s = 30$–80 MPa. The values of ξ_{av} and ξ_m were found to be $\xi_{av} = 10^{-5}$–10^{-3} s^{-1}; $\xi_m = 0.05$–0.3 s^{-1}. The results of the calculations based on the above equations are given in Fig. 5.5: solid lines denote the dependence of the force P (kN) on the roll feed Δh (mm/rev) for different temperatures (numbers near the curves). Horizontal dashed line indicates the upper limit of the available equipment; in this example $P_{max} = 30$ kN. The dependence of the force on the roll on the angular velocity of the disc is presented in Fig. 5.6. It was assumed in the calculations that $\xi_{av} = 0.002$ s^{-1}; $\xi_m = 1$. The duration of the process was 8 minutes.

The comparison of the theoretical predictions with the corresponding experimental data enables one to conclude that the procedure involved allows a first order approximation with reasonable accuracy for engineering applications, i.e., parameters like the contact area, the axial, radial and tangential components of the force on the roll, the moments applied to the roll and the workpiece, the average values of the strain and the strain rate and the power requirement to rotate the roll and the workpiece can be determined.

5.1.7 Clutching

During the compression of axi-symmetric parts using dies of complex shape, the metal may expand from a smaller section into a larger cavity. In classical theory of metal working, this is prohibited as it leads to an irremediable defect known as clutching [497, 498]. In this case, the points of the material originally on the surface reach the interior of the formed part and the defects present on the original surface are now found in the product. Incomplete die filling may also be present. These defects cannot usually be eliminated in subsequent operations. It was believed that clutching was a common problem in all materials. However, practical experience has shown that under certain conditions isothermal superplastic forging is not accompanied by clutching and die filling is complete even when metal flow takes place from a smaller section into a larger cavity. Some rules of thumb have also been evolved to select the conditions under which this kind of flow will be present. This problem is of interest in mathematical modelling but as of now the procedures for setting up the boundary value problem and solving them are not developed. Efforts should be directed towards this, as successful modelling would reduce costly experimentation. This facility to break convention allows the development of new technologies, e.g., the well-known scheme of 'sandy clock' is discussed at the end of this chapter.

5.2 Modelling of Sheet Metal Processes

The most remarkable applications have been with respect to superplastic forming (SPF) of sheet materials [2–4, 10–12, 18–29]. With the advent of this technology, complex articles can now be produced in single or two stage operations. The principal method of production is blow/pressure forming, in which gas pressure is applied on one side of a sheet. The sheet then flows into a die of predetermined shape and geometry located on the other side of the sheet [499–508]. It is believed, see e.g., [14, 509], that superplastic gas pressure forming of fine-grained sheet materials has already reached 'a state of maturity'.

SPF processes are widely used in industry. A number of aluminium alloys have been successfully formed by this route. These include the Al–Cu series (e.g., Supral 100 or 150), the Al–Mg series (e.g., 4913), the Al–Zn–Mg series (e.g., 7075 and 7475) and the Al–Li series (e.g., 2090 and 8090). Titanium sheet alloys, most importantly Ti–6Al–4V, are widely used in the aerospace industry. Primary reasons for the use of titanium alloys are [463]: weight saving (*vis–a–vis* steel); space saving (compared with Al alloys); increased operating temperatures (compared with steel and Al alloys); improved corrosion resistance (compared with Al alloys and low alloy steels) and composite compatibility (*vis–a–vis* Al alloys).

Fig. 5.7. SPF/DB produced parts (courtesy R.V. Safiullin)

In combination with diffusion bonding, superplastic forming (SPF/DB) has led to the development of new types of bonded structures, namely, sandwich (cellular) panels [4, 26, 510–518]. This has resulted in the further development of the technology and has, as a consequence, promoted new approaches to the design of complex profile structures with high stiffness that are of special interest to the aerospace industry. This technology offers substantial cost savings by reducing the number of detail parts and mechanical fasteners, by achieving more efficient load carrying structures, and by increasing the fly-to-buy ratio of titanium alloys. The SPF/DB process has also been successfully employed with Al alloys, although the relevance of this technology for Ti alloys is definitely greater. Presently, the SPF and SPF/DB processes are commonly accepted as advanced technologies for forming aerospace components (Fig. 5.7). As superplasticity has been established more recently in such advanced materials as intermetallic compounds, ceramics and composites and that too at high strain rates, there is an increased need to model these processes.

Modelling of SPF processes can be done at various levels: macroscopic (structural level), mesoscopic (grain level) and microscopic (atomic level). In this section, macroscopic level modelling is considered. Mathematical modelling of the SPF and SPF/DB processes includes the analysis of a number of simplified design schemes of deformation of the sheet materials. Two such idealised schemes are often used in the literature: the long rectangular thin sheet and the circular membrane. Both of them are considered below. Some problems associated with the construction of mathematical models are also discussed.

This section can be considered as an exception from the common rule accepted in this book (elsewhere only infinitesimally small strains are considered). In this section, finite strain behaviour of superplastics is considered as an example of such an approach. Detailed considerations of the finite strain behaviour of superplastics are given in Appendixes A and B.

5.2.1 Simplifications in Modelling SPF and SPF/DB Processes

The simplifying assumptions used while setting up the boundary value problems in SPF and SPF/DB processes are seldom stated explicitly. In this section, the commonly used simplifications are discussed.

Conventional superplasticity is present over a strain rate range of ~ 10^{-5} to 10^{-1} s^{-1}. Therefore, the inertia forces in the equations of motion of the continuum can be neglected. Besides, the influence of the mass forces (due to gravity) is neglected. Thus, one can consider the equilibrium equations instead of the equations of motion while setting up the boundary value problem. For high strain rate superplasticity the upper limit of strain rate may go up to 1-10 s^{-1} (see, e.g., [6, 519–523]); impact superplasticity can be observed in consolidated alloys of amorphous or nanocrystalline powders (e.g., [519]). The influence of inertia forces may be important for such materials. But, these are not considered here.

Superplasticity is usually observed at elevated temperatures of the order of half the absolute melting temperature. When the grain size is in the sub-micrometer or nanometre range, the temperature of forming can decrease considerably. In all cases, however, the optimal temperature interval will be rather narrow. Therefore, one can consider the isothermal boundary value problem, i.e., the temperature boundary value problem is not usually considered.

Titanium alloy sheets used in SPF and SPF/DB processes are not isotropic; in particular, the properties measured along the rolling direction are distinctly different from those measured transverse to the rolling direction. The difference in yield stress, for example, may be as high as 20%. But, this is not usually taken into consideration while modelling the SPF or SPF/DB processes. The material to be formed is usually assumed to be isotropic and incompressible.

The sheet to be formed may be rigidly clamped along its periphery or, alternatively, left free. The following two limiting boundary conditions are usually considered at the contact between the sheet and the die: (i) free sliding of the shell over the die, or (ii) full adherence of the shell to the die in the contact zone. It is possible that an intermediate condition is realised in practice. However, the use of such a boundary condition will lead to considerable complications in the boundary value problem. The thickness of the sheet is usually small compared with the typical dimensions of the die. Therefore, bending and shear effects are negligible. Then, one can use the membrane theory to describe the stress state in the deformed sheet. The fundamentals of this theory were presented in Sect. 4.2.3.

Superplastic deformation is characterised by extremely large tensile ductility. Therefore, the elastic strains can be neglected in comparison with the plastic strains. Besides, SPF and SPF/DB processes are characterised by smooth changes in strain rate and neglecting the elastic strain rates will pose no serious problems in this case. In view of the above, the following assumptions are made in further analysis:

(a) the material is uniform, isotropic and incompressible;
(b) average grain size d is small as compared to sheet thickness and width;

(c) elastic strains are negligible;
(d) the strain rates are sufficiently small so that the problem can be considered to be quasi-static;
(e) isothermal conditions prevail during forming;
(f) the die is absolutely rigid so that its geometry does not change (e.g., the radius of the (cylindrical) mould R_0 = constant);
(g) the specimen is rigidly clamped along its periphery;
(h) the initial thickness of the sheet is sufficiently small so that bending and shearing effects are negligible.

If any of the above assumptions are omitted in a particular analysis, the same will be mentioned.

5.2.2 Main Challenges in Modelling SPF and SPF/DB Processes

The following problems are to be addressed while exploiting the SPF and SPF/DB processes. Optimal conditions of superplastic forming should be known. As the optimal strain rates in many commercial superplastic alloys lie in the range 10^{-4}–10^{-2} s^{-1}, typical superplastic forming times will be about 10^2–10^4s. If SPF is carried out under constant pressure, the strain rate may vary widely [2, 3, 384] and the forming conditions may even fall outside the permissible range. Therefore, determination of the pressure–time cycle to maintain the strain rate in the optimal range is of considerable practical importance. This requirement assumes greater significance because visual control of the process is not possible. Mathematically, the optimal condition can be written as

$$\sigma_e = \sigma_{opt} = constant \quad or \quad \xi_e = \xi_{opt} = constant \tag{5.38}$$

where σ_{opt}, ξ_{opt} are the co-ordinates of the point of inflection of the sigmoidal $\log\sigma - \log\xi$ curve. Polycrystalline materials in a superplastic state are characterised by a strong dependence of stress on strain rate; the strain dependence of stress in contrast is rather weak. Therefore, it is often assumed that the stress is a single-valued function of strain rate.

Usually, some strain rate distribution is present during real SPF or SPF/DB, i.e., ξ_e will vary between a minimum $\xi_e^{\min}$ and a maximum $\xi_e^{\max}$. The following modification to Eq. (5.38) has been suggested:

$$min\left(\xi_e^{max} - \xi_{opt}\right)^2 \tag{5.39a}$$

It is also possible to modify Eq. (5.38) as

$$\xi_e^{\max} \equiv \xi_{opt} = constant \tag{5.39b}$$

Equation (5.39a) is convenient for use in FEM-modelling of SPF and SPF/DB processes, while Eq. (5.39b) is more relevant for analytical modelling of these processes. In particular, condition (5.39b) can be used to obtain an analytical expression for the pressure time cycle, as it was done in [57, 95, 282, 384, 524].

Experimental investigations show that, usually, the thickness distribution in parts made by SPF or SPF/DB is not uniform, see, e.g., [2, 3, 87, 282, 384, 517, 525–531]. The extent of non-uniformity is inversely related to the value of m. In practice, thickness non-uniformity is a very serious problem in the manufacture of sandwich cellular structures. In fact, maximum permissible thinning in a component is one of the limiting factors in exploiting SPF/DB technology [517]. Thus, it is important to estimate numerically the degree of exhaustion of the plasticity resource. Development of effective ways of reducing differences in thickness across a section is of interest as well. Recently, an engineering method of determining the accumulated strain has been suggested [87]. The use of this method facilitates the design of SPF/DB parts in a rational way. Further investigations, which may be based on estimating the damage accumulation (see Sect. 4.1.3), could be of interest.

The microstructure of the material may change significantly during SPF or SPF/DB so that the mechanical properties may vary notably as well. In particular, grain growth is present in titanium alloys at relatively low strain rates. This causes (strain) flow hardening and an increase in the flow stress. When this effect is significant, localised non-optimal deformation and failure can result. The effects of a change in grain size [532–534] should be taken into account while modelling SPF processes, see, e.g., [282]. SPF processes for Al-based alloys have to consider in addition the growth and interlinkage of internally nucleated voids [535–541]. However, the application of hydrostatic pressure effectively eliminates cavitation.

Thus, as correctly emphasised in [434], the main challenges in the mathematical modelling of SPF and SPF/DB processes involve the calculation of the pressure-time cycle and the prediction of thickness distribution and maximum thinning in a part. In addition, the possibility of numerically estimating the degree of exhaustion of the plasticity resource will lead to a rational design of SPF and SPF/DB parts.

5.2.3 SPF of Hemispherical Domes

There are many reports in the literature, e.g., [54, 282, 384, 460, 524–530, 542–550], in which the superplastic forming of a thin circular diaphragm into a cylindrical or hemispherical die is considered. The assumptions (a)–(h) listed at the end of Sect. 5.2.1 are taken to be valid for all the models. In this section, a number of models for bulge forming of hemispheres, which also use some additional hypotheses, are considered. The membrane theory is used to describe the stress state. It is necessary to note that some problems may arise in denoting the principal stresses and strain rates. As in Sects. 4.2.3 and 4.3.3.1, let the notations σ_t, σ_c and, ξ_t , ξ_c be used for the tangential and circumferential (hoop) stress and strain rate,

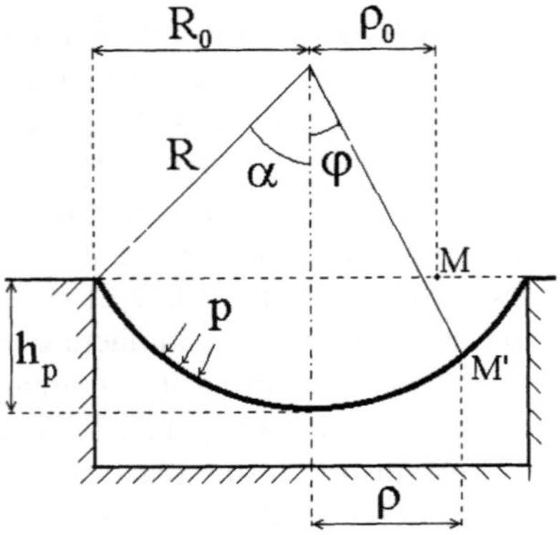

Fig. 5.8. Bulge profile (Schematic)

respectively. Finite strain behaviour of a sheet material during deformation is considered below. Some background information is given in Appendixes A and B.

5.2.3.1 Finite Strain Behaviour

Finite strain behaviour is presently analysed using the above mentioned assumptions, standard procedures of mechanics of solids [401] and further the terminology used in [452].

One can use Lagrangian (x^i) and Eulerian (x^i) co-ordinates for analysing SPF. The position at a given moment of time t of a point M with Lagrangian co-ordinates x^i can be given by the radius-vector **r** (x^i, t) which is determined with respect to a fixed Cartesian system of co-ordinates. The basis vector $\vec{b}_i$ and the metric tensor g_{ij} at time t are given by definition as

$$\vec{b}_i = \frac{\partial \vec{r}}{\partial x^i};\ g_{ij} = \vec{b}_i \cdot \vec{b}_j \tag{5.40}$$

The velocity vector of a particle can then be decomposed as $V = V_i \vec{b}_i = V_j \vec{b}_j$ where $\vec{b}_j$ denotes the reciprocal basis vector, subscript and superscript denote covariant and contravariant components, respectively, and a repeated index means summation. The strain rate tensor V_{ij} is given by

$$V_{ij} = \frac{1}{2}\dot{g}_{ij} = \frac{1}{2}\left(V_{i,j} + V_{j,i}\right) \tag{5.41}$$

where the dot means a time derivative and the comma signifies covariant differentiation with respect to the metric tensor g_{ij} at time t. The equivalent strain rate ξ_e is defined by

$$\xi_e = \sqrt{\frac{2}{3}\xi^i_j \xi^j_i} \tag{5.42a}$$

where

$$\xi^i_j = \hat{g}^{is}\xi_{sj}\ ;\ \xi_{ij} = V_{ij} - \left(\frac{1}{3}\right)V_{kk}\ ;\ \hat{g}^{is} = \vec{b}_i.\vec{b}_s \tag{5.42b}$$

Here ξ^i_j and ξ_{ij} are the mixed and the covariant components of the strain rate deviator, respectively. Representing the Caushy stresses as σ^{ij} and deviatoric Caushy stress as S^{ij}, the equivalent stress σ_e is defined as

$$\sigma_e = \sqrt{\frac{3}{2}g_{ik}g_{jl}S^{ij}S^{kl}} \tag{5.43a}$$

where

$$S^{ij} = \sigma^{ij} - \frac{1}{3}g^{ij}\left(g_{kl}\sigma^{kl}\right) \tag{5.43b}$$

The relationship Eq. (4.3) is then generalised for the case of finite strain behaviour as

$$S^{ij} = \frac{2\sigma_e}{3\xi_e}g^{ik}g^{jl}\xi_{kl} \tag{5.44}$$

Further details can be found in text books, e.g., [420].

5.2.3.2
Jovane's Model

Jovane [384] assumed that the geometry of the deforming shell at any moment of time is part of a sphere of radius R and half arc angle α (Fig. 5.8). Besides, he also assumed that at any instant, the thickness of the dome is uniform along its profile. The volume of the membrane can be calculated as the triple integral in a spherical system of co-ordinates. As a result, the following relation can be found with the assumption that h = constant (h/R<<1):

$$V = 2\pi R^2(1-\cos\alpha)h\left(\frac{1-h}{R}+\frac{h^2}{3R^2}\right) \cong 2\pi R^2(1-\cos\alpha)h \tag{5.45}$$

Taking into account the incompressibility condition $V = V_0 = \pi(R_0)^2 h_0$

$$h = h_{id} = h_0 \cos^2\left(\frac{\alpha}{2}\right) = \frac{h_0}{1+H^2} \tag{5.46}$$

where $H = h_p/R_0 = \tan(\alpha/2)$ is the relative dome height; index '*id*' denotes ideal thinning corresponding to uniform thickness distribution.

The metric tensor g_{ij} will then be given as

$$\{g_{ij}\} = \begin{bmatrix} 1+H^2 & 0 & 0 \\ 0 & 1+H^2 & 0 \\ 0 & 0 & \dfrac{1}{(1+H^2)^2} \end{bmatrix} \tag{5.47}$$

Covariant components of the strain rate tensor can be found using relationships Eq.(5.41) and the mixed components in accordance with Eq. (5.42b) (the components g^{ij} of the metric tensor of the reciprocal basis can be found as the components of the matrix, which is reciprocal to that of Eq.(5.47)). Eventually one arrives at

$$\{\xi^i_j\} = \begin{bmatrix} \dfrac{H\dot{H}}{1+H^2} & 0 & 0 \\ 0 & \dfrac{H\dot{H}}{1+H^2} & 0 \\ 0 & 0 & \dfrac{-2H\dot{H}}{1+H^2} \end{bmatrix} \tag{5.48}$$

Now, one can calculate the equivalent strain rate by substituting the strain rate components from Eq. (5.48) in Eq. (5.42a).

$$\xi_c = \xi_t = \frac{H\dot{H}}{1+H^2}; \quad \xi_e = \frac{2H\dot{H}}{1+H^2} \tag{5.49}$$

Substituting Eq. (5.49) in Eq. (5.38) one can obtain the following time dependence for the dome height H for the constant strain rate regime of loading.

$$H = \sqrt{\exp(\xi_{opt} t) - 1}\ , \quad 0 \le t \le t_1 \tag{5.50}$$

where $t_1 = \ln 2/\xi_{opt}$ is the blowing time for a hemisphere (see also [384]). In accordance with the principal equations of the membrane theory one can derive for a spherical shell the following equations (see Sect. 4.2.3)

$$\sigma_e = \sigma_c = \sigma_t = \frac{pR}{(2h)} \tag{5.51}$$

where σ_t is the tangential (meridian) principal stress, σ_c is the circumferential (hoop) principal stress.

Pressure–time cycle can be found from Eqs. (5.51) and (5.46) as

$$\frac{p}{\sigma_{opt}} = \frac{\frac{4h_0}{R_0}H}{1+H^2}, \quad 0 \le H \le 1 \tag{5.52}$$

where H (t) is from Eq. (5.50). It should be noted that the above analysis differs somewhat from that presented by Jovane [384]. But, the results are the same– the above equations extend up to the initial moment of time $t = 0$.

Sheets of eutectic lead–tin alloy were used in [384] to validate the results. The results of experiments in forming superplastic diaphragms provide some justification for the assumptions made in the analysis and, generally, support the validity of the predicted time–height relationship for a bulging process that follows a stepwise loading path.

Jovane's model has been reproduced many times in the literature. It has been noted in many publications that since the die does not deform during forming (R_0 = constant) deformation along the periphery is negligible, see, e.g., [57, 282, 524, 525, 546]. Some consequences are presently examined. It follows from Eq. (5.49) that the circumferential strain rate $\xi_c = 0$. If one could calculate the main circumferential (hoop) strain ε_c as

$$\varepsilon_c = \int_0^{t_1} \xi_t dt = \int_0^1 \frac{H_p dH_p}{1+H_p^2} = \frac{1}{2}\ln 2 \tag{5.53}$$

then, one can conclude that at the periphery $\varepsilon_c = ln(2\pi R^b/2\pi R_0) = ln\ (R^b/R_0)$. Therefore, according to Jovane a diaphragm of initial radius R_0 will be transformed into a hemisphere of radius $R^b = R_0\sqrt{2}$, that is more than 1.4 times larger than the die radius! Thus, the assumption of thickness uniformity during forming leads to a serious contradiction. From the viewpoint of mechanics, Jovane's model as well as its later modifications e.g., [54, 544], do not satisfy the obvious boundary condition: $R^b = R_0 =$ constant, which is to be satisfied if the die does not deform and the diaphragm is rigidly clamped along its periphery.

5.2.3.3 Geometric /Kinematic Models

The following hypothesis has been suggested recently [57]: a meridian passing through the dome apex is uniformly stretched at any moment of time. This hypothesis leads to a simple analytical expression for the thickness distribution, which is in agreement with experimental data. Using this hypothesis, the kinematics of bulge forming of dome and the finite strain behaviour of superplastics are presently analysed.

Let point M in the starting diaphragm be at a distance $\rho_0 = vR_0$ from the point O (which coincides with the center of the initial disc) (Fig. 5.8) at $t = 0$. Here v^* is a number that characterises the initial position of the point in the dome; $0 \le v^* \le 1$. At some moment of time $t > 0$, the point M moves to M', and O to O '. Let φ be the angle between the symmetry axis and the dome radius drawn to point M'. φ defines the position of the point in the dome.

Let $\vec{b}_1, \vec{b}_2, \vec{b}_3$ be the basis of the convective co-ordinate system x^i at the initial moment of time $t = 0$. Assuming it to be orthonormalised one obtains that metric tensor $\{g_{ij}^0\}$ is a unit matrix. At some moment of time $t > 0$, the basis of the convective co-ordinate system at point M' $\vec{b}_1, \vec{b}_2, \vec{b}_3$ is not, in general, orthonormalised. In accordance with the above hypothesis at any time all meridians passing through the apex are stretched according to the relation $R\alpha/R_0 = \alpha/\sin\alpha$. Also, from this hypothesis

$$v^* = \frac{R\varphi}{R\alpha} = \frac{\varphi}{\alpha} = \frac{\rho_0}{R_0}$$

In contrast, a latitude passing through point M' is stretched by ρ/ρ_0 times, or (taking into account that $\rho = R\sin\varphi$, $\rho_0 = vR_0 = R_0\varphi_0/\alpha$) $\alpha\sin\varphi / [\varphi\sin\alpha]$ times. Then, taking that the initial basis vector has unit length one can obtain

$$\left|\vec{b}_1\right| = \frac{\alpha}{\sin\alpha};\ \left|\vec{b}_2\right| = \left[\frac{\alpha}{\sin\alpha}\right]\cdot\left[\frac{\varphi}{\sin\varphi}\right];\ \left|\vec{b}_3\right| = \frac{h}{h_0} \qquad (5.54)$$

Thus, $\vec{b}_1, \vec{b}_2, \vec{b}_3$ are an orthogonal basis but not an orthonormalised basis. The metric tensor will then be found from Eqs. (5.40) as

$$\{\hat{g}_{ij}\} = \begin{bmatrix} \left[\frac{\alpha}{\sin\alpha}\right]^2 & 0 & 0 \\ 0 & \left[\frac{\alpha}{\sin\alpha}\right]^2\left[\frac{\varphi}{\sin\varphi}\right]^{-2} & 0 \\ 0 & 0 & \left[\frac{h}{h_0}\right]^2 \end{bmatrix} \qquad (5.55)$$

Writing the incompressibility condition using Lagrangian variables

$$\det g_{ij} = \det g_{ij}^0 = 1 \qquad (5.56)$$

Substituting Eq. (5.55) in Eq. (5.56), one obtains that

$$h(\varphi,\alpha)=h_0\left[\frac{\sin\alpha}{\alpha}\right]^2\cdot\frac{\varphi}{\sin\varphi} \tag{5.57}$$

Equation (5.57) is identical to those given in [57, 524]. It is necessary to emphasise that in accordance with Eq. (5.57) thickness distribution does not depend on the properties of the material; in particular on the strain rate sensitivity index *m*. Therefore, this model can be regarded as being geometrical, i.e., thickness distribution does not depend on material properties but is strictly defined by the geometry of the shell. This conclusion is not in agreement with a number of well-known experimental results, e.g., [2, 3, 462, 525, 530].

The procedure to calculate the pressure–time cycle is described in detail in [57]. A comparison of the results of the theoretical analysis with the experimental data for high values of *m* reveals the usefulness of the model in predicting the thickness distribution present (Table 5.1). (It is to be noted that the values of *K* and *m* in the constitutive equation $\sigma = K\xi^m$ were determined in a constant gas pressure forming experiment [57].)

Covariant components of the strain rate tensor can be found in accordance with Eqs. (5.41) and the mixed components from Eq. (5.42b). The component g^{ij} of the metric tensor of the reciprocal basis can be found as the components of the matrix, which is the reciprocal of Eq. (5.55). It will include only diagonal elements so that the strain rate components are

$$\begin{aligned}&\xi_c=\frac{\dot\alpha}{\alpha}-\dot\alpha\cot\alpha;\xi_t=\dot\varphi\cot\varphi-\dot\alpha\cot\alpha\\&\xi_n=\frac{\dot h}{h}=-\xi_t-\xi_c\end{aligned} \tag{5.58}$$

Equivalent strain rate is given by Eq. (5.42a). In particular, the following relations are valid at the dome apex:

Table 5.1 Comparison of the predicted and the experimentally measured duration for constant strain rate superplastic forming of hemispherical domes of Ti–6Al–4V alloy [57]

Strain rate (s^{-1})	Flow stress (MPa)	Duration of forming		Error %
		Predicted	Experimental	
2.4×10^{-4}	11.4	62 min 43 s	45 min 45 s	37
4×10^{-4}	14.0	37 min 36 s	39 min 56 s	6
8×10^{-4}	19.0	18 min 48 s	19 min 50 s	5
1.2×10^{-3}	23.0	12 min 30 s	14 min 00 s	11
4×10^{-3}	38.0	3 min 46 s	3 min 15 s	16

$$\xi_t^a = \xi_c^a = \frac{\dot{\alpha}}{\alpha} - \dot{\alpha}\cot\alpha;\ \ \xi_e^a = 2\left[\frac{\dot{\alpha}}{\alpha} - \dot{\alpha}\cot\alpha\right] \tag{5.59}$$

At the dome periphery,

$$\xi_t^b = \frac{\dot{\alpha}}{\alpha} - \dot{\alpha}\cot\alpha,\ \xi_c^b = 0, \xi_e^b = \left(\frac{2}{\sqrt{3}}\right)\left[\frac{\dot{\alpha}}{\alpha} - \dot{\alpha}\cot\alpha\right] \tag{5.60}$$

Then, from the equations of equilibrium at the dome apex

$$\sigma_t^a = \sigma_c^a = \sigma_e^a \tag{5.61}$$

S_c= ($2\sigma_c$–σ_t). Therefore, at the dome periphery (taking into account $\xi_c^b = 0$ from Eq. (5.60))

$$\sigma_t^b = 2\sigma_c^b \tag{5.62}$$

It is easy to see that Eq. (5.61) is in good agreement with the principal equation of the membrane theory Eq. (4.127) while Eq. (5.62) contradicts it. This problem has already been considered in Sect. 4.3.3 while considering the model of Ghosh and Hamilton [282].

It turns out to be impossible to satisfy simultaneously the equilibrium Eqs. (4.123) and the proportionality between the stress and strain rate deviators described by Eq. (4.3) within the framework of the geometric model [57]. The principal equations of the membrane theory fail close to the edge of the dome where the shear strains are considerable and can not be neglected. From the point of view of mechanics, this problem (also found with the model of Ghosh and Hamilton [282] and others) is due to the fact that close to the edge of the shell the simplified equilibrium Eqs. (4.123) are not valid. Thus, the way to modify these models is to consider the more adequate equilibrium Eq. (4.2). Notwithstanding these contradictions, good agreement between the experimental results and the theoretical models is reported [57, 282] concerning technologically important parameters. Whether this is fortuitous or there are other reasons for this agreement is not clear yet.

Chandra and Kannan [546] have suggested another geometric model. Both uniform and non-uniform thinning have been considered assuming plane strain condition Eq. (5.62) at the dome periphery and balanced biaxial condition Eq. (5.61) at the pole. Then, similar to Ghosh and Hamilton [282], using one of the equations of the membrane theory, the relationship between the thickness at the pole and the edge is obtained using the equation

$$\frac{h_a}{h_b} = \frac{\sigma_t^b}{\sigma_t^a} = \left(\frac{2}{\sqrt{3}}\right)^{m+1} \cdot \left(\frac{\xi_n^b}{\xi_n^a}\right)^m \tag{5.63}$$

It is further assumed [546] that the thickness increases linearly with the arc length starting from the pole, i.e.,

$$h(\varphi,\alpha) = h_a + \left(\frac{\varphi}{\alpha}\right)\cdot(h_b - h_a) \tag{5.64}$$

so that at $\varphi = 0$, h = h_a and $h = h_b$ at $\varphi = \alpha$.

The following comments are made. According to this analysis, as $m \to 0$, from Eq. (5.63) $h_a > h_b$, i.e., the thickness at the pole is greater than at the periphery. This is not discussed in [546]. Also, in that paper the symbol θ in Figs. 3 and 5 represent different concepts. The theory is said to agree well with the experimental results on Al-alloys 7475 and 8090 in so far as the thickness distribution is concerned. But, no validation has been made with respect to the pressure–time cycle suggested in [546].

5.2.3.4 Model of Cornfield–Johnson and its Modifications

Cornfield and Johnson [525] and Guo et al. [528–530] have developed a mathematical model for SPF of hemispherical domes, which takes into account thickness non-uniformity. This model was improved later [547, 548]. Apart from the standard hypothesis (a)–(h) listed at the end of Sect. 5.2.1, a common assumption in all these approaches is that the geometry of the median plane is part of a sphere at any instant of deformation. The initial disc is divided into a number of rings and the equilibrium of each ring is then considered based on the principal equations of the membrane theory Eq. (4.123). The resulting system of equations is solved numerically. Cornfield and Johnson [525] have suggested that the force tangent to a meridian is constant along the dome profile, i.e.,

$$\sigma_t h = \frac{pR}{2} \tag{5.65}$$

Ghosh and Hamilton [282] have adopted the same condition and Eq. (5.63) also means the same.

Standard power relation $\sigma = K\xi^m$ is used in [525, 528–530, 547, 548] as a scalar constitutive equation. It is extended to the 3D case in accordance with Eq, (4.3). It is shown [525] that as m decreases, the thickness non-uniformity increases. Based on careful experiments involving the use of IMI 318 (Ti–6Al–4V) up to relative heights $H = 0.76$ and 1.0, it was observed that the shape of the dome was always very close to that of a part of a sphere and that good agreement between the predictions and the experimental results existed so far as thickness distribution was concerned. The conclusion that thickness non-uniformity increased with a decrease in m was later confirmed by Smirnov [2] for a number of alloys.

More recently, Guo et al. [528–530] have performed a refined experiment: two circular sheets of Al-alloy 7475 were rigidly clamped along their common periphery (R_0 = 20 mm). A number of holes of diameter 500 μm were drilled in one of the sheets. This aggregate was superplastically formed at 499°C. It was found that the ratio of the hole dimensions in the meridian and the circumferential directions increased logarithmically with the relative height of the dome. This experimental fact was used as a basis for developing a mathematical model for the process. Usugi et al. [547] improved this model using the incremental strain theory of plasticity and the presence of non-equal biaxial stress states in the deformed shell. They also replaced the experimentally found ratio of circumferential stress to tangential stress, σ_c/σ_t, by a theoretical prediction [529, 530].

As already mentioned, the assumption about the spherical shape of the deformed shell leads to a conclusion that is consistent with the principal equations of the membrane theory, viz., $\sigma_t = \sigma_c = pR/2s$. Cornfield and Johnson [525] have also noted that the equilibrium equations of the membrane theory are not valid close to the edge, as shown by Timoshenko [549] quite sometime ago.

5.2.3.5 *Holt's Model and its Modifications*

Holt [542] seems to be the first investigator who rejected the hypothesis of a spherical shape so far as superplastic bulge forming of domes was concerned. He assumed that the stress and the strain rate states were 'balanced biaxial', that is, $\varepsilon_t = \varepsilon_c$ and $\sigma_t = \sigma_c = pR/2h$, $R = \rho/\sin\varphi$ (see Fig.5.8 for definitions of ρ and φ). Then, taking into account that $\sigma_e = \sigma_t$ and $\xi_e = 2\xi_t = 2\dot{\rho}/\rho$

$$\sigma_e = \sigma_t = \frac{p\rho}{2h\sin\varphi} = K(\xi_e)^m = K\left(\frac{2\dot{\rho}}{\rho}\right)^m \tag{5.66}$$

From the incompressibility condition $\dot{h}/h + 2\dot{\rho}/\rho = 0$ one can find after integration that $h/h_0 = (\rho_0/\rho)^2$. Then, Eq. (5.66) can be rewritten as

$$\frac{\dot{\rho}}{R_0} = \frac{1}{2}\left(\frac{pR_0}{2Kh_0}\right)^{\frac{1}{m}}\left(\frac{\rho}{R_0}\right)^{\frac{3}{m}+1}\left(\frac{\rho_0}{R_0}\right)^{-\frac{2}{m}}(\sin\varphi)^{-\frac{1}{m}} \tag{5.67}$$

This equation was used to follow the development of a bulge under constant pressure. The sheet was divided into ten annuli, and then Eq. (5.67) was solved for each one of them. The results of the calculations show that the thickness non-uniformity increased with decreasing value of m. A specific feature of the model of Holt is the prediction of the absence of deformation near the edge. This result contradicts numerous experimental findings; it is now commonly recognised that notable thinning is present in the vicinity of the edge during the SPF of a circular diaphragm. Thus, one of the main assumptions of Holt's analysis ($h_t = h_c$) is not valid close to the edge.

Table 5.2 Thickness along a dome cross-section formed at 927°C and $p = 0.4$MPa [544]

	Thickness of cross section, (mm)					
	Pole					Edge
Experimental	0.651	0.691	0.752	0.807	0.937	1.021
Predicted	0.643	0.655	0.744	0.816	1.170	1.370

Yang and Mukherjee [460] assumed the shape of the deformed shell to be a part of an ellipsoid. Standard power relation $\sigma = K\xi^{m}$ was used to describe the scalar properties. To illustrate the influence of strain rate sensitivity on the shape of the formed domes, experiments were done on superplastic Zn–22Al, AA 4913 and AA 7475 alloy sheets. It was shown that when m was low, a conical rather than a spherical bulge was obtained.

Holt's model was later modified [527] to predict the thickness distribution, the thickness strain and the thinning factor. The material constants for the superplastic alloy Ti–6Al–4V (K and m in $\sigma = K\xi^{m}$) were obtained as a function of strain rate by a polynomial regression analysis described in [317]. A good correlation between the predicted and the experimentally measured values was found (Table 5.2). Superplastic forming of Ti–6Al–4V and Sn–Pb eutectic alloy sheets were investigated experimentally in [523]. It was shown that when m was in the superplastic range the bulge profile could be described as an arc of a circle and true hemispheres could be formed out of sheets of both the alloys (Table 5.3).

Table 5.3 Calculated and measured values of the radius of curvature of the bulge (base radius of the dome = 80 mm) [526]

Material	Bulge Height (mm)	Radius of Curvature (mm)	
		Calculated	Measured
Ti–6Al–4V	9.0	360.1	361.0
	26.8	132.8	133.0
	32.6	114.5	115.0
	40.0	100.0	100.0
	63.0	82.3	83.0
	80.0	80.0	80.0
Sn–Pb	35.0	108.6	109.0
	45.0	93.6	94.0
	53.0	86.8	87.0
	68.0	81.1	81.5
	78.0	80.0	80.5
	80.0	80.0	80.0

5.2.4
Free Forming of Spherical Vessels

5.2.4.1
Description of the Process

Laycock [499] has pointed out that certain superplastic sheet components can be produced without tooling or with only specific regions of the final shape defined by tools. British Aerospace [550] have produced ducts and tanks by free blowing of edge welded titanium blanks. Later, this process was used by Kruglov et al. [453, 551, 552] to produce spherical and ellipsoidal tanks for aerospace and oil industries (Fig. 5.9). A typical feature of this process is that the equatorial diameter of the preform decreases during forming. It was experimentally established [552] that the radius of the spherical vessel R_s was related to the preform radius R_0 as $R_0/R_s = 1.20–1.25$. Therefore, it is of practical importance to predict theoretically such a decrease and to calculate the pressure–time cycle and the thickness distribution.

5.2.4.2
Mathematical Model

A simple engineering model for the free forming of spherical vessels was suggested in [524]. It is based on the same assumptions as made in the geometric model for a circular membrane (see Sect. 5.2.3.3). The difference lies in the boundary conditions: it is assumed in [524] that thickness in the vicinity of the edge does not vary during blowing. Taking into account the symmetry of an edge welded preform of a circular membrane of initial radius R_0 being bulged (Fig. 5.10), the problem can be analysed using the convective co-ordinate system x^i

Fig. 5.9. Spherical vessel and its preform (courtesy: A. Kruglov)

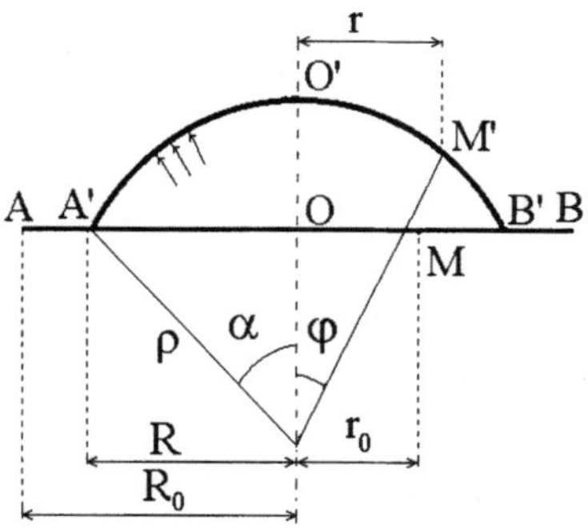

Fig. 5.10. Free forming of a spherical vessel (Schematic)

(Sect. 5.2.3.1). (This system coincides with the Cartesian system of co-ordinates when $t = 0$.) With time, points M, O, A and B move to points M', O', A' and B' respectively. Since the co-ordinates of particles in the continuum M in the convective system of co-ordinates does not change with time, the angles φ and α are related as $\varphi = \nu\alpha$. Taking into account that $r_0 = \nu R_0 = R_0\varphi/\alpha$ and $\rho = R/sin\alpha$, the metric tensor is obtained as

$$\hat{g}_{ij} = \begin{bmatrix} \left(\frac{R}{R_0}\right)^2\left(\frac{\alpha}{\sin\alpha}\right)^2 & 0 & 0 \\ 0 & \left(\frac{R}{R_0}\right)^2\left(\frac{\alpha}{\sin\alpha}\right)^2\left(\frac{\sin\varphi}{\varphi}\right)^2 & 0 \\ 0 & 0 & \left(\frac{h}{h_0}\right)^2 \end{bmatrix} \tag{5.68}$$

Thickness distribution can be found from the incompressibility condition, which can be written, in the Lagrangian approach as Eq. (5.56) [401].

$$h(\varphi,\alpha,R) = h_0\left[\frac{R_0}{R}\right]^2\left[\frac{\sin\alpha}{\alpha}\right]^2\frac{\varphi}{\sin\varphi} \tag{5.69}$$

If in Eq. (5.69) R is taken as equal to R_0, Eq. (5.57) that corresponds to rigid clamping of the circular membrane along its edge is obtained. For free forming, from Eq. (5.69) and using the condition $h = h_0$ *at* $\varphi = \alpha$ it can be shown that

$$R = R_0\sqrt{\frac{sin\,\alpha}{\alpha}} \tag{5.70}$$

and for thickness distribution

$$h(\varphi,\alpha) = h_0 \left(\frac{\sin\alpha}{\alpha} \right) \frac{\varphi}{\sin\varphi} \tag{5.71}$$

The following pressure–time cycle is derived [555].

$$\frac{p}{\sigma_{opt}} \cong \frac{2h_0}{R_0} \sqrt{6\xi_{opt} t} \cdot \exp\left(-\frac{3\xi_{opt}}{2} \right) \tag{5.72}$$

The theoretical relationship between the radii of the spherical vessel and the preform is obtained from Eq. (5.70) at $\alpha = \pi/2$ as $R_0 = R_s\sqrt{\pi/2} \cong 1.25R_s$, which is in good agreement with the experimental result [552]: $R_0/R_s = 1.20$–1.25. Thickness distributions calculated using Eqs. (5.57) and (5.71) are presented in Fig. 5.11. For comparison, the corresponding experimental data for Ti–6Al–4V alloy sheet are also presented.

5.2.4.3 Wrinkling in Superplastic Forming

During the free forming of spherical or ellipsoidal shells, the equatorial ring containing the welded joint often loses its circular shape. Many wrinkles/folds arise in the equatorial zone in the early stages of forming. If the initial sheet thickness is small and the quality of the weld is good, then these wrinkles may not form till the end of bulging. If thick sheets and welds of coarse microstructure are used, wrinkles are present in the formed part; this detracts from the appearance and can lead to rejection. Thus, elimination of wrinkling is of practical interest. In conventional materials, in contrast, folds form in thin walled shells (see, for example, [549]). This is a classical problem (see, for example, [553]). For this case a circular ring changes shape into an ellipse and gives rise to a limited number of wrinkles.

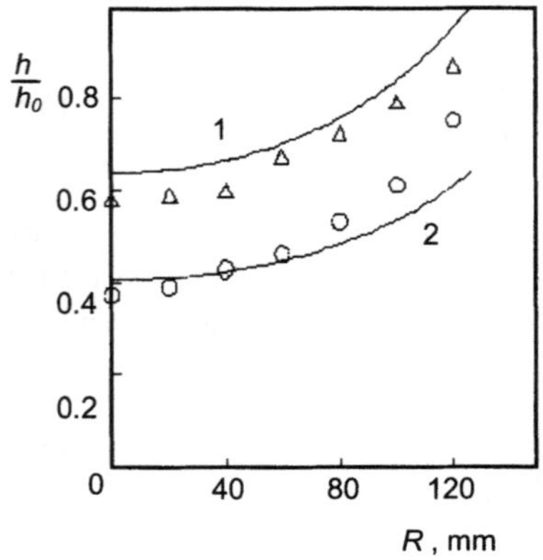

Fig. 5.11. Dependence of normalised thickness *h/ho* on distance from the pole [98]. Solid lines – theoretical predictions. Curve 1 and triangles (experimental) correspond to free forming while curve 2 and circles (experimental) to forming with rigidly clamped edges. Strain rate $\xi = \xi_{opt} = 2\cdot10^{-4}$ s^{-1}, $h_0 = 5$ mm, $R_0 = 100$ mm, $R_0/R_s = 1.2$

Wrinkling in the early stages of free superplastic forming may be explained as follows. Let the weld seam be regarded as a rigid ring on which two thin membranes are drawn. Such an assembly is regarded as an elastic ring gripping two relatively soft circular membranes. Then, the equilibrium of the elastic ring under a uniform load q can be considered. This leads to an expression for the minimum value of the critical load, q_{cr} [553] (which corresponds to the so-called 'first tone') as

$$q_{cr} = \frac{3EJ}{R^3} \tag{5.73}$$

where E is Young's modulus and J is the moment of inertia of the weld seam. If the distributed load $q < q_{cr}$, then the ring retains its circular shape; once the condition $q > q_{cr}$ is satisfied, the shape of the ring gets distorted. To estimate the distributed load q, the equation of the membrane theory Eq. (4.123) is used. It is easy to show that

$$q = \frac{p\left(R^2 - h_p^2\right)}{2h_p} \tag{5.74}$$

where h_p is the current dome height. Substituting Eq. (5.74) in Eq. (5.73)

$$p_{cr} = \frac{2h_p}{R^2 - h_p^2}\frac{3EJ}{R^3} \tag{5.75}$$

In the early stages of forming $h_p << R$ and p_{cr} is small (p_{cr} is proportional to h_p). But, the load q is large (q is proportional to $1/h_p$). Therefore, in the very early stages of SPF, the load applied at the weld seam starts from its maximum value and decreases to zero. And so, at least in the early stages $q > q_{cr}$ and wrinkles will form. Another conclusion of practical importance is that in the early stages, p_{cr} changes as $1/R^5$. Shells tested in [552] had a typical diameter of 100 mm which is an order of magnitude larger than those tested earlier. A shell firmly clamped along its periphery can be regarded in the model as a ring with a very large stiffness EJ. Then, from Eqs. (5.73) and (5.75), it follows that q_{cr} and p_{cr} can be very large and wrinkling may not be observed in many superplastic forming experiments. Thus, the model can explain some features of wrinkling during the SPF of a spherical shell. In reality, however, the problem is far more complicated. One of the predictions is that the diameter of the equatorial part decreases notably during forming so that the ratio of the diameters R_0 and R_s is about 1.25. Thus, the weld seam decreases approximately 20% in length; this is far from being a problem in elasticity as treated in the model. In general, one should consider the problem of shape distortion for an elasto-plastic ring. Also, in experiments unlike in the model, the weld seam shape distortion may be outside the equatorial plane. Finally, an alternative explanation is possible: if it is assumed that the weld seam does not change in total length, the 20% decrease in diameter after forming can cause shape distortion.

At a first glance, the wrinkling phenomenon may appear to be exotic. But, similar phenomena have been observed in formed parts of ellipsoidal shape. Also, shape distortion can proceed according to another deformation mechanism known in the theory of shells as 'mirror reflection mechanism'. It would be interesting to formulate the corresponding boundary value problems in order to solve them.

5.2.5 SPF of a Long Rectangular Membrane

An analysis of the existing methods of titanium cellular structure manufacture has facilitated decision making on the main types of structures and methods of their production [516, 517]. The structures produced by the SPF/DB method are distinguished by the number of sheets used and may consist of 2, 3 or 4 sheets. The geometry of the sandwich structure is characterised by a repeating rectangular profile. Therefore, it is enough to study just one cell. As the rectangular form is common in many parts, it is thoroughly investigated, see, e.g., [55, 184, 282, 382, 383, 554–557]. In this section, superplastic forming of a long rectangular sheet into a rectangular die is considered. Principal equations of the membrane theory and the hypotheses (a)–(h) listed at the end of Sect. 5.2.1 are assumed to be valid. Constant pressure forming of a long rectangular membrane has already been considered in Sect. 3.6.2. Here, two important technological characteristics, viz., thickness distribution and pressure–time cycle will be examined.

5.2.5.1 Thickness Distribution

Ghosh and Hamilton [55, 556] have considered the bulging of a long rectangular sheet into a mould. They have analysed free forming as well as the final stages of deformation into a die (after contact between sheet and die). For the latter case the following boundary condition is used: it is assumed that the sheet sticks to the die and deformation ceases in the contacting part of the sheet. Later, this condition was modified to take into account possible sliding of the sheet along the die (see, e.g., [383]).

Cellular structures consist of outer sheets called casings and inner sheets that give rise to the required profile, known as the core. Core and casings are connected by solid state bonding and together they form a stiff integral structure. The sheets of the core are initially connected by discontinuous welding in such a way that the weld pattern is at right angles to the lines defining the rectangular cells. The discontinuity of the weld is necessary to create the gaps during SPF, which allows the forming gas to pass through all cells. Seam resistance welding is a convenient way of connecting the core sheets. It provides the needed strength at the forming temperature with minimal weld width and heat affected zones. The configuration of the framework (to be distinguished from interaction between core and casing) proceeds with friction at the contact zone [558]. Mathematically this means that one can assume the boundary condition of free sliding along the vertical walls of the die. Full adhesion is assumed at the bottom of the die. In general, the following three situations can be considered:

1. Adhesion at the bottom as well as at the walls (a–a);
2. Adhesion at the bottom and sliding along the walls (a–s);
3. Free sliding along all contact surfaces (s–s).

The case (s–s) is the simplest from a mathematical point of view. In this case, the thickness of the deformed part (a rectangular box having the dimensions $D \times W$) is uniform along its profile.

$$h_f = \frac{h_0}{1+\dfrac{D}{W}} = \frac{h_0}{1+\overline{D}} = constant \tag{5.76}$$

where the subscript f denotes final thickness and $\overline{D}=D/W$ is relative die depth. It is easy to show that for the case (a–a), the minimal thickness (at the corner of the box) will tend to zero. In practice, however, the upper limit of gas pressure will be below p_{max}, so that deformation will stop at a certain stage. The minimum thickness can be calculated from the expressions

$$\frac{h_{\min}}{h_0} = \left(\frac{2}{\sqrt{3}} \cdot \frac{\sigma_0 h_0}{p_{\max} W}\right)^{\frac{4-\pi}{2\pi-4}} \left[\frac{8}{\pi^2} \cdot \frac{\tan^{-1}\overline{D}}{\overline{D}}\right]^{\frac{\pi}{2\pi-4}} \quad if\ \ D \le W \qquad (a-a)$$

$$\frac{h_{\min}}{h_0} = \left(\frac{2}{\sqrt{3}} \cdot \frac{\sigma_0 h_0}{p_{\max} W}\right)^{\frac{4-\pi}{2\pi-4}} \left\{\frac{2}{\pi} \cdot \exp\left[-\frac{2}{\pi}\left(\overline{D}-1\right)\right]\right\}^{\frac{\pi}{2\pi-4}} \quad if\ \ D \ge W \quad (a-a) \tag{5.77a,b}$$

The maximum thickness will be at the centre of the bottom portion (if $D \le W$) or at the upper part of the wall (if $D \ge W$), i.e.,

$$\frac{h_{\max}}{h_0} = \frac{1}{1+\overline{D}^2} \cdot \frac{\overline{D}}{\tan^{-1}\overline{D}} \quad if\ D \le W \quad (a-a)$$

$$h_{\max} = \left(\frac{2}{\pi}\right) h_0 \qquad if\ D \ge W \quad (a-a) \tag{5.78a,b}$$

For the case (a–s), the maximum thickness is determined by Eq. (5.78a) [558] and

$$h_{max} = \frac{h_0}{\left[\dfrac{\pi}{2} - 1 + \overline{D}\right]} \quad if\ D \ge W \quad (a-s) \tag{5.79}$$

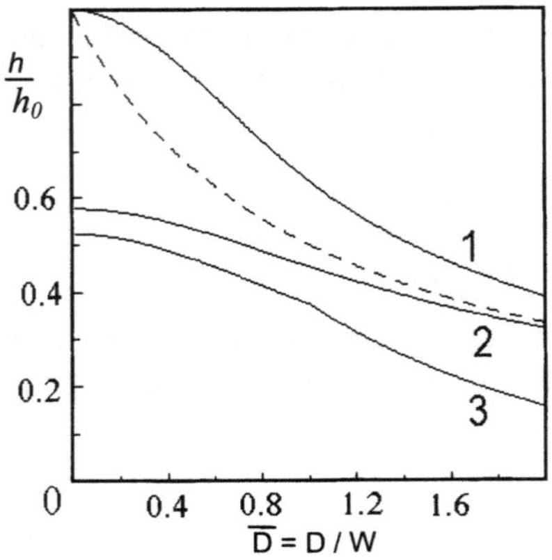

Fig. 5.12. Dependence of normalised thickness, h/h_0, on relative depth $\overline{D} = D/W$: 1 – maximum thickness h_{max}, calculated accordingly to Eq. (5.78); 2 – minimum thickness calculated using Eq. (5.80); 3 – minimum thickness calculated using Eq. (5.77); σ_0 =10 MPa, h_0 = 1 mm; D =15 mm and p_{max} = 20 atm. Dashed line indicates the solution for the ideal case described by Eq. (5.76)

The minimum thickness can be calculated from

$$\frac{h_{min}}{h_0} = \frac{8}{\pi^2}\cdot\left(\frac{2}{\pi}\right)^{\frac{4-\pi}{\pi-2}}\cdot\frac{tan^{-1}\overline{D}}{\overline{D}} \cong 0.5772\frac{tan^{-1}\overline{D}}{\overline{D}} \quad if\ D \le W \quad (a-s)$$

$$\frac{h_{min}}{h_0} = \frac{1}{\frac{\pi}{2}-1+\overline{D}}\cdot\left[1-\frac{\pi-2}{\pi-2+2\overline{D}}\right]^{\frac{4-\pi}{\pi-2}} \quad if\ D \ge W \quad (a-s) \tag{5.80a,b}$$

In Fig. 5.12, the theoretical dependence calculated in accordance with Eqs. (5.76)–(5.80) are presented. Dashed line corresponds to the ideal case (no friction) where in thickness distribution is uniform along the box profile. If adhesion is present in the contact zone, thickness non-uniformity develops.

5.2.5.2 Pressure –Time Cycle

Substituting the expression for equivalent strain rate in Eq. (5.38) and integrating one obtains

$$\frac{\alpha}{\sin\alpha} = \exp\left(\frac{\xi_{opt}t\sqrt{3}}{2}\right) \tag{5.81}$$

which gives (in an implicit form) the time dependence of $\alpha(t)$. The following approximate solution of Eq. (5.81) may be recommended in practice:

$$\alpha(t) \approx \sqrt{3\sqrt{3}\xi_{opt}t} \cong 2.28\sqrt{\xi_{opt}t} \tag{5.81a}$$

The inaccuracy of this solution does not exceed 5% for $0 \le \alpha \le, \pi/2$. Using the principal equations of the membrane theory one obtains $p = 2h\sigma_{opt}/(R\sqrt{3})$. Then, as $R \sin\alpha = W$

$$p(t) = \frac{2h_0}{W\sqrt{3}}\sigma_{opt}\alpha \exp\left(-\frac{\sqrt{3}\xi_{opt}t}{2}\right) \tag{5.82}$$

where $\alpha(t)$ is from Eq. (5.81). The results of calculations based on Eqs. (5.81) and (5.82) are presented in Fig. 5.13. For comparison, results obtained when Eq. (5.81a) is used instead of Eq. (5.81) are also shown by dashed lines. Evidently the approximate Eq. (5.81a) can be used in practice. The deforming shell will touch the die at time $t = t_k$, defined by the condition $H(t_k) = D$. The duration of free forming is given by

$$\tau_k = \xi_{opt}t_k = \frac{2}{\sqrt{3}}\ln\frac{\alpha_k}{\sin\alpha_k} \tag{5.83}$$

where $\alpha_k = 2\arctan(D/W)$. In [556, 559], an iterative procedure for calculating the pressure–time cycle for a long rectangular box section is given. Using the algorithm suggested in [559] and Eq. (5.82), good agreement between theory and experimental results was established. Malinin [184] has used two limiting boundary conditions at the contact between the sheet and the die: (i) free sliding of the shell over the die; (ii) full adhesion of the shell to the die in the contact zone. It is easy to show that the procedure described above is applicable to both these cases. But, Malinin [184] did not consider the superplastic forming regime. Therefore, the pressure–time cycle was not determined.

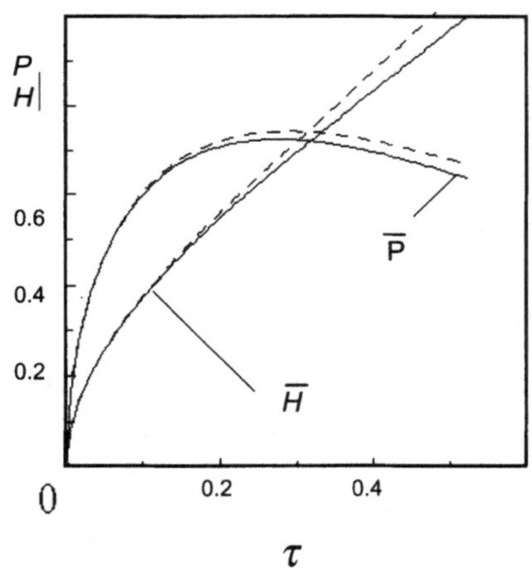

Fig. 5.13. The dependence of the normalised pressure (Eq. (5.82)) and the normalised height on the normalised time $\tau = \xi_{opt}t$, calculated according to Eqs. (5.81), (5.81a) respectively (solid and dashed lines).

5.2.5.3 Comparison with Experimental Results

The following types of experiments have been used to validate the above analysis [87, 93]: (i) constant pressure forming; (ii) constant strain rate forming; (iii) uniaxial constant cross head velocity tests; (iv) uniaxial load relaxation tests. All tests were done at the same temperature on Ti–6Al–4V alloy specimens. Experimental details are available in [57].

Constant pressure tests were done till the shell contacted the die bottom. Results are presented in Table 5.4. Values of p_i and t_i (i =1, 2, 3, 4, 5) were chosen in pairs in order to calculate m (Table 5.5). K was calculated using the well-known two parameter power law (Table 5.6).

The considerable scatter seen in Tables 5.5 and 5.6 reveal the inadequacy of the two parameter (power law) constitutive equation used. In the last column of these tables the average values of m and K are given which may be used if the present results are to be understood in terms of the above analysis. Alternatively m and K may be evaluated using the experimental data given in Table 5.4. As

Table 5.4 Constant gas pressure tests at T = 900°C (D denotes the depth of die)

i	p_i, (atm)	t_i, time taken to contact the die (s)	
		D=10 mm	D=15 mm
1	6	935	2550
2	8	524	1290
3	10	313	940
4	12	215	594
5	14	171	400[a]

[a]specimen failed

Table 5.5 The calculated values of m

D (mm)	Calculated m for each $p_i - t_i$										Average of m
	p_1–t_2	p_1–t_3	p_2–t_3	p_1–t_4	p_2–t_4	p_3–t_4	p_1–t_5	p_2–t_5	p_3–t_5	p_4–t_5	
10	0.497	0.467	0.433	0.472	0.455	0.485	0.499	0.500	0.557	0.673	0.487
15	0.422	0.495	0.705	0.476	0.523	0.397	0.457	0.478	0.394	0.390	0.475

Remark: In this and the following tables and the values of m have only two reliable digits after the decimal point. Similar remark is applicable for the value of K in the following table where the digit after the decimal point is without significance.

Table 5.6 Material constant K (MPa·s^m) calculated from the power law

m	D, mm	J_m[a]	Calculated value of K for each p_i, MPa·s^m					Average K MPa·s^m
			6 atm	8 atm	10 atm	12 atm	14 atm	
0.487	10	0.08980	767.6	764.4	737.0	731.8	760.8	752
0.475	15	0.18415	674.9	651.0	700.2	675.7	653.3	671

[a]The value of J_m is calculated according to procedure given in Chap. 3

$$\frac{t_i}{t_0} = \left(\frac{p_0}{p_i}\right)^m \tag{5.84}$$

where t_0, p_0 is an arbitrary pair of experimentally measured t and p values (reference point). For a least squares fit

$$\Phi(m) = \sum_{i=1}^{N}\left(\ln\frac{p_i}{p_0} - m\ln\frac{t_0}{t_i}\right)^2 \to \min$$
$$m = \left\{\sum_{i=1}^{N}\ln\frac{p_i}{p_0}\ln\frac{t_0}{t_i}\right\}\cdot\left\{\sum_{i=1}^{N}\left[\ln\frac{t_o}{t_i}\right]^2\right\}^{-1} \tag{5.85}$$

The value of m, determined using Eq. (5.85) with p_i, t_i, listed in Table 5.4, is presented in Table 5.7. Here, the scatter is not so large as in Table 5.5. The value of K can be found as follows:

$$\Psi(K) = \sum_{i=1}^{N}\left\{p_i - K\frac{2}{\sqrt{3}}\cdot\frac{h_0}{W}\left[\frac{2J_m(\alpha)}{t_i\sqrt{3}}\right]^m\right\}^2 \to min$$

Therefore,

$$K = \left\{\sum_{i=1}^{N}\frac{p_i}{t_i^m}\right\}\cdot\left\{\frac{2}{\sqrt{3}}\cdot\frac{h_0}{W}\cdot\left[\frac{2J_m(\alpha)}{\sqrt{3}}\right]^m\cdot\sum_{i=1}^{N}\frac{1}{t_i^{2m}}\right\}^{-1} \tag{5.86}$$

Table 5.7 The value of m, for the p_i–t_i corresponding to Table 5.4 determined using Eq. (5.85)

D, mm	Calculated value of m for each p_i–t_i pair					Average m
	p_1–t_1	p_2–t_2	p_3–t_3	p_4–t_4	p_5–t_5	
10	0.4839	0.4793	0.4802	0.4714	0.4887	0.484
15	0.4684	0.4873	0.4672	0.4752	0.4523	0.470

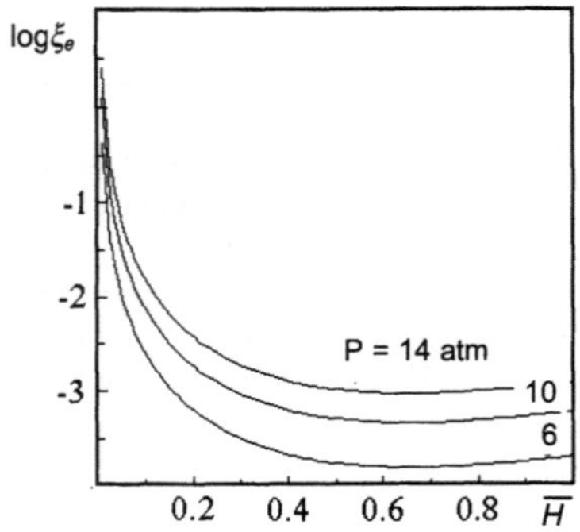

Fig. 5.14. Dependence of the equivalent strain rate ξ_e (s^{-1}) on the relative height $\overline{H}$ calculated with $m = 0.47$, $K = 645$ MPa·s^m. Constant pressures employed are indicated

K was calculated from the experimental data, obtained at various die depths, e.g., for $D = 10$ mm, $m = 0.484$; $K = 652$ MPa·s^m. For $D = 15$ mm, we have $m = 0.470$; $K = 645$ MPa·s^m. To estimate numerically the strain rate intervals corresponding to the constant pressure tests, Eq. (3.150) can be written as

$$\log\xi_e = \frac{1}{m}\cdot\log\left[\frac{\sqrt{3}}{2}\cdot\frac{p_0 W}{K h_0}\cdot\frac{\alpha}{\sin^2\alpha}\right] \tag{5.87}$$

The dependence of the equivalent strain rate on the relative height $\overline{H}$ is presented in Fig. 5.14. It is seen that ξ_e can vary from 10^{-1} to $10^{-4}s^{-1}$.

To verify the results theoretical time dependence H (t) was found using the above values of K and m. From Table 5.8 the accuracy of modelling is shown to be better than 5% for D =10 mm and better than 10 % for D =15 mm. For comparison, a number of constant cross head velocity tests were carried out in uniaxial tension in vacuum. Pairs of σ_i, ξ_i were chosen from Fig.1.1 and analysed in terms of Eq. (3.4). Findings are given in Table 5.9

Table 5.8 Theoretical and experimental values of the duration of constant gas pressure forming

P_i, atm	Duration of constant gas pressure forming, s					
	D =10 mm, m = 0.484, K = 652 MPa·s^m			D =15 mm, m = 0.470, K = 645 MPa·s^m		
	Expt.	Calc.	Error %	Expt.	Calc.	Error %
6	935	938	0.3	2550	2536	0.5
8	524	518	1.1	1290	1375	6.6
10	313	326	4.2	940	855	9.0
12	215	224	4.2	594	580	2.4
14	171	163	4.7	400*	418	4.5

Table 5.9 The values of material constants K (MPa·s^m) and m, calculated from Fig. 1.1

	Percentage elongation (e, %)								
	20	30	40	50	60	70	80	90	100
m	0.671	0.664	0.643	0.624	0.599	0.576	0.558	0.537	0.524
K	1333	1363	1243	1118	984	870	797	711	665

It is clear that m and K decrease monotonically with e. (A similar behaviour has also been reported in other materials [92, 547].) To compare the results obtained for m from the constant gas pressure tests the corresponding equivalent engineering strain e for the SPF process was evaluated by integrating the equivalent strain rate with respect to time [87]. From Eq. (3.148)

$$s = \int_0^t \xi_e dt = \int_0^\alpha \frac{2}{\sqrt{3}}\left(\frac{1}{\alpha} - \cot\alpha\right) d\alpha = \frac{2}{\sqrt{3}} \ln\frac{\alpha}{\sin\alpha} \tag{5.88}$$

Then, the equivalent tensile engineering strain [87][3] e^{eqv} is

$$e^{eqv} = \left[\left(\frac{\alpha_k}{\sin\alpha_k}\right)^{\frac{2}{\sqrt{3}}} - 1\right] \cdot 100\% \tag{5.89}$$

where $\alpha_k = 2\tan^{-1}(D/W)$. The results are presented in Table 5.10. The last set involved load relaxation experiments analysed following two specimens standard procedures [72]. The results are displayed in Table 5.11.

Theoretical dependence of H (t) calculated using the values of m and K listed in Table 5.9 is compared with the experimental results in Table 5.12. In all cases, the theoretically predicated duration was less than half the time of deformation required experimentally. Thus, the use of uniaxial experimental data can lead to large errors in the prediction of time of forming.

Table 5.10 Equivalent tensile strain [87], calculated according to Eq. (5.89)

D, mm	$\overline{H} = H/W$	α_k, rad	ε_{tr}	e^{eqv}, %
10	0.667	1.176	0.278	32.3
15	1	1.571	0.521	68.4

[3] This parameter corresponds to the same value of the accumulated strain (Odqvist's parameter) in a uniaxial tensile test

Table 5.11 Experimental results of load relaxation tests [72]

v,mm/ min[a]	σ_{in}, MPa	t_1, s	t_2, s	M
1.4	27.6	6.8	21.5	0.473
1.2	24.2	5.6	17.4	0.482

[a] v is the cross head velocity immediately before arrest. σ_{in} is stress at the beginning of relaxation. t_1, t_2 are the parameters of the load relaxation curve: $\sigma(t_1) = \sigma_{in} / 2$; $\sigma(t_2) = \sigma_{in} /4$; $M = ln2/ln[t_2 (t_2 - t_1)/t_1]$

It is interesting to note that the value of m determined in the constant pressure tests (last column of Table 5.7) are practically the same as the value of m determined from the load relaxation tests (last column of Table 5.11). But, the behaviour of superplastic materials in a load relaxation test is quite different from that in a uniaxial tensile test or in constant gas pressure forming (unsteady versus steady state conditions, differences in the role of grain boundary sliding) [11, 72, 147]. It is not clear if such a close agreement is just fortuitous. Additional experiments are necessary for an unequivocal conclusion. It is also relevant that in the experiments of Hamilton and Ghosh [560] m value similar to what was obtained in the above gas pressure forming experiments has been reported in uniaxial tension over two decades of strain rate.

Table 5.12 Theoretical predictions made on the basis of uniaxial tests

P_i, atm	Duration of forming, s					
	D =10 mm, m = 0.664, K = 1363 MPa·s^m			D =15 mm, m = 0.576, K = 870 MPa·s^m		
	Expt.	Calc.	t_{iexp} / t_{icalc}	Expt.	Calc.	t_{iexp} / t_{icalc}
6	935	307	3.0	2550	885	2.9
8	524	199	2.6	1290	537	2.4
10	313	142	2.2	940	365	2.6
12	215	108	2.0	594	266	2.2
14	171	87	2.0	400*	203	2.0

Table 5.13 Results of tests at constant strain rate

ξ_i, s^{-1}	σ_i, MPa	H, mm	Duration of Test, s		Error %
			Expt.	Calc.	
1×10^{-3}	25.1	14.2	434	482	11
8×10^{-4}	22.6	14.4	491	609	21
4×10^{-4}	16.3	14.5	949	1242	31

Comparison between the theoretical predictions and the experimental observations in constant strain rate tests is made in Table 5.13. (Material constants used for calculating the pressure–time cycle were taken from the results of constant gas pressure forming: $m = 0.47$; $K = 645$ MPa·s^m.) Reasonable agreement is seen. The following general conclusion can be drawn. The use of non-standard tensile specimens can lead to large errors if material constants obtained from these tests are used to estimate technological parameters.

5.2.6 Estimating Strain in SPF and SPF/DB Processes

In general, to estimate the accumulated strain (Odqvist parameter) at a point it is necessary to solve the corresponding boundary value problem, derive the velocity field and the strain rate tensor, and integrate its second invariant along the streamline passing through that point. As this is rather difficult, it appears attractive to evaluate the strain from the original thickness and the thickness at the given point. This method is explained below, along with a procedure for converting the accumulated strain into an equivalent strain.

When L/W >>1, plane strain condition prevails (see Sect. 5.1.5) and $\xi_z = 0$. From incompressibility condition, $\xi_t = -\xi_n$ where ξ_t and ξ_n are, respectively, the tangential and the normal strain rate. It is evident that all non-diagonal components of the strain rate tensor in this case are equal to 0. Thus, the effective strain rate is $\xi_e = 2|\xi_n|/\sqrt{3}$. Taking into account $\xi_n = -(1/h)\, dh/dt$ (h is the thickness), the Odqvist parameter is found to be

$$s^{plane} = \int_0^t \xi_e dt' = \frac{2}{\sqrt{3}} \ln\left(\frac{h}{h_0}\right) \cong 1.155 \ln\left(\frac{h}{h_0}\right) \tag{5.90}$$

where h_0 is the initial sheet thickness.

Superplastic forming of axi-symmetric parts is characterised by a biaxial strain state. The tangential (ξ_t) and the circumferential (ξ_c) strain rate tensor components being equal at the pole (see Sect. 5.2.3). From the incompressibility condition it follows then that the normal strain rate is given by $\xi_n = -2\xi_t = -2\xi_c$. In this case, $\xi_e = |\xi_n|$ and so

$$s^{biaxial} = \int_0^t \xi_e dt = \ln\left(\frac{h}{h_0}\right) \tag{5.91}$$

Since $\xi_c = 0$ at the edge, plane strain condition is present here (see discussion in Sect. 5.2.3). Then the accumulated strain near the edge is calculated according to Eq. (5.90). Thus, at an arbitrary point between the pole and the edge, the accumulated strain is

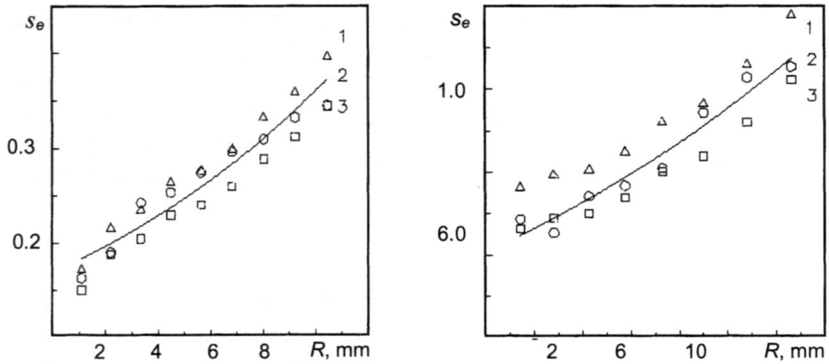

Fig. 5.15. The dependence of accumulated strain s on the distance from the pole R for an axisymmetric part of height **a** 5 mm and **b** 10 mm [87]: Δ, O and □ are calculated according to Eqs. (5.90), (5.94) and (5.91), respectively

$$s = (1.000 - 1.155)\ln\left(\frac{h_0}{h}\right) \tag{5.92}$$

For comparing deformation capabilities in different types of loading, the concept of equivalent strain is useful (Sect. 5.3). Experimental validation of Eq. (5.92) was attempted in [87] using Ti–6Al–4V alloy sheets. The accumulated strain was determined by two independent methods. The first was based on Eq. (5.92). In the second method, an estimate was obtained by measuring the change in distance between pre-marked positions. For this, a net of concentric circles of 1 mm diameter was used. Distance between the circles after deformation and the thickness distribution were measured using a microscope to an accuracy ±0.001 mm. If r_i and R_i are respectively the initial and the final radius of the i-th circle (i = 1, 2, N), then the main logarithmic strains are given by [394]

$$s_1 = ln\left(\frac{R_i - R_{i-1}}{r_i - r_{i-1}}\right), \quad s_2 = ln\left(\frac{R_i}{r_i}\right), i = 1,2,...,N \tag{5.93}$$

Then the value of the strain intensity s_e (assuming incompressibility condition) is

$$s_e = \frac{2}{\sqrt{3}}\sqrt{\varepsilon_1^2 + \varepsilon_2^2 + \varepsilon_1\varepsilon_2} \tag{5.94}$$

where s_1 and s_2 are from Eqs (5.93). Accumulated strain versus distance from the symmetry axis relationship is presented in Fig. 5.15. Evidently, the accumulated strain increases with distance from the central line. Curve 2 was plotted using the marker net lying between curves 1 and 3. (Curves 1 and 3 correspond to the upper

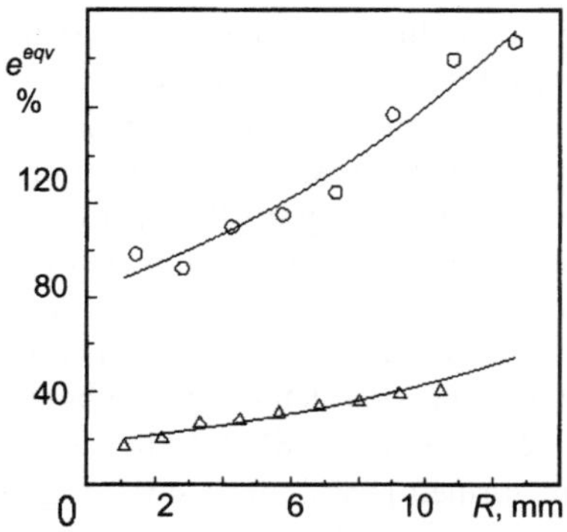

Fig. 5.16. Dependence of equivalent tensile strain, on distance from the axis of symmetry *R* for domes of 5 mm height (triangles) and 10 mm height (circles). Solid lines are the experimental data

and the lower estimates of the accumulated strain.) Thus, the method can be used to estimate the accumulated strain. Equivalent tensile strain (%) e^{eqv} , versus distance from the symmetry axis *R* (in mm) plots are shown in Fig. 5.16. Thus, it is possible to convert the thickness distribution into strain distribution by this method. As the limiting strain in tension can be found by simple experimentation, the limit of formability of a component of a given geometry can be known and this will lead to rational design.

5.3 Deformation Processing of Materials

5.3.1 General Notes

To obtain a grain size conducive to superplastic flow (typically 10–15 μm in commercial alloys) thermomechanical treatments are resorted to. Too high a temperature will lead to coarse grains in the processed material, while too low a temperature or too high a plastic strain at a given temperature will cause cracking. Thus, it is essential to choose an optimal combination of temperature and deformation for obtaining the best results in the final product. Many methods are available that exploit phase or structural transformations in the solid state and these are of special significance to highly alloyed and/or difficult to form materials. A major requirement is that the microstructure should be uniform throughout the volume being processed. This becomes an acute problem when sub-microcrystalline or nanocrystalline structures are to be obtained. But, this area has acquired considerable importance in recent times. Some of the interesting techniques used to produce submicrocrystalline structures are as follows: saturation of Ti-alloys with hydrogen [561], employing (cryo) cooling –heating cycles in stainless steel to take advantage of the phase transformation present [561], multi-stage upsetting of the intermetallic TiAl under gradually decreasing temperature [106, 564]. These are innovative empirical procedures developed by metallurgists/materials scientists. However, they are yet to evolve into standard metal working operations.

Many investigations [565–579] have shown that large plastic deformation at low temperatures ($T < 0.3T_m$) results in rather similar microstructures in different materials. Formation of cells, their transformation into subgrain boundaries and with increasing mis-orientation between subgrains their conversion into small nuclei of new grains are all part of standard literature on grain refinement, recovery and recrystallisation. Moreover, structural evolution and final grain size depend on the thermomechanical processing steps, sequence, and step details, e.g. subgrains form in molybdenum following heavy plastic deformation [568], but after torsion under pressure fresh grains of size 0.1 μm are formed [580]. In a steel, a grain size of about 20 nm could be produced by torsion [572]. Equal channel angular (ECA) extrusion is yet another way of producing an ultra-fine grain size [570, 571]). These methods are yet to be employed on an industrial scale. Their main features are examined below so that a better quantitative understanding becomes available in the future.

5.3.2 Torsion under Pressure and ECA Extrusion

The schemes of torsion under pressure and ECA extrusion are presented, respectively, in Figs. 5.17 and 5.18. These methods have been successively used to produce ultra fine grain sizes [570–578, 581–583]. During torsion under pressure plastic deformation is due to the combined action of a torque $\bar{\tau}$ and an axial force P transferred to the specimen by a mobile upper cross head. Unlike in simple torsion under pressure, the axial force is active during ECA extrusion; a shear momentum is produced by the reaction of the intersecting walls of channels of same size and cross-section. Also, ECA extrusion is a multi-cycle process: the specimen is repeatedly extruded through the intersecting channels (usually 5–14 cycles). In the case of torsion at constant specimen height (the so-called 'fixed-end torsion') deformation will correspond to simple shear (see Appendix). It has been shown [581–583] that simple shear is encountered in ECA extrusion also. Consequently, these schemes of deformation possess specific features of simple shear, namely, directed development of macro-deformation; and a possibility of accumulating large plastic deformation by cyclic alteration of the shear direction. In other words, material flow is non-monotonic.

An important feature in torsion under pressure is that after a certain number of rotations the changes in structure-sensitive properties, e.g., hardness, with further deformation are rather small. This is reflected in transmission electron microscopic observations where changes in microstructure are not significant [571, 584]. Such an observation was present after ECA extrusion also [581–583] although the average grain size after this deformation schedule was larger than in the former case [571, 584]. The reasons for this kind of mechanical response are not clear although broadly speaking they may be connected with the differences in the velocity fields and the magnitudes of the flow and force parameters.

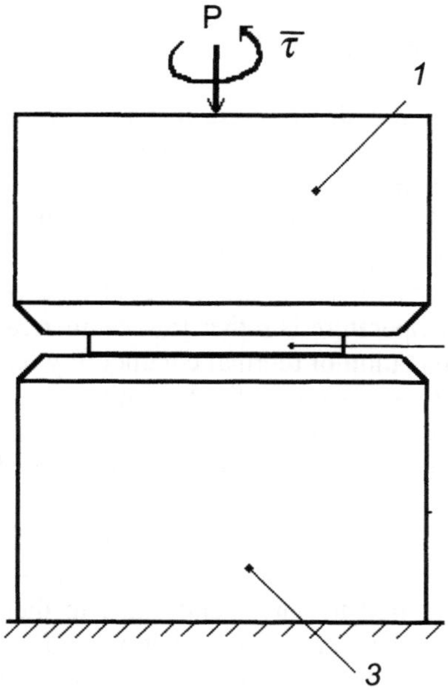

Fig. 5.17. Torsion under pressure:
P – axial force, $\overline{\tau}$ – torque *1* – upper anvil *2* – specimen *3* – lower anvil

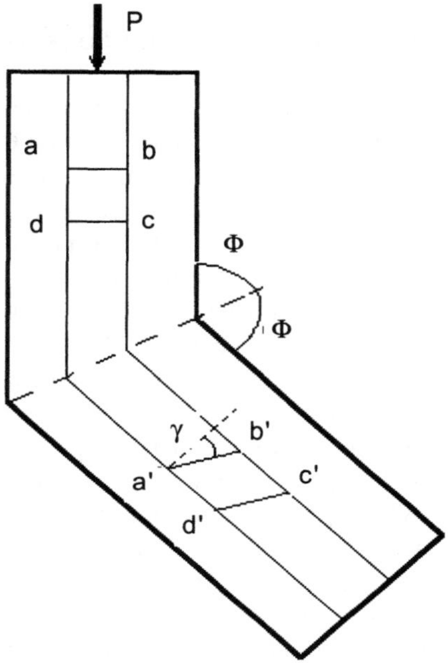

Fig. 5.18. Equal channel angular (ECA) extrusion: P – axial force, 2Φ – intersection channel angle, γ – shear angle

A practical difficulty with both the processes, in particular ECA extrusion, is the finite strength of the equipment, which has to withstand the large deformation forces required. But, as it is seen in Sect. 5.1.3, compression with simultaneous torsion reduces the axial force needed to obtain a given axial upset. Then, for a given axial force more accumulated strain can be obtained using this process than by simple compression or torsion. Evidently, this will lead to greater grain refinement.

The main problem with the use of torsion under pressure (Bridgman anvils) lies in material slip along the instrument with serious problems associated with the determination of the sliding rate as well as the accumulated strain (which differs significantly from the values estimated from the number of revolutions). Another limitation arises from the smallness of the specimen that can be used (0.1–1.0 mm in height and 1–10 mm in diameter). Also, it is well-known that buckling will preclude the use of torsion under compression when the height to diameter ratio is greater than about three. But, from an industrial point of view it is desirable to handle specimens of height to diameter ratio of about 10. Thus the method is to be regarded as a laboratory device to assess formability which is of potential use with more massive specimens.

An acceptable industrial design, however, is lacking [313]. But it is satisfying to note that this method, with some modifications, has been successful in producing on small scale materials with submicrometrer and nanometre range grain sizes.

5.3.3 Thermomechanical Conditions for Grain Refinement

Experiments have revealed that in pure metals and single phase materials, a decrease in the deformation temperature leads to a decrease in the average grain size. To obtain a sub-microcrystalline grain size in metals of commercial purity the deformation temperature should not exceed (0.25–0.30) times the absolute melting temperature, T_m [571, 584]. During ECA extrusion, local temperature rise, due to mechanical energy dissipation in a narrow zone of deformation should also be considered, even when the overall specimen temperature is rather low. Numerical estimates have been made in [407] where the equation of thermal conductivity has been used with the following simplifications: (i) temperature field is uniform and (ii) work of plastic deformation is assumed to be transformed entirely into heat. For example, for strain hardening it was calculated that $\Delta T \approx 110$ K for a strain of $\varepsilon_e = 1$. Thus, in principle, local heating can be considerable.

When specimens are thin, temperature rise will not be significant due to heat dissipation through massive cross heads/platens. Dispersion strengthened alloys can be processed at higher temperatures (compared to pure metals) due to their grain coarsening resistance and this will reduce the work of deformation. However, this would require the use of materials of greater heat resistance for the manufacture of equipment. Attention is now turned to hydrostatic pressure. Higher the hydrostatic pressure, greater is the plasticity and lower the diffusivity. Thus, the recovery processes get impeded [585]. The magnitude of the hydrostatic pressure depends on the material, the scheme of deformation and the accumulated strain. (During torsion under pressure, axial stress is usually about 3000–4000 MPa, and in ECA extrusion it is about 1500–2000 MPa.)

With respect to the role of initial structure and loading conditions (values of strain and strain rate) it has also been shown recently [586] that in ECA extrusion as well as torsion under pressure the texture becomes weaker when the strains are large. In some studies (see, e.g., [570, 584]) the initial grain size and texture did not affect the final grain size significantly if the plastic deformation during processing was large. In general, a deformation-induced change in the orientation distribution of polycrystals influences the flow stress and resultant strain if the grain size is comparable to the specimen size. But, if the average grain size is very small in comparison with specimen size, as a first approximation, the material may be assumed to be an isotropic continuum and standard methods of continuum mechanics used to describe the strain state in the deformed material. However, even when one uses the methods of mechanics of continua for describing the deformed state there are some difficulties, e.g., description of finite strain behaviour.

The efficiencies of different types of metalworking can be compared by studying the kinetics of grain refinement [170, 171]. Thus, it is necessary to develop a concept of equivalent strain states in different processes. For infinitesimally small strains, there is only one strain tensor whose components have obvious meaning: the three diagonal components of the strain matrix describe linear deformations, while the non-diagonal components describe shear strains. Also, in most metal working processes the change in metal volume is negligible and so, the sum of the diagonal components is zero (incompressibility condition). Hence, the strain tensor is identical to its deviator. For finite strains one can introduce various strain tensors (see Appendix). Each strain tensor allows the unequivocal description of the strain state but the components of all the tensors do not have a clear meaning as in the case of infinitesimally small strains. Thus, it is difficult to compare any two deformed states. The problem is more complicated if one wishes to compare two different deformed states that are characterised by six numbers (components of the strain tensor). Then, one has 6 parameters to be controlled. If one considers only the invariants of the strain tensor one should control three independent numbers. Evidently, it is desirable to have only one parameter to control with respect to the kinetics of the process. In practice, the value of the Odqvist parameter is often used to estimate the accumulated strain under different kinds of metalworking. This parameter is often used in mechanics of solids as 'fictitious time' (ordering parameter of the process). A similar approach has been used in [407] for comparing different kinds of deformation treatments.

5.3.4 On Some Principles of Structure Refinement

It is of interest to derive some general principles for obtaining a uniform ultra-fine-grained structure. It is well-known that the final grain size d_s obtained by severe plastic deformation is determined by the value of the Zener–Hollomon parameter $Z = \xi \, exp \, (Q/RT)$ [129], where Q is the activation energy, R the gas constant, T the absolute temperature of deformation and ξ is the strain rate. Thus, in principle, it is possible to obtain the desired value of grain size by plastic deformation under an appropriate combination of T and ξ. This is the key principle used while obtaining ultra-fine grained structures by severe plastic deformation [590, 591]. But obtaining a homogeneous structure throughout the volume of a part is very difficult. In practice, the accumulated strain is chosen to be sufficiently high which causes structural non-uniformity. If one compares the different methods of deformation treatments using Odqvist's parameter (see Sect.5.3.3) one would obtain the results presented in Table 5.14. From this Table, it is clear that loading schemes that employ two components are very effective. For example, in compression with torsion one can easily achieve $s \sim 30$. But to obtain $s \sim 16.1$ in a tensile test, one has to achieve an elongation of about 10^7 without necking. Such values are not practically achievable. In principle one can use die-drawing [565, 566], rolling [567] or ordinary extrusion achieve $s \sim 16.1$. However, the geometry of the part will be changed considerably after such a treatment.

In order to reduce damage accumulation and consequent failure, one has to use deformation modes that employ high values of hydrostatic pressure. This problem is of special significance at low temperatures and large deformation forces often used to produce submicro-crystalline and nano-crystalline microstructures.

Torsion under pressure (Fig.5.17) is often used to achieve large plastic strains. In this case due to contact friction forces the specimen is subjected largely to shear deformation. It has been established that nano-crystalline and submicro-crystal-

Table 5.14 Odqvist's parameter s_e calculated for different deformation processes [407]

ECA Extrusion		Torsion		Tension / Compression		Compression + Torsion	
cycles	s_e	N	s_e	$L_k/L_0 = H_0/H_k$	s_e	N^2	s_e
10	11.5	1	18.2	5	1.6	1	30.2
12	13.9	2	33.4	10	2.3	2	60.3
14	13.2	3	54.6	1 000	3.9	3	90.4
16	18.5	4	72.8	1 000 000	13.8	4	121
18	20.8	5	91.0	10 000 000	13.1	5	151

line microstructures can be obtained by this deformation mode in copper, nickel, iron, steels, intermetallic compounds, etc. (But, this method is confined to the laboratory only as the dimensions of the specimen used are very small.)

Bulk (3 dimensional) forging is the obvious way of achieving considerable plastic deformation. By this technique, structural non-uniformities can be drastically reduced and submicro-crystalline structures obtained [567–569]. However, development of plastic instabilities/buckling is a problem due to grain refinement and consequent strain softening. Also, for example, the strength of two-phase titanium alloys becomes 1/2–1/3 the original value with increasing strain [591]. Then after first upsetting, the yield stress in the deformed regions is considerably less than that of the dead metal zones (which are found near the area of contact with the equipment). Further upsetting along other directions does not lead to any notable refinement of the dead metal zones because they become floating solid bodies of greater strength within a soft mantle (regions of refined grain structure). There are two possible ways of overcoming this problem. (1) One can increase the strain rate and/or decrease the temperature to raise the level of the stress–strain curve. (This will reduce the plastic instabilities.) However, this method will increase the deformation force/energy and the probability of specimen failure due to decreased formability with increased strain rate/decreased forming temperature. (2) Using many relatively small upsettings so that the flow stress difference in the different zones in successive upsettings is not large and the deformation force and strain per cycle are small. In this case, the processing will be time consuming. Multi-stage forging is often used in practice because there is no better way currently available for grain refinement in large billets.

In ECA extrusion (Fig.5.18) a long billet is pushed through the channels as many times as necessary to refine the grain size (typically, 12–14 times). The drawback of ECA extrusion is the low value of accumulated strain per pass, which is determined by the angle between the channels Φ.

One more deformation mode used to refine the grain structure is known as 'sandy clock' [592, 593]. In this case the billet *1* is placed between rods *2* and *3* inside die *4* (which has an axi-symmetric constriction), see Fig. 5.19. Then, one moves the rods *2, 3* and the die *4* in such a way that the billet *1* is repeatedly pushed through the narrow portion of the die. As a result, a large accumulated strain can be achieved. This deformation mode has the same drawback as ECA

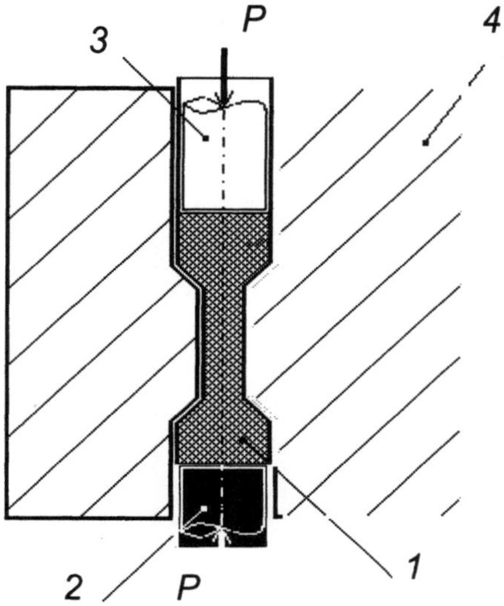

Fig. 5.19. Schematic of the deformation process known as 'sandy clock': *1* – material to be deformed *2*, *3*– dies; *4* –container

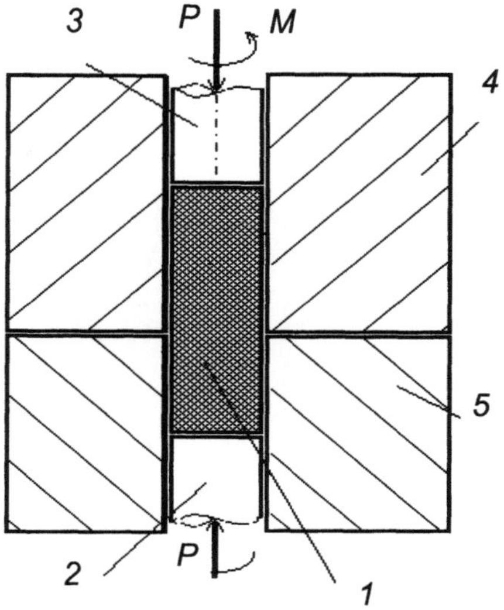

Fig. 5.20. Method to twist an axi-symmetric preform under pressure: *1* – material processed. *2*, *3* - dies; *4*, *5* – container [93]

extrusion: the accumulated strain per cycle is low and is determined by the configuration of the die. Thus, many deformation modes can be used to obtain an ultra-fine grained structure. Torsion under pressure, ECA extrusion and 'sandy clock' modes all allow large strains to be developed while retaining the same shape and dimensions of the billet. *The shape and dimensions of the billet in these cases will be the same before and after each cycle of treatment.* ECA extrusion and 'sandy clock' mode eliminate the dead metal zones and help achieve a uniform structure. The principle is to control the local deformation zones so as to displace the dead metal zones throughout the billet. As a result, the entire volume acquires a refined structure.

Based on the above principles, a new method for obtaining an ultra-fine grained structure has been suggested [594]. Here, the die is divided into two parts *4* and *5*, which can rotate relative to each other (Fig.5.20). Such a die design allows the localisation of the deformation zone in the vicinity of the contact plane between parts *4* and *5*. Cylindrical rods *2* and *3* can be displaced along the vertical axis, which will allow all the layers of the billet to be worked.

In this method, the relative kinematic characteristics of parts *4*, *5* and rods *2*, *3* can be changed. In addition, the geometry of the channel can be changed to combine this method with the 'sandy clock' deformation mode. Alternatively, ECA extrusion and this method can be combined. In this case, one has to use channels of elliptical cross-section, eccentricity of which will be unambiguously determined by the angle Φ between the channels. Experimental results [170] do suggest that combinations of deformation modes lead to more efficient grain refinement than simple types of loading.

6 Problems and Perspectives

In this chapter, some problems of and perspectives on investigations in superplasticity are presented. As elsewhere in the book, only aspects of phenomenology and mechanics are considered.

From the viewpoint of mechanics, the phenomenology of superplasticity can be considered as a domain of creep of materials where the strains are large. This would imply that investigations on the phenomenology of superplasticity would benefit not only by the common methods of mechanics of solids but also from the methodologies of well-developed theories of creep, viscoplasticity and thermo-elasto-viscoplasticity.

Towards this goal, two directions of work can be identified–one for specialists in superplasticity and the other for those in mechanics.

1. The integration of the methods and approaches of mechanics of solids into the methods of superplasticity research (by specialists in superplasticity).
2. Finding solutions to such problems of superplasticity, which lie within the framework of mechanics and mathematics, e.g., mathematical models, stability of flow (by specialists in mechanics).

The first group may take up, among others, the following tasks:

- Use methods of theory of similarity and dimensional analysis to simulate material behaviour and technological processes.
- Use the theory of constitutive equations.
- Develop standard procedures for experimental investigations.
- Use the theory of experiments for the inhomogeneous stress–strain state.
- Use the theory and methods of development of constitutive equations in experiments (determination of the constants of the constitutive equations).
- Adapt the main ideas of the theory of elasto-plastic processes and use them for the classification of forming processes.

The second group may attempt at completing the following investigations:

- Developing phenomenological constitutive equations (per boundary value problems) based on physical models.
- Experimenting on superplastic materials under complex loading conditions.
- Formulating constitutive equations for superplastics in tensor form when large deformation is present.
- Developing methods for including structure/microstructure in constitutive equations in explicit forms (as internal state variables).

- Developing the theory of experiments to include inhomogeneous stress–strain states and large deformation.
- Evolving new methods of identifying constitutive equations, including those where inhomogeneous stress–strain states.
- Working out the phenomenological models of damage accumulation and incorporating them into constitutive equations.
- Developing methods to analyse the stability of plastic flow in technological processes.
- Developing new approaches to establish adequate boundary conditions at contact surfaces in technological boundary value problems.
- Evolving new methods of physical modelling to describe: (a) the behavior of materials in the superplastic state; (b) the technological processes that exploit superplastic flow.

The combined skills of specialists in superplasticity and mechanics will also be required to solve certain types of problems. This is necessitated by the fact that mechanical properties of superplastics are strongly influenced by microstructure. Then, the development of constitutive equations that explicitly convey the importance of microstructure is essential, i.e., the model should clearly reveal that mechanical properties of a material of a given chemical composition but different microstructures will be different. In such a situation, either the constants of the constitutive equation or the form of the constitutive equation will be different. Therefore, the constitutive equations should contain internal variables of clear physical meaning that will signify the importance of microstructure. Specialists in superplasticity will help identify such important microstructural parameters.

It is well-known in mechanics that mechanical response depends on the history of loading, particularly at low homologous temperatures of deformation. In the theory of elastic-plastic flow, definite classification of the deformation processes depending on the degree of complexity is possible (see Chap. 2). Recommendations for choosing correctly the constitutive equations for each case are also available. Thus, there is scope for the materials scientist to identify in consultation with the specialist in mechanics, the type of the deformation processes that would represent the microstructural evolution accompanying a given thermomechanical treatment/loading history of a material in the superplastic state.

Materials technologists and specialists in mechanics may combine to develop mathematical models for technological processes, obtain correct and as far as possible adequate relationships for the boundary conditions, simplify the formulation of boundary value problems and choose methods of solution for the boundary value problems. In this third group of activites, in the immediate future, the following problems may be focussed on.

- Establishing the micro- and the meso-structure parameters, which are most sensitive to temperature and history of deformation.
- Creating databases, including data on the relationship between thermomechanical treatment and microstructural evolution.

- Identifying the necessary conditions for superplastic flow from a wider class of mechanical parameters.
- Collating and analysing experimental data concerning damage accumulation in superplastic flow processes.
- Understanding unstable flow under conditions of superplasticity, both experimentally and analytically.

A systematic study of this kind will allow the manufacture of components/ products of pre-defined mechanical properties. However, this will demand the starting microstructure to be specified accurately throughout the workpiece.

Specialists in superplasticity will benefit from a knowledge of the theory of elasto-plastic processes and its effective application in the problems outlined above. In this theory the following two basic ideas are involved: process trajectories (time dependence of strain tensor and temperature) and types of deformation trajectories. These concepts should be effectively used in experiments as well as in constructing and applying the constitutive equations for studying microstructural evolution and in the mathematical modelling of technological processes. Since all technological processes are characterised by multiaxial stress states, superplastic behaviour should be investigated under multiaxial loading conditions as well. That is, microstructural evolution, damage accumulation and mechanical properties should be evaluated under multiaxial loading.

6.1 Influence of Strain History on Evolution of Structure

For both fundamental and practical reasons, a study of the kinetic changes in structure that depend on the deformation path is of interest. In such an investigation, the principles of grain refinement by large plastic strain and new methods of obtaining superplastic microstructures will get established.

Presently, thermal and/or thermomechanical processing is used to obtain the desirable microstructures. The thermal treatments are well documented, e.g., treatments in vacuum, air, water, etc. In contrast, the effects of the history of mechanical loading are documented rather poorly, e.g., just a mention that a material was rolled or upset by a certain amount. Ideally, one should record the history of all the components of the strain tensor (or stress tensor).

It is convenient to use the theory of elasto-plastic processes to describe the features of mechanical loading. This theory was briefly outlined in Chap. 2. Then, it becomes possible to incorporate the experience accumulated by specialists in mechanics of solids with the knowledge gained by materials scientists. Such a synthesis will be useful in studying the mechanical response of materials in which the structure changes during deformation. Two examples of such collaborative works are considered below.

With reference to multiaxial loading, a materials scientist focuses on a study of technological processes. In these processes the material does not experience the same stress and strain throughout. Therefore, often, one does not know the exact deformation history at any point in the material.

In their investigations, materials scientists attempt to evolve a relationship between microstructural evolution at fixed points in a material and the scheme of deformation or the parameters of the technological process (total load, tool movement, velocities, pressure etc.). But, this relationship is not single valued as it depends not only on material properties but also on the technological process itself. This means that after a change in a process procedure, one needs to look at the relationship once again.

From the viewpoint of solid mechanics, this problem (establishment of a correlation between a process and the microstructural parameters) is divided into two parts. Firstly, one must understand how a change in microstructure is related to the parameters of the known deformation path at a given point. This relationship represents the properties of the material. If the same deformation path is gained at any point in another technological process, the behaviour of the material at that point will be evaluated from the same relationship. The second part involves the establishment at a connection between a technological process and the deformation path at each point in the material. This part results from a solution of the boundary value problem, the technique of which is well developed (see Chap. 4).

It is clear that a required microstructure can be gained in many ways following different deformation paths. If one knows the above mentioned set of relationships between the microstructure and the deformation parameters one will be able to state and solve the problem of optimization. Consequently, one can try to optimize the technological process also. This would mean that one should develop a technological process in such way that different points on the material go through the required deformation paths so that, at these points, the desired microstructure is present.

It is to be remembered that the deformation history (see Chap.2) is described by the deformation trajectory/path. This curve can be characterized by geometric parameters (curvatures) and a point on it can be specified either by the length of the curve up to that point (Odqvist parameter) or by the distance from the origin of that point (strain intensity). For monotonic loading the two parameters are equal. For all other cases they are different.

The above points will be illustrated by examples. The transformation of a coarse-grained lamellar ($\alpha+\beta$) morphology into an equiaxed one (including a submicrometer grain size) in titanium alloys is usually effected by large plastic deformation in the ($\alpha+\beta$) phase field, e.g., by multi-stage forging [595, 596]. Such a treatment, in general, is time consuming. But, refinement of structure is an essential prerequisite for superplasticity. Besides, the fine equiaxed structures promise some improvements in mechanical properties. Therefore, the development of fine grain microstructures using different kinds of mechanical working is of interest. As a result, this aspect has been thoroughly investigated to identify the highly effective thermomechanical treatments [597–603]. The influence of strain rate was studied in [599]. It was established that forging reduction should at least be 50%. In contrast, approximately 25% rolling reduction leads to a similar structural change [597]. Thus, the degree of plastic deformation needed to transform a microstructure strongly depends on the scheme of deformation.

It is important to ensure that the deformed structure is uniform. Towards this end multiaxial forging is often resorted to [601–603]. But, such processes are non-monotonic in nature. (The deformation path is not smooth, but is a broken line.) The influence of this factor on structural transformation remains obscure. In two recent papers [170, 171], different ways of transforming a microstructure was studied by examining in a Ti-alloy the kinetics of transformation of a lamellar microstructure into an equiaxed one.

The first problem was to identify a suitable parameter to evaluate quantitatively the degree of structural transformation. The efficiency of the process of transformation following different treatments of a lamellar structure into an equiaxed one was evaluated by comparing the equiaxed α-particles, n, visible in a selected area of specimen cross-section corresponding to a fixed accumulated strain, s, strain rate ξ and temperature T. The ratio of the number of equiaxed α-particles to their total number in the selected specimen cross-section, n was calculated (an α-particle was considered equiaxed if its aspect ratio was less than two) and estimated as an average of 12 field views (total number of grains examined was greater than 1000). The confidence level was greater than 95%.

The second problem was concerned with the need to establish the identity of strain rate and strain for different schemes of deformation. It is well-known that following large plastic deformation the strain may be calculated in different ways. Deformation paths/trajectories of different types that have been realised by two-component tension/torsion testing are shown in Fig. 6.1. They may be characterised as follows. Paths a and b represent simple, single component, monotonic processes. The difference between them lies in the stress state. Path c represents a simple, two component, process. Path d represents a single component, non-monotonic process. Finally, path e represents a complex, two component, non-monotonic process, having 90° breaks on the deformation path.

The material used was a (Russian) titanium alloy VT9 (Ti-6.6%Al-0.5%Mo-1.7% Zr-0.27%Si). The alloy was deformed in the (α+β) region. The samples were subjected to the following heat treatment: 30 minutes annealing at 1030°C (β region) followed by air cooling; 1 hour annealing at 850°C and then water quenching. (Annealing at 1030°C followed by air-cooling led to the formation of relatively small matrix β-grains and rather thin lamellae. Additional annealing at 850°C and water quenching gave rise to a non-equilibrium phase composition.) Subsequent deformation at 960°C promoted the transformation of the lamellae [604]. Based on tensile testing at 960°C, it was established that maximum elongation could be obtained if the strain rate was around $10^{-3}s^{-1}$.

It is emphasised that all the deformation paths represented in Fig. 6.1a–e is with respect to a 2D space. Therefore, it is of interest to study more complex deformation paths, which are characterised by three-, four- or five-dimensional curves in the Iljushin deformation space (see Sect. 2.4.3). Unfortunately, it is impossible to

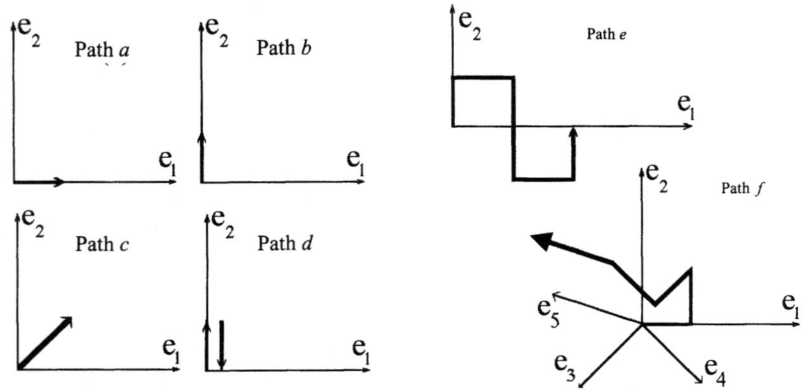

Fig. 6.1. Deformation paths in the deformation space $\text{э}_1\text{э}_2$, where э_1 corresponds to the tensile component while э_2 corresponds to the torsion component [13]: Path *a* – tension, $v = 0.08$ mms^{-1}; Path *b* – torsion, $\omega = 0.016$ s^{-1}; Path *c* – tension with simultaneous torsion, $v = 0.046$ mms^{-1}; $\omega = 0.014$ s^{-1}; Path *d* –reverse torsion $\omega = 0.017$ s^{-1}; Path *e* – tension alternating with torsion $v = 0.08$ mms^{-1}; $\omega = 0.017$ s^{-1}; Path *f* – the special process of multi-stage forging

achieve such loading using standard tension/torsion scheme of deformation. It is easy to show that usual multiaxial forging of a cube is also characterised by a flat deformation path. Therefore, Iljushin proposed a deformation path in which the projections on each of the 5 axes in the deformation space are approximately of equal value. Dmitriev [605] has calculated such a deformation path in detail.

In this scheme, a cube (50×50×50 mm) with a lamellar microstructure was upset 18 times along its three edges in a pre-determined sequence. The forging reduction was 20% for each of the 18 upsettings. After this, the workpiece was quenched in water. Then, a new cube was cut from the deformed and quenched material maintaining pre-calculated angles [605]. This procedure was repeated so that the total number of upsetting was 54. Load was applied using a hydraulic press. The temperature was 960 ± 20°C, and the strain rate about 10^{-3} s^{-1}. The microstructures were studied at the corners of all the three cubes and at the centre of the last cube.

The results are presented in Fig. 6.2. It is seen that as the accumulated strain increases, the proportion of equiaxed particles increases rapidly to start with but the rate falls subsequently. The gradually increasing part of the *n*–*s* curve was absent when necking led to failure at $s > 0.8$. Under tension with simultaneous torsion, a high degree of refinement of structure with $n = 70\%$ could be obtained at $s = 0.8$ itself. In broken deformation paths (reverse torsion and tension alternating with torsion), in contrast, *n* did not exceed 40% even when *s* reached 1.6. Five-dimensional loading path which had considerably more breaks for the same accumulated strain was still less efficient: $n \cong 20\%$ at $s = 1.6$. Even when *s* reached a value of 10, *n* did not exceed 50%. Thus, it is seen that the kind of loading determines the magnitude of *n*(*s*). This fact itself confirms the necessity to include the characteristics of complex loading among the parameters of thermomechanical

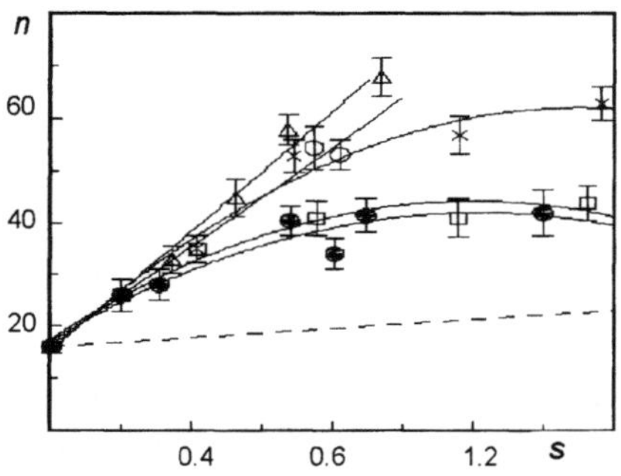

Fig. 6.2. The fraction of equiaxed particles, *n*%, formed as a function of accumulated strain *s* in titanium alloy VT9 of an initial lamellar structure. T = 960°C and $\xi \approx 10^{-3}\, s^{-1}$ ooo–tension; xxx–torsion; ΔΔΔ–tension with simultaneous torsion; □□□–reverse torsion; ••• -tension alternating with torsion. Dashed line corresponds to multi-stage forging (Path 6–*f* in Fig.6.1)

treatment. However, the set of experimental data in [170] is poor. Therefore, one has to consider with caution the following statement contained in [170]: 'monotonic loading is more efficient than non-monotonic loading, so far as grain refinement is concerned'. This statement can not be considered to be general, because the specific ways of complex loading should be indicated while making such a statement. For example, the result obtained in [171] contradicts this statement (where cyclic torsion under fixed axial strain was used). Thus, the following conclusion seems to be more acceptable: there are many complex deformation trajectories and so the problem of finding the optimal (with respect to the structure aimed at) deformation trajectory should be investigated further.

Additional experiments on the same material under similar experimental condition [171] are pictorially depicted in Fig. 6.3. The microstructures were examined in a section perpendicular to the specimen axis. The following parameters were determined: mean size of α-phase particles ($\overline{L}$), the grain size distribution function, a mean value of the ratio of the maximum to the minimum grain size ($\overline{K}$) and volume fraction of equiaxed α-particles (*n* %). No less than 1000 grains/particles were examined. The results are presented in Table 6.1.

Two initial structural states were examined: lamellar and globular structures. It was found that the globular microstructure did not change significantly after deformation by paths *1–4* and only a weak tendency for grain growth was seen. In contrast, the lamellar microstructure changed after tests along all the loading paths; α-phase on β-boundaries disappeared, plates of α-phase inside β-grains changed shape (became thicker and curved) eventually approaching a globular shape.

X-ray studies showed that in the samples deformed by paths *1* and *4* the starting texture of both types of (lamellar and globular) microstructures was replaced by a complex multi-component texture with different distributions of intensity maxima. The distribution depended on the path of loading.

Thus, the results obtained in [170, 171] lead to the conclusion that the loading path, in addition to the accumulated strain (Odqvist parameter), influences the kinetics of structural changes significantly. These are first efforts in which specialists in mechanics and materials science have worked together to derive conclusions

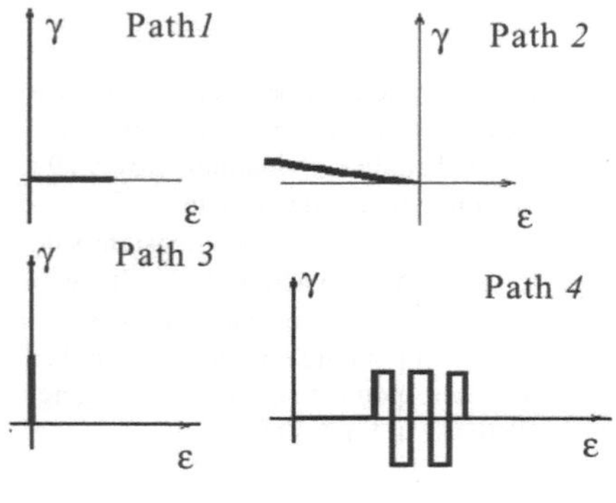

Fig. 6.3. Experimental programme of [171]: Path *1* – tension, Path *2* – compression + shear, Path *3* – torsion, Path *4* – tension + reversible torsion

of practical relevance. Evidently, such works are very important and should be pursued in the future.

6.2 Constitutive Equations including Structural Parameters

Investigations of the mechanical properties of materials which strongly depend on the instantaneous microstructure are important. To evolve adequate constitutive equations in these cases the initial microstructure as well as its evolution with deformation should be considered.

In the literature on superplasticity, this problem is often solved by constructing physical models based on micro-mechanisms of deformation. Such models include both micro- and macro-parameters (micro-parameters are structural characteristics while macro-parameters characterise the macroscopic stress–-strain state). It is assumed that the micro-parameters can be found from investigations at atomic/microscopic level, e.g., Burgers vector, diffusion coefficient, etc. In principle, all physical models are assumed to be amenable to a description of the macro-

Table. 6.1 Microstructural parameters and deformation Path [171]

Deformation path	$\overline{L}$, (μm)		$\overline{K}$		n, (%)		ε_e
	Lamellar	Globular	Lamellar	Globular	Lamelar	Globular	
Initial	21	4	6.3	1.9	16	70	0
1	10	4.5	4.1	1.9	41	72	1.6
2	12	4.5	5.1	2.0	31	69	0.8
3	14	4.5	5.3	1.8	30	70	0.6
4	7	4.5	3.1	2.0	48	75	1.1

scopic state, that is, the values of σ, ε, ξ in physical equations are regarded as macroscopic parameters. In contrast, the internal variable approach is widely used in mechanics of solids (see Chaps 2,3). Such an approach is extremely useful for constructing constitutive equations taking into account structural changes present during deformation. There are two different approaches so far as the procedure of constructing such models is concerned. One of them is traditional: suggesting a specific mechanical model, along with one or more internal variables, to account for a given set of experimental data. Here, no effort is made to attribute physical meaning to any of the variables and the choice of the model will be based on experience. (Thus, constitutive equations containing internal variables can be considered as a functional of strain state, i.e., the value of the stress depends not only on the current strain state but also on the history of variation of strain state.) Another way is to introduce internal variables with physical meaning, e.g., average grain size, dislocation density, etc. Although the second method appears to be more complicated, it allows the use of the vast knowledge of physicists/materials scientists to obtain a comprehensive description of the mechanical properties of materials. As it was already mentioned in Chap.2, the scheme of introducing internal variables was suggested by Rabotnov [272]. However, as yet there is no effective algorithm in practice. The following scheme to construct a constitutive equation having a physical basis can be suggested.

Let a physical model be considered as a conventional CE assuming (as usual in mechanics) that *all* material constants are determined from *macro-experiments*. Thus, all material constants including the physical ones are to be considered as fitting coefficients, the physical sense of the (physical) constants being ignored for the time being. The following axiom is to be adopted while doing this. If the physical model captures the main features of the phenomenon (here superplastic flow), it will contain only those parameters, which directly influence macro-behaviour; all other parameters are to be excluded from the constitutive equation. (More precisely, these parameters will be taken into account indirectly in material functions of the CE.) Then at least in principle, it should be possible to determine all the parameters of a constitutive equation from *macro-experiments*. This will also enable the materials scientists to estimate the magnitudes of the micro-parameters (or some combinations of them) from macro-experiments and thus provide convincing evidence for the soundness of the physical model by comparing these values with estimates obtained from micro-experiments or physical theory.

Some examples of such an approach are presently considered. Some models of superplasticity (e.g., the one due to Padmanabhan and Schlipf) do not include implicitly the structural characteristics as functions or functionals of the process of deformation. The physical basis of such models is reflected by micro-constants. The model of Padmanabhan and Schlipf for optimal superplastic flow was considered in Chap.3. This model is the result of a detailed physical description. Yet the method used to determine the material constants *a, b* and *c* is based on macro-measurements only. In Sect. 3.2, a procedure for estimating the physical constants of this model was presented. The results have shown that the values are close to what are obtained from micro-experiments/theoretical considerations.

A wide range of physical models known in the literature can be represented as

$$\xi = f(\sigma, T, d) \tag{6.1}$$

$$\dot{d} = \varphi(d, \sigma, T), d\big|_{t=0} = d_0 \tag{6.2}$$

(See also Sect. 2.4.4.) One has to take the following steps to 'transform' a physical model into a CE:

1. All micro-physical constants (or their combinations) should be considered as material constants.
2. It is necessary to analyse formally the equations obtained (system of equations should be closed, the estimations of the orders of different variables should be made, the formal mathematical analysis of the system of equations is to be carried out, etc.).
3. The analysis of the system of equations is to be carried out proceeding from the common requirements of the theory of constitutive equations (see Chaps. 2, 3, 4). The methods of determining the material constants are to be developed.

One can use empirical data accumulated by materials scientists in order to construct a CE. Materials scientists have compiled vast empirical knowledge on mechanical properties in the form of reference books and databases. In particular, significant studies have been undertaken with respect to structure–mechanical property correlations in polycrystalline materials. These empirical correlations include the effects of ξ, σ and T (see, e.g., [210, 606–608]), as well as structural characteristics. The Zener–Hollomon parameter [129] is often used to describe the temperature–strain rate combination in deformation, i.e.,

$$Z = \xi \cdot \exp\left(\frac{Q}{RT}\right) \tag{6.3}$$

where R is the gas constant and Q is the activation energy. Many experimental studies have revealed that during high temperature plastic deformation the flow stress reaches a value σ_{ss} that depends on the recrystallised grain size d_R and Z. The following universal relationships are often used [210, 606, 607]:

$$\sigma_{ss} = C_1 \cdot (d_R)^{-k} \tag{6.4}$$

$$Z = C_2 \cdot (d_R)^{-n} \tag{6.5}$$

where C_1, C_2, k, n are material constants, $3 < n < 9$ and $0.7 < k < 1.1$. Such empirical results help identify a common pattern for a physical theory of plastic deformation of polycrystalline materials.

Since in mechanical tests different combinations of d and Z can be employed, it is convenient to use the co-ordinate plane d-Z suggested by Sakai and Jonas [591, 609] (Fig. 6.4). At small d and Z, dynamic grain growth is present, but at large d and Z grain refinement is normally observed. In the latter case, strain softening, which contradicts Drucker's postulate [219] is seen. The dynamically stable grain size d_s is nearly independent of the initial grain size d_0 and is determined by the value of Z

$$d_s = f(Z) \tag{6.6}$$

Higher the Z value, finer will be the equilibrium grain size. Equation (6.6) represents the boundary between the two domains shown in Fig. 6.4. Thus, structural changes during hot deformation of polycrystalline materials are of great importance and so they should be incorporated into constitutive equations.

An example of constructing a CE for superplastic flow based on the principles of plastic deformation has been considered in [136, 137, 610]. In [137] the optimal conditions for superplastic flow are determined and in [610] the constitutive equations are derived. It is assumed that the presence of defects in polycrystalline materials introduces additional energy $L = \sum L_{ie}$, where L_{ie} is the average energy per unit volume due to the presence of defects of the i-th type. The rate of change in L is assumed to be determined by a balance between the rate of generation of defects $\dot{L}^+$ and the rate of their annihilation, $\dot{L}^-$, i.e., $\dot{L} = \dot{L}^+ - \dot{L}^-$. Only two structural features are considered in [610]: dislocations and grain boundaries. The vector properties are described by Saint–Venant type equations:

$$\sigma_{ij} = A\xi_{ij} \tag{6.7}$$

and scalar properties by

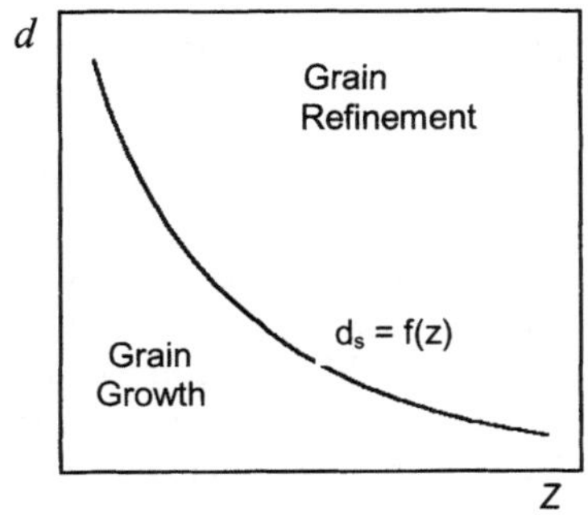

Fig. 6.4. General pattern of structure-mechanical behaviour relationship in polycrystalline materials

$$\sigma_e = \hat{C}_1 \sqrt{L_\rho} + \hat{C}_2 L_D \tag{6.8}$$

Kinetic equations for structural characteristics are given by

$$\begin{aligned} \dot{L}_\rho &= C_1 \xi \sqrt{L_\rho} - C_4 \sqrt{L_\rho} L_D^{n-1} \exp\left(-\frac{q}{RT}\right) \\ \dot{L}_D &= C_2 \xi L_D - C_4 \sqrt{L_\rho} L_D^{n-1} \exp\left(-\frac{q}{RT}\right) - \\ &\quad C_3 L_D^n \exp\left(-\frac{q}{RT}\right) - C_5 \sqrt{L_\rho} L_D^{n-1} \exp\left(-\frac{q}{RT}\right) \end{aligned} \tag{6.9}$$

Initial conditions are described as

$$L_D(0) = L_{D_0}, \quad L_\rho(0) = L_{\rho_0} \tag{6.10}$$

where $C_1 = (3\alpha/2)\hat{C}_1$, $C_2 = (3\alpha/2)\hat{C}_2$, C_3, C_4, C_5, α, n and q are material constants.

It is shown in [31] that in the steady state region when $\dot{L}_\rho = 0$ and $\dot{L}_D = 0$ one can derive Eqs. (6.4) and (6.5) from Eqs. (6.9). System of equations (6.8)–(6.10) are solved numerically using standard Runge–Kutta method. Comparison of the results of the theoretical analysis with the experimental data reveals that the hypothesis $\alpha =$ constant is valid. It has been concluded that the model reflects qualitatively the main features of the structure–mechanical behaviour relationship in polycrystalline materials during near steady state loading. Moreover, kinetic equations (6.9) allow some useful predictions concerning hot deformation and annealing, e.g., when $\xi = 0$. From Eq. (6.9), the following well-known grain growth relationship $d^p - d_0^p = Ct$, where p and C are constants, is obtained. Thus, this approach is another example of a useful collaboration between specialists in mechanics and materials science.

6.3 The Concept of Database 'TMT–Structure–Properties'

The interdependence between thermomechanical treatments and structural changes is a fundamental problem in materials science. A large volume of information on this aspect is available concerning metals and alloys, in particular steels and cast irons. Considerable amount of work has also been done to correlate the structure of polycrystalline materials with their mechanical properties. Many recommendations with regard to regimes of thermal/thermomechanical treatments have been made. Much of this information is yet to be included in computer databases. Materials scientists have acquired considerable skills in alloy design, i.e., control of mechanical properties through a manipulation of microstructure/alloy chemistry. Much of this knowledge, however, is empirical and yet to be quantified. The constitutive equations developed are far from complete. Creation of a scientific database to include information on thermomechanical treatments, resulting microstructures, mechanical properties and constitutive equations is an urgent requirement.

In the literature, heat treatments are well described but the descriptions of mechanical treatments are somewhat sketchy. But, as shown in Sect. 6.1 the deformation path/history is also very important. Therefore, the experience gained in mechanics of solids should be effectively used. For example, in the theory of elastoplastic processes due to Iljushin, mechanical treatments are classified based on the curvature of the deformation trajectory. Incorporation of such ideas with the efforts of materials scientists can lead to a deeper understanding of the nature of superplastic flow and the development of new technologies for structure preparation and metal working.

It is of great practical importance to model the plasticity and the strength resources of a material or a component. This involves two stages: (i) selection of a microstructure, assigning it an appropriate level of property resource; and (ii) development of a technological process to obtain an article/component having such a microstructure. This is based on one of the cardinal principles of materials science, viz., structure determines properties. For example, in two phase titanium alloys this principle has been established with numerous experimental data. To obtain a component of specified properties, the following stages are necessary: (a) identification of the mechanical attributes that determine the plasticity resource, e.g., yield strength, fatigue limit etc.; (b) establishing a relationship between the initial microstructure, its behaviour during processing and the final structure/properties in the formed component; (c) finding out the effects of mechanical history/deformation trajectories typical of metal working processes on the structure of the material.

Items (i) and (ii) mentioned above will now be considered. Step (i) includes the creation of a database 'initial microstructure–plasticity resource of material'. Using this database, along with the recommendations of technologists and materials scientists, one can solve the following inverse problem: Selection of the microstructure of a component, which will provide a predetermined level of resource. Mostly, only uniaxial experiments are used to construct such databases. Such databases can provide only standardised characteristics of the resource (fatigue properties, long-time strength, etc.). In practice, the parts and components are subjected to complex (non-uniaxial) loading conditions. Therefore, it is necessary to add to such databases the corresponding information concerning the magnitude of the resource under non-uniaxial loading. In practice, such information is sometimes obtained from the experiments on the parts and components themselves under loading conditions that are close to service conditions. (It may also be obtained from field data.) In such cases, the problem is to deduce the stress–strain state and so this information can not be used in the databases. Therefore, one has to carry out experiments under controlled programmes with respect to strain or stress (i.e., complex loading of specimens in accordance with programmes that are close to service conditions).

Step (ii) involves the technology to produce a part leading to a predetermined microstructure. In practice, this problem is usually solved by technologists on the basis of uniaxial experiments, e.g., tension/compression. However, a real technological process is characterised by multiaxial loading conditions. Thus, uniaxial experiments are not enough to solve problem (ii) and so one has to create a database 'thermomechanical treatment – microstructure' for complex loading conditions as well (see Sect. 6.1). As in the above case, this database will be used to solve the following inverse problem: selecting a regime of thermomechanical treatment to obtain a predetermined microstructure.

Now, the problem is considered as a whole: one has to obtain an article (part, component) having a predetermined property resource. As was outlined above, a solution to this problem can be obtained by using two databases: 'initial microstructure–property resource of material' and 'thermomechanical treatment–microstructure'. In practice, one is interested in the relationship between the thermomechanical treatment and the property resource. Therefore, one needs to have only the database 'thermomechanical treatment–resource'. Evidently, the microstructure can then be considered as an 'intermediate' (internal) parameter. This situation is similar to that which arises while constructing constitutive equations that include internal variables (Chap. 2).

The boundary value problems that arise in connection with the above mentioned situations (i) and (ii) are now considered. Formulation of the boundary value problem for a technological process that gives rise to a specified microstructure: Using the database microstructure of material – property resource one chooses a microstructure distribution throughout the part that would result in the specified property resource in the product (component). Then the database thermomechanical treatment – microstructure of material is used to decide the thermomechanical process (in particular, the deformation trajectories) that will give rise to an initial microstructure of the required kind. Towards this end, the new class of optimisation boundary value problems should be studied in such a way that one can select the technological parameters of a process correctly to obtain at any point in a workpiece the required temperature and deformation trajectories.

The main problem can be stated in the following way: one has constitutive equations, equations of motion, boundary conditions and initial conditions. One needs to find the laws that determine the technological parameters (e.g., loads, feeds, etc.,) that will ensure that the deformation trajectory and geometrical characteristics are very close to those needed. Once the optimization parameters (rate of tool movement, load etc.,) are specified, the problem can be mathematically stated as follows: equalities (from the equation system of the problem) and inequalities (from the boundary and initial conditions of the problem). One needs to minimise some functional, which characterises the deviation throughout the material of the deformation trajectory from the requisite one.

This problem appears to be very complex and difficult. But in practice, the situation is somewhat clearer, because each microstructure type can be obtained not by just one thermomechanical process, but by a whole class of thermomechanical processes. Thus the difficulty of obtaining at each point in a workpiece the requisite deformation path is not formidable. One only needs this path to belong to a class of trajectories that leads to the structure required.

6.4 Challenges in Mechanics of Superplasticity

In this section the main problems to be solved in mechanics of superplasticity are stated in greater detail than was done at the beginning of this chapter.

6.4.1 Experimental Superplasticity

Most experiments to date dealing with the rheological behaviour of superplastics have employed uniaxial testing at constant temperature, average grain size and strain rate. Flow stress, strain rate sensitivity index and elongation to fracture are usually reported. But, such studies are of limited significance for technological processes that involve complex loading schemes. Effects of two or three dimensional stress/strain states, non-proportional and/or non-monotonic loading have seldom been examined. This lacuna should be made up. Further, knowledge of behaviour of superplastics under transient regimes of loading is important from both practical and theoretical points of view. Even in uniaxial testing, estimation of non-uniformity in deformation and direct measurement of strain using an extensometer are necessary. With regard to non-uniaxial testing, it should be noted that thin-walled tube specimens, often used with other classes of materials, are not suitable for studying superplastic flow (due to the problem of buckling/unstable collapse). Consequently, one may have to design some new type of specimen, testing equipment and methods of analysis. It should also be noted that a similar problem arises in analysing even the results of uniaxial testing if a prominent neck develops during deformation.

The following experimental programs, along with adequate methods of analysis, could be of interest.

1. Systematic experiments with proportional deformation to establish the existence of 'unique curves' for superplastic materials. (A 'unique curve' is a universal relationship among the invariants of stress, strain and strain rate tensors for all proportional loading conditions.) In fact, in case of finite strains, choosing appropriate invariants is not easy because the solution is not unique. In fact, for superplastic materials one should first of all establish a domain of validity for the relationship $\sigma_e = f(\xi_e)$.

2. Experiments in which the deformation trajectories of industrial significance are simulated to establish the validity of the hypothesis concerning the proportionality of stress and strain rate tensors would be highly relevant. The domain of applicability of this hypothesis should be determined with sufficient reliability.
3. Experimental construction of the 'superplasticity surface': This surface will be constructed in a special space of the parameters of the deformation process, e.g., invariants of stress, strain, strain rate tensors, temperature, structural characteristics. This surface will be confined to the region of superplastic behaviour.
4. Experimental determination of the material constants for different constitutive laws and the development of databases containing information on the mechanical properties of superplastics.
5. Experimental determination of the field of applicability of different constitutive equations.
6. Database for damage accumulation and plasticity resource in superplastic materials corresponding to different strain and temperature histories that are typical of metalworking.
7. Development of the database 'history of thermomechanical treatment – evolution of microstructure' (includes strain histories typical of metal working processes).
8. Development of efficient methods for preparing microstructures most conducive to superplastic forming using the mechanical theory of inelastic processes and the rules of thumb prescribed by materials scientists. As one of these methods is severe plastic deformation, use of special complex loading in preliminary deformation can make the technique more effective.
9. Investigation of the role of hydrostatic pressure in superplastic forming including its influence on damage accumulation.
10. Approximate methods to determine the conditions at the contact surface between an equipment and the workpiece: These conditions can not be found directly and so one has to determine them indirectly (a non-linear inverse mathematical problem that often defies unambiguous solution). Reasonable and substantiated restrictions to facilitate a solution should be derived from technological experience.
11. Accumulation of experimental data concerning the transient regimes of superplastic flow.
12. Development of methods of physical (mechanical) modelling of technological processes of superplastic metalworking. In particular, development of methods of modelling the rheological behaviour of superplastics using model materials (see Appendix C).
13. Use of results of group analysis of boundary value problems for understanding experimental data (see Appendix D).
14. Investigation of mechanical anisotropy in the as received condition of superplastics, as well as after superplastic deformation (see, e.g., [611]).

6.4.2 Constitutive Equations

It is evident from many passages in this book that from the points of view of both materials science and mechanics of solids superplasticity may be regarded as a domain of creep. The important feature of this type of creep is that deformation is large or very large and, in general, the mechanisms of deformation are quite different from what is seen during conventional creep. Appropriate macro-constitutive equations can be taken from the theory of creep and modified to suit finite strains. There is also a theory of damage accumulation in the theory of creep. However, it is not yet clear if it is possible to use these approaches directly to analyse superplastic flow. Kinetics of damage accumulation that depends on the history of straining should be investigated for a superplastic material. It is already known that this is important in conventional (non-superplastic) materials.

An analysis of the constitutive equations available in the literature on superplasticity (Chaps.2 and 3) points to the following main challenges.

1. One has to develop the theory of finite strains to describe non-uniaxial superplastic flow. In case of finite strains, it is not easy to choose the most appropriate measure to describe a deformed state, along with the corresponding measure of the stress state (problem of non-uniqueness). The application of the commonly used logarithmic measure meets with major difficulties if the main axes of the metric tensor rotate during deformation.
2. It is necessary to find out experimentally the applicability of the standard hypothesis concerning the direct proportionality between the stress and strain rate tensors for the set of deformation trajectories corresponding to technological processes, e.g. special experiments that simulate industrially relevant deformation trajectories. The field of adequacy of this hypothesis should be determined with sufficient accuracy. If this hypothesis turns out to be invalid the constitutive law in the tensor form should be written as the so-called three-terms equation representing a linear relationship among the stress, strain and strain rate tensors. Then, the vector properties are given in a way similar to what obtains in the theory of elasto-plastic processes.
3. One has to state clearly the concept of a 'unique curve' for superplastic flow (the universal relationship among the invariants of stress, strain and strain rate tensors under certain loading conditions). In this regard, the measures to be used and the invariants to be taken into consideration are yet to be decided.
4. It is of great importance to determine experimentally *all* the material constants of a constitutive law. Theoretical methods to estimate them from experimental results should be developed and tested. In particular, it seems reasonable to determine the characteristics of superplastic flow from technological experiments (e.g., see [57]). Theoretical aspects of such investigations represent a non-linear inverse problem, which is yet to be solved.

This problem gets very complicated if the constitutive model contains many material constants.

5. It is of interest to develop a theory and new experimental methods to investigate constitutive equations under multiaxial loading.
6. The physical constitutive relations are to be used to construct the phenomenological constitutive laws that describe the scalar properties. It is emphasised that the meaning of the specific material constants and parameters entering the physical relationships (e.g., average grain size, grain boundary diffusion coefficient) is of no consequence in the constitutive relations. These material constants should be treated in a formal way, considering them to be some internal variables. That is, these are to be regarded as fitting coefficients of the constitutive equations.
7. Fields of applicability of different constitutive equations should be defined, including those of the constitutive equations already well-known.
8. A criterion to distinguish between superplastic and conventional behaviour under multiaxial loading should be developed. The boundaries of superplastic flow are to be given in terms of tensor invariants (stress tensor, strain rate tensor and temperature are to be considered). For example, it is known that the limits of superplasticity in a uniaxial tensile test will change when the mode of testing is changed.
9. The constitutive equations intended for use in practical applications should be applicable not only for superplastic flow but also for the neighbouring regions of high temperature creep and short time plasticity. It is obvious that these constitutive equations may have different mathematical forms in different regions. In that case, they should be related at the boundaries, as defined in Sect. 1.3.
10. It is desirable to use the constitutive equations of non-linear viscoelasticity along with the above mentioned equations of creep. In particular, one can use the following idea, which is used in the theory of complex mixtures: the contributions of different mechanisms to the net stress are summed (rather than the contributions of different mechanisms to the net strain rate) – see, e.g., [612].
11. One has to develop the internal variable approach for constructing the constitutive equations for superplastics [613].
12. Only those constitutive equations should be selected for practical calculations that employ the minimum number of material constants, see. e.g., [105, 389, 394].
13. It is necessary to apply the thermodynamical approach while constructing and analysing the constitutive equations as also while analysing the experimental data.
14. The boundary value problem should be clearly formulated. Methods of numerically solving them using only those constitutive equations that are already substantiated experimentally should be developed.

Appendix A: Finite Strain Kinematics of Solids

A.1 Basic Concepts

A movement is always defined with respect to some system of co-ordinates. Every point in 3D space is identified by three numbers x^1, x^2, x^3 (its co-ordinates). Lines along which two co-ordinates are constant in value are termed co-ordinate lines; if these lines are straight, the system of co-ordinates is termed rectangular.

Motion of a given point (particle) in a continuum with respect to the system of co-ordinates x^1, x^2, x^3 can be described by three time dependencies $x^i \dot{L} = f^i(t)$. Let some particle O in the continuum be characterised by co-ordinates at $t = 0$. Then, one can write the law of motion for this point as

$$x^i = x^i\left(\zeta_1, \zeta_2, \zeta_3, t\right), \ i = 1,2,3 \tag{A.1}$$

Functions (A.1) are referred to as the laws of motion of a point in a continuum. Let $\overline{D}_0$ be the region occupied by a body at $t = 0$, $(\zeta) \in \overline{D}_0$. Then, $x^i = x^i(\zeta^i, t)$ define the region $\overline{D}$, occupied by the body at $t > 0$: $\overline{D} = \{x^i = x^i(\zeta^i, t) \mid (\zeta^i) \in \overline{D}_0\}$. Co-ordinates $\zeta^1, \zeta^2, \zeta^3$ which individualise the particles in a continuum are referred to as Lagrangian or material co-ordinates while the spatial co-ordinates x^1, x^2, x^3 are termed the Eulerian or spatial co-ordinates (see also Chap. 2). The system of co-ordinates $\zeta^1, \zeta^2, \zeta^3$ is referred to as the material or convective system of co-ordinates. This system is 'frozen' into a continuum and deforms with it.

Let the co-ordinate line x^i be drawn through some point O. Let F be a neighbouring point on this line. It is evident that the vector $\overrightarrow{OF} = \vec{r}_F - \vec{r}_O = \Delta\vec{r}$. The basis vectors $\vec{e}_i$ are defined as

$$\vec{e}_i = \lim_{\Delta x^i \to 0} \frac{\Delta\vec{r}}{\Delta x^i} = \frac{\partial\vec{r}}{\partial x^i}, i = 1,2,3 \tag{A.2}$$

Similarly, the basis vectors of a convective co-ordinate system are introduced as

$$\hat{\vec{e}}_i = \frac{\partial \vec{r}}{\partial \zeta^i}, \quad i = 1,2,3 \tag{A.3}$$

If the co-ordinate system x^1, x^2, x^3 is Cartesian, then $\vec{e}_1 = \vec{i}$, $\vec{e}_2 = \vec{j}$, $\vec{e}_3 = \vec{k}$ where $\vec{i}$, $\vec{j}$, $\vec{k}$ are mutually orthogonal unit vectors. A co-ordinate system is introduced by the observer. Therefore the laws of motion should be independent of the choice of the co-ordinate system. This circumstance leads to some restrictions on the mathematical functional forms used in (A.1).

It will be assumed further that functions $x^i(\zeta^1, \zeta^2, \zeta^3, t)$ are sufficiently smooth; the relation between vectors $\vec{x}$ (x^1, x^2, x^3) and $\zeta(\zeta^1, \zeta^2, \zeta^3)$ is unambiguous. Considering the increments in $\vec{x}$ and $\vec{\zeta}$ one can write

$$\begin{aligned}
dx^1 &= \left(\frac{\partial x^1}{\partial \zeta^1}\right) d\zeta^1 + \left(\frac{\partial x^1}{\partial \zeta^2}\right) d\zeta^2 + \left(\frac{\partial x^1}{\partial \zeta^3}\right) d\zeta^3 \\
dx^2 &= \left(\frac{\partial x^2}{\partial \zeta^1}\right) d\zeta^1 + \left(\frac{\partial x^2}{\partial \zeta^2}\right) d\zeta^2 + \left(\frac{\partial x^2}{\partial \zeta^3}\right) d\zeta^3 \\
dx^3 &= \left(\frac{\partial x^3}{\partial \zeta^1}\right) d\zeta^1 + \left(\frac{\partial x^3}{\partial \zeta^2}\right) d\zeta^2 + \left(\frac{\partial x^3}{\partial \zeta^3}\right) d\zeta^3 \\
or \quad dx^i &= \left(\frac{\partial x^i}{\partial \zeta^j}\right) d\zeta^j \quad ; i, j = 1,2,3
\end{aligned} \tag{A.4}$$

Similarly one can write

$$d\zeta^i = \left(\frac{\partial \zeta^i}{\partial x^j}\right) dx^j \tag{A.5}$$

Decomposing the increments using Eqs. (A.2) and (A.3) one obtains

$$d\vec{x} = dx^1\vec{e}_1 + dx^2\vec{e}_2 + dx^3\vec{e}_3 = dx^i\vec{e}_i \tag{A.6a}$$

$$d\vec{\zeta} = d\zeta^1\hat{\vec{e}}_1 + d\zeta^2\hat{\vec{e}}_2 + d\zeta^3\hat{\vec{e}}_3 = d\zeta^i\hat{\vec{e}}_i \tag{A.6b}$$

At this stage the concept of space metric which is necessary to understand the concept of length in this space is introduced. Let

$$\hat{g}_{ij} = \vec{e}_i \cdot \vec{e}_j = \frac{\partial \vec{r}}{\partial x^i} \cdot \frac{\partial \vec{r}}{\partial x^j} = \frac{\partial \zeta^k}{\partial x^i} \cdot \frac{\partial \zeta^k}{\partial x^j} \tag{A.7a}$$

$$\hat{g}_{ij} = \hat{\vec{e}}_i \cdot \hat{\vec{e}}_j = \frac{\partial \vec{r}}{\partial \zeta^i} \cdot \frac{\partial \vec{r}}{\partial \zeta^j} = \frac{\partial x^k}{\partial \zeta^i} \cdot \frac{\partial x^k}{\partial \zeta^j} \tag{A.7b}$$

Then, the square of the length of some vector is defined as

$$|d\vec{x}|^2 = d\vec{x} \cdot d\vec{x} = dx^i dx^j \vec{e}_i \vec{e}_j = dx^i dx^j g_{ij} \tag{A.8a}$$

$$|d\zeta|^2 = d\zeta^i \cdot d\zeta = d\zeta^i d\zeta^j \hat{\vec{e}}_i \hat{\vec{e}}_j = d\zeta^i d\zeta\zeta^j \hat{g}_{ij} \tag{A.8b}$$

Keeping in mind that the length is an invariant, one can find from Eqs. (A.5) and (A.8) that

$$|d\vec{x}|^2 = dx^i dx^j g_{ij} = d\zeta^i d\zeta^j \hat{g}_{ij} = \frac{\partial \zeta^i}{\partial x^k} dx^k \frac{\partial \zeta^i}{\partial x^m} dx^m \hat{g}_{ij} \tag{A.9}$$

Consequently, the equations for transformation of g_{ij} are

$$g_{km} = \frac{\partial \zeta^i}{\partial x^k} \frac{\partial \zeta^j}{\partial x^m} \hat{g}_{ij} \tag{A.10}$$

From (A.10) it follows that the values g_{ij} are to be considered as the covariant components of some tensor $\tilde{g} = g_{ij} \vec{e}^i \vec{e}^j$. This tensor is referred to as the fundamental metric tensor. This tensor is symmetric, since by definition, $g_{ij} = g_{ji}$. Let the matrix $\{g^{ij}\}$ be the reciprocal of $\{g_{ij}\}$. The reciprocal basis vectors $\vec{e}^j = g^{ji} \vec{e}_i$ can be found from

$$\vec{e}^1 = \frac{\vec{e}_2 \times \vec{e}_3}{\vec{e}_1 \cdot (\vec{e}_2 \times \vec{e}_3)}; \vec{e}^2 = \frac{\vec{e}_3 \times \vec{e}_1}{\vec{e}_1 \cdot (\vec{e}_2 \times \vec{e}_3)}; \vec{e}^3 = \frac{\vec{e}_1 \times \vec{e}_2}{\vec{e}_1 \cdot (\vec{e}_2 \times \vec{e}_3)} \tag{A.11}$$

It is pertinent to note that the concept of reciprocal basis is closely related to the concept of reciprocal basis of a crystalline lattice, commonly used by materials scientists. In Cartesian system of co-ordinates $\vec{e}^i = \vec{e}_i$. Consequently the reciprocal basis coincides with the primary one. Therefore, the covariant and the contravariant components are identical. The metric tensor is widely used to juggle the indexes; it enables one to obtain the contravariant and the mixed components from known covariant components.

A.2
Theory of Deformations

A.2.1
Strain Tensors

Consider two neighbouring particles in a continuum, O and F. Let $d\vec{r}$ be the vector connecting these points at $t = 0$, its length ds_0 being equal to

$$ds_0^2 = d\vec{r} \cdot d\vec{r} = d\zeta^i d\zeta^j \vec{e}_i \cdot \vec{e}_j = g_{ij} d\zeta^i d\zeta^j \tag{A.12}$$

At some arbitrary moment of time $t > 0$ points O and F transfer to O' and F', the distance ds between them being equal to

$$ds^2 = d\vec{r}' \cdot d\vec{r}' = d\zeta^k d\zeta^m \hat{\vec{e}}_k \cdot \hat{\vec{e}}_m = \hat{g}_{km} d\zeta^k d\zeta^m \tag{A.13}$$

$d\vec{r}'$ is the vector connecting O' and F '. Then, one can write

$$ds^2 - ds_0^2 = 2\varepsilon_{ij} d\zeta^i d\zeta^j, \tag{A.14}$$

where

$$\varepsilon_{ij} = \frac{1}{2}\left(\hat{g}_{ij} - g_{ij}\right) = \frac{1}{2}\left(\frac{\partial x^k}{\partial \zeta^i} \cdot \frac{\partial x^m}{\partial \zeta^j} g_{km} - g_{ij}\right) \tag{A.15}$$

Eq. (A.15) gives the Lagrangian strain tensor (also known as Green strain tensor). Another measure of strain can be obtained if one uses Eulerian co-ordinates. In this case,

$$ds^2 - ds_0^2 = 2\varepsilon_{km} dx^k dx^m, \tag{A.16}$$

where

$$\varepsilon_{km} = \frac{1}{2}\left(g_{km} - \frac{\partial \zeta^i}{\partial x^k} \cdot \frac{\partial \zeta^j}{\partial x^m} g_{ij}\right) \tag{A.17}$$

Eq. (A.17) gives the Eulerian strain tensor (also referred to as Almansi strain tensor). The covariant components of Green and Almansi tensors coincide. Mixed and contravariant components of Green and Almansi tensors can be found using $\hat{g}_{ij}$ and g_{ij}, respectively.

Sometimes Caushy strain is introduced as follows

$$E_{ij} = \frac{\partial \zeta^k}{\partial x^i} \cdot \frac{\partial \zeta^k}{\partial x^j}, \quad L_{ij} = \frac{\partial x^k}{\partial \zeta^i} \cdot \frac{\partial x^k}{\partial \zeta^j} \tag{A.18}$$

where E_{ij} and L_{ij} are the components of the Caushy strain in Eulerian and Lagrangian representations, respectively.

In the literature, an approach based on a consideration of displacements of material particles is used. In this case, the components of strain tensors can be found from the known field of displacements. For a particle in a continuum defined by $\zeta^1, \zeta^2, \zeta^3$, the law of motion is $\vec{x} = \vec{x}(\vec{\zeta}, t)$, where the vector $\vec{\zeta}$ individualises the particle. The displacement vector, by definition, is $\vec{u} = \vec{x} - \vec{\zeta}$ For the case when the metric of the reference configuration (at $t = 0$) is Cartesian one can find from Eq. (A.15) that

$$\varepsilon_{ij} = \frac{1}{2}\left(\frac{\partial x^k}{\partial \zeta^i} \cdot \frac{\partial x^k}{\partial \zeta^j} - \delta_{ij}\right) = \frac{1}{2}\left(\frac{\partial u^i}{\partial \zeta^j} + \frac{\partial u^j}{\partial \zeta^i} + \frac{\partial u^k}{\partial \zeta^i} \cdot \frac{\partial u^k}{\partial \zeta^j}\right) \tag{A.15a}$$

A.2.2 Geometrical Sense of Strain Tensor Components

Writing

$$\hat{g}_{ij} = \hat{\vec{e}}_i \cdot \hat{\vec{e}}_j = \left|\hat{\vec{e}}_i\right| \cdot \left|\hat{\vec{e}}_j\right| \cdot \cos\hat{\psi}_{ij} \tag{A.19}$$

$$\overset{0}{g}_{ij} = \overset{0}{\vec{e}}_i \cdot \overset{0}{\vec{e}}_j = \left|\overset{0}{\vec{e}}_i\right| \cdot \left|\overset{0}{\vec{e}}_j\right| \cdot \cos\overset{0}{\psi}_{ij} \tag{A.20}$$

one finds that

$$\frac{\left|\hat{\vec{e}}_i\right|}{\left|\overset{o}{\vec{e}}_i\right|} = \frac{\left|\dfrac{\partial \vec{r}}{\partial \zeta^i}\right|}{\left|\dfrac{\partial \vec{r}_0}{\partial \zeta^i}\right|} = \frac{\left|d\vec{r}_i\right|}{\left|d\vec{r}_{0i}\right|} = \frac{ds_i}{ds_{0i}} = 1 + l_i \, (no\, summation) \tag{A.21}$$

where $l_i = (ds_i - ds_{0i})/ds_{0i}$ are coefficients of relative extension along the directions ζ^i. Substituting Eqs. (A.19)–(A.21) in Eq. (A.15) one obtains

$$2\varepsilon_{ij} = \left[(1 + l_i)(1 + l_j)\cos\hat{\psi}_{ij} - \cos\overset{0}{\psi}_{ij}\right]\left|\overset{0}{\vec{e}}_i\right| \cdot \left|\overset{0}{\vec{e}}_j\right| \tag{A.22}$$

From Eq. (A.22) with $i = j$

$$2\varepsilon_{ii} = \left[(1 + l_i)^2 - 1\right]\overset{0}{g}_{ii} \quad (no\, summation) \tag{A.23}$$

As $\overset{0}{g}_{ii} = 1$ $(i = 1,2,3)$,

$$l_i = \sqrt{1 + 2\varepsilon_{ii}} - 1 \quad (no\ summation) \tag{A.24}$$

For infinitesimally small strains, from Eq. (A.24) $l_j \cong \varepsilon_{jj}$. Thus, for small strains the diagonal terms of the strain tensor are the same as the coefficients of relative extension along the axes of the Cartesian system of co-ordinates.

Considering non-diagonal elements $(i \neq j)$ and taking into account that $\overset{0}{\psi}_{ij} = \pi/2$ and setting $\hat{\psi}_{ij} = \pi/2 - \chi_{ij}$ one can obtain from Eqs. (A.15), (A.19) and (A.20) that

$$2\varepsilon_{ij} = \left|\hat{\vec{e}}_i\right| \cdot \left|\hat{\vec{e}}_j\right| \sin \chi_{ij}, \; i \neq j \tag{A.25}$$

or

$$\sin \chi_{ij} = \frac{2\varepsilon_{ij}}{\sqrt{\hat{g}_{ii}\hat{g}_{jj}}} \quad (no\ summation) \tag{A.26}$$

Thus, right angles are distorted during deformation and the covariant components ε_{ij} $(i \neq j)$ characterise the distortions. If the strains are small, then $\chi_{ij} \cong 2\varepsilon_{ij}$.

A.2.3 Method of Determining the Principal Components of a Strain Tensor

One can find below a formal procedure that allows the determination using any second rank tensor, e.g., Green strain tensor, Almansi strain tensor. The principal components of a second rank tensor equal its eigenvalues while the eigenvectors give the directions of its main axes. The eigenvector of any linear operator A in a linear vector space is defined as

$$A\vec{y} = \lambda\vec{y}, \qquad \vec{y} \neq 0 \tag{A.27}$$

where λ is a scalar value referred to as the eigenvalue of A. Rewrite Eq. (A.27) as

$$(A - \lambda E)\vec{y} = 0 \tag{A.27a}$$

where E is a unit matrix. From vector Eq. (A.27a) in a scalar form, e.g., using basis Eq.(A.3), one obtains a system of three linear equations for the three unknown components of $\vec{y} = \{y^j, j = 1,2,3\}$ as

$$\left(A^i_j - \lambda\delta^i_j\right)y^j = 0, \quad i, j = 1,2,3 \tag{A.28}$$

As $\vec{y} \neq 0$, system (A.28) has non-zero solutions only if the determinant $|A-\lambda E| = 0$, or

$$\begin{vmatrix} a_1^1 - \lambda & a_2^1 & a_3^1 \\ a_1^2 & a_2^2 - \lambda & a_3^2 \\ a_1^3 & a_2^3 & a_3^3 \end{vmatrix} = 0 \tag{A.29}$$

One can derive from Eq. (A.29) the following characteristic equation

$$\lambda^3 - I_1\lambda^2 + I_2\lambda - I_3 = 0 \tag{A.30}$$

where

$$\begin{aligned} I_1 &= a_1^1 + a_2^2 + a_3^3; \\ I_2 &= a_1^1 a_2^2 + a_2^2 a_3^3 + a_3^3 a_1^1 - a_2^1 a_1^2 - a_3^2 a_2^3 - a_1^3 a_3^1 \\ I_3 &= \det a_j^i \end{aligned} \tag{A.31}$$

The solutions λ_1, λ_2, λ_3 of Eq. (A.30) are the principal components a_1, a_2, a_3 for operator A. If the matrix of the operator A is symmetric, then all the three roots λ_1, λ_2, λ_3 are real. The eigenvectors corresponding to the roots λ_1, λ_2, λ_3 can be found by substitution in Eq. (A.28). Sometimes, the following additional condition is assumed in order to obtain a unique set of eigenvectors: $|\vec{y}| = 1$. It can be shown that the eigenvectors corresponding to different eigenvalues are orthogonal, i.e., they are linearly independent. These vectors give the directions of the main axes where the matrix A has a diagonal form

$$A = \begin{bmatrix} a_1 & 0 & 0 \\ 0 & a_2 & 0 \\ 0 & 0 & a_3 \end{bmatrix} \tag{A.32}$$

Characteristic equation (A.30) in principal axes is

$$(a_1 - \lambda)\cdot(a_2 - \lambda)\cdot(a_3 - \lambda) = 0 \tag{A.33}$$

The values of I_1, I_2 and I_3 in Eqs. (A.31) do not change on going from one system of co-ordinates to another, i.e., they are invariants.

$$I_1 = a_1 + a_2 + a_3; \quad I_2 = a_1a_2 + a_2a_3 + a_3a_1; \quad I_3 = a_1a_2a_3 \tag{A.34}$$

Green and Almansi strain tensors introduced above have different eigenvalues for the same deformed state. Actually, writing Eqs. (A.15) and (A.17) for direction $d\vec{r}_i$ along i-th principal axis

$$ds_i^2 - ds_{0i}^2 = 2\hat{\varepsilon}_i ds_i^2 = 2\overset{o}{\varepsilon}_i\, ds_{0i}^2 \tag{A.35}$$

Hence,

$$\overset{o}{\varepsilon}_i = \frac{\hat{\varepsilon}_i}{1-2\hat{\varepsilon}_i}, \quad \hat{\varepsilon}_i = \frac{\overset{o}{\varepsilon}_i}{1+2\overset{o}{\varepsilon}_i}, \quad i=1,2,3 \tag{A.36}$$

The invariants I_1, I_2, I_3 of Almansi tensor are denoted as $\overset{0}{I}_1$, $\overset{0}{I}_2$, $\overset{0}{I}_3$ while those of Green tensor as $\hat{I}_1$, $\hat{I}_2$, $\hat{I}_3$. These invariants are not the same. It can be shown that

$$\begin{aligned}
\hat{I}_1 &= \frac{\overset{0}{I}_1 + 4\overset{0}{I}_2 + 12\overset{0}{I}_3}{1+2\overset{0}{I}_1 + 4\overset{0}{I}_2 + 8\overset{0}{I}_3}; \\
\hat{I}_2 &= \frac{\overset{0}{I}_2 + 6\overset{0}{I}_3}{1+2\overset{0}{I}_1 + 4\overset{0}{I}_2 + 8\overset{0}{I}_3}; \\
\hat{I}_3 &= \frac{\overset{0}{I}_3}{1+2\overset{0}{I}_1 + 4\overset{0}{I}_2 + 8\overset{0}{I}_3}
\end{aligned} \tag{A.37}$$

The coefficient of volume expansion $\theta = (dV - dV_0)/dV_0$ can be expressed through the invariants as

$$\Theta = \sqrt{1-2\overset{0}{I}_1+4\overset{0}{I}_2-8\overset{0}{I}_3} - 1 = \frac{1}{\sqrt{1-2\hat{I}_1+4\hat{I}_2-8\hat{I}_3} - 1} \tag{A.38}$$

A.2.4 Volumetric and Deviatoric Parts of Strain Tensors

Strain tensor or any other tensor may be decomposed as

$$T_\varepsilon = \varepsilon_m U + D_\varepsilon \tag{A.39}$$

where $\varepsilon_m = I_1/3$, U is unit matrix; D_ε is the so-called strain deviator tensor.

$$D_\varepsilon = \begin{bmatrix} \varepsilon_{11}-\varepsilon_m & \varepsilon_{12} & \varepsilon_{13} \\ \varepsilon_{21} & \varepsilon_{22}-\varepsilon_m & \varepsilon_{23} \\ \varepsilon_{31} & \varepsilon_{32} & \varepsilon_{33}-\varepsilon_m \end{bmatrix} \tag{A.40}$$

The first invariant of any deviator is always equal to zero. Often, the most interesting is the second invariant of the strain deviator; one of its forms is

$$I_2(D_\varepsilon) = -\frac{1}{2}\left(\varepsilon_j^i\varepsilon_i^j - 3\varepsilon_m{}^2\right) = -\frac{1}{6}\left[\left(\varepsilon_1^1 - \varepsilon_2^2\right)^2 + \left(\varepsilon_2^2 - \varepsilon_3^3\right)^2 + \left(\varepsilon_3^3 - \varepsilon_2^2\right)^2 + 6\left(\varepsilon_2^1\varepsilon_1^2 + \varepsilon_3^2\varepsilon_2^3 + \varepsilon_1^3\varepsilon_3^1\right)\right] \quad (A.41)$$

The third invariant is equal to the determinant

$$I_3(D_\varepsilon) = \det\left(D_j^i\right) = \det\left[\varepsilon_j^i - \left(\frac{1}{3}\right)\varepsilon_k^k\right] \quad (A.42)$$

A.3 Strain Rate Tensor

A.3.1 Covariant Components of Strain Tensor

The main feature of superplastics is their high strain rate sensitivity. Therefore, the concept of strain rate tensor is of special importance in considering superplastic flow.

Consider the change in the length of some material fibre $d\vec{R}$ in the vicinity of some point O during an infinitesimally small time interval dt. Let O and O' be the positions of point O at moments of time $t = t_0$ and $t = t_0 + dt$ respectively. The basis of the convective system of co-ordinates $\zeta^1, \zeta^2, \zeta^3$ will be $\hat{\vec{e}}_1$, $\hat{\vec{e}}_2$, $\hat{\vec{e}}_3$ at $t = t_0$ and $\hat{\vec{e}}_1'$, $\hat{\vec{e}}_2'$, $\hat{\vec{e}}_3'$ at $t = t_0 + dt$. Repeating the considerations invoked in Sect. A2.1 one obtains

$$\frac{d}{dt}\left(dR'^2 - dR^2\right) = 2\dot{\varepsilon}_{ij}\,d\zeta^i d\zeta^j \quad (A.43)$$

where $\dot{\varepsilon}_{ij}$ are the covariant components of the strain rate tensor. While considering the deformed state different strain tensors are introduced, e.g., Green strain tensor, Almansi strain tensor, etc. (Sect. A2.1). Often two different deformed states are compared (usually the initial and the current deformed states). The difference between the two deformed states may be significant. But, while considering strain rate states two very close deformed states are compared. Consequently, the deformations can be regarded as infinitesimally small. Therefore, only strain rate tensor is conventionally considered. This tensor characterises the current kinematic state in a continuum. The mixed components of the strain rate tensor are determined unambiguously by using the metric tensor $\{\hat{g}_{ij}\}$. Comparing Eqs. (A.14), (A.15) and (A.43), one can infer (assuming the functions $\hat{g}_{ij}$ to be differentiable) that

$$\dot{\varepsilon}_{ij} = \frac{1}{2} \cdot \frac{d\hat{g}_{ij}}{dt} \qquad \text{(A.44)}$$

Equation (A.44) can also be derived by differentiating Eq.(A.15a). Comparing Eqs. (A.15) and (A.44) one can conclude that

$$\varepsilon_{ij} = \int_{t_0}^{t} \dot{\varepsilon}_{ij} dt. \qquad \text{(A.45)}$$

where the integration should be done along the path of a particle (convective integration).

Kinematic meaning of the strain rate components are determined by the sense of the strain tensor components: covariant components of strain rate tensor having repeated indexes $\dot{\varepsilon}_{ii}$ characterise the rates of the relative elongation of fibres lying along the i-th axis, while the components $\dot{\varepsilon}_{ij}$ $(i \neq j)$ characterise the rate of distortion of right angles formed by material fibres.

A.3.2 Distortion and Spin Tensors

It is known that the motion of an absolutely rigid body can be presented as the superposition of translational and rotational movement. When the body can deform during movement, the situation becomes more complicated. Conventionally the rigid movement of a continuum is excluded from consideration. If one uses the above described approach based on a consideration of the convective system of co-ordinates, it is implied that rigid motion is excluded automatically. However, the concept of field velocities is often introduced while considering motion in a continuum.

$$\vec{V} = \frac{\partial \vec{r}}{\partial t} = \frac{\partial x^i}{\partial t}\vec{e}_i = V^i \vec{e}_i \ , \qquad \text{(A.46)}$$

where $\vec{r}$ is the radius vector of a particle, V^i are the components of their velocity in the frame of reference with basis $\vec{e}_i$. The derivative is to be obtained for an individual point in a continuum, i.e., at fixed $\zeta^1, \zeta^2, \zeta^3$.

Let $\Delta\zeta^i$ be the relative co-ordinates of particles in a continuum in the vicinity of some point O. The velocity field is assumed to be continuous and differentiable. Let $\vec{V}$ be the velocity of O and $\vec{V}'$ the velocity of O' close to O. Using Taylor series expansion one obtains that

$$\vec{V}' = \vec{V} + \frac{\partial \vec{V}}{\partial \zeta^i}\Delta\zeta^i = \vec{V} + \frac{\partial\left(V_j \vec{e}^j\right)}{\partial \zeta^i}\Delta\zeta^i = \vec{V} + \left(\nabla_i V_j\right)\Delta\zeta^i \vec{e}^j \tag{A.47}$$

where

$$\nabla_i V_j \equiv \frac{\partial V_j}{\partial \zeta^i} - \Gamma^k_{ji} V_k \tag{A.48}$$

$$\Gamma^k_{ji} = \frac{1}{2}\hat{g}^{ks}\left(\frac{\partial \hat{g}_{js}}{\partial \zeta^i} + \frac{\partial \hat{g}_{is}}{\partial \zeta^j} - \frac{\partial \hat{g}_{ji}}{\partial \zeta^s}\right) \tag{A.49}$$

Here $\nabla_i V_j$ is the so called covariant derivative of the covariant components of the velocity vector while Γ^k_{ji} are Kristoffel symbols. $\nabla_i V_j$ is referred to as the distortion tensor. Evidently, it includes the rigid movement of a continuum as well as its deformation. In order to separate them one can rewrite Eq. (A.47) as

$$\vec{V}' = \vec{V} + \frac{1}{2}\left(\nabla_i V_j + \nabla_j V_i\right)\Delta\zeta^i \vec{e}^j + \frac{1}{2}\left(\nabla_i V_j - \nabla_j V_i\right)\Delta\zeta^i \vec{e}^j \tag{A.50}$$

where the symmetric part gives the strain rate tensor

$$\dot{\varepsilon}_{ij} = \frac{1}{2}\cdot\frac{d\varepsilon_{ij}}{dt} = \frac{1}{2}\cdot\left(\nabla_i V_j + \nabla_j V_i\right) \tag{A.51}$$

The antisymmetric part gives the spin tensor

$$\omega_{ij} = \frac{1}{2}\cdot\left(\nabla_i V_j - \nabla_j V_i\right) \tag{A.52}$$

A.3.3 Strain Rate Tensor Invariants

Strain rate tensor has three eigenvectors corresponding to its eigenvalues $\dot{\varepsilon}_1$, $\dot{\varepsilon}_2$, $\dot{\varepsilon}_3$ (conventionally it is assumed that $\dot{\varepsilon}_1 \geq \dot{\varepsilon}_2 \geq \dot{\varepsilon}_3$). Let η^1, η^2, η^3 be the principal axes for the strain rate tensor. Then,

$$\dot{\varepsilon}_{ij} = \begin{bmatrix} \dot{\varepsilon}_1 & 0 & 0 \\ 0 & \dot{\varepsilon}_2 & 0 \\ 0 & 0 & \dot{\varepsilon}_3 \end{bmatrix} \tag{A.53}$$

The characteristic equation is

$$\lambda^3 - \dot{I}_1\lambda^2 + \dot{I}_2\lambda - \dot{I}_3 = 0 \tag{A.54}$$

where the strain rate invariants are

$$\begin{aligned} \dot{I}_1 &= \dot{\varepsilon}_1^1 + \dot{\varepsilon}_2^2 + \dot{\varepsilon}_3^3 = \dot{\varepsilon}_k^k = \dot{\varepsilon}_1 + \dot{\varepsilon}_2 + \dot{\varepsilon}_3; \\ \dot{I}_2 &= \frac{1}{2}\left[\left(\dot{\varepsilon}_i^i\right)^2 - \dot{\varepsilon}_j^i \dot{\varepsilon}_i^j\right] = \dot{\varepsilon}_1\dot{\varepsilon}_2 + \dot{\varepsilon}_2\dot{\varepsilon}_3 + \dot{\varepsilon}_3\dot{\varepsilon}_1; \\ \dot{I}_3 &= \det\left(\dot{\varepsilon}_j^i\right) = \dot{\varepsilon}_1\dot{\varepsilon}_2\dot{\varepsilon}_3 \end{aligned} \tag{A.55}$$

The mixed components of the strain rate tensor are to be found using the metric tensor as follows.

$$\dot{\varepsilon}_j^i = \hat{g}^{is}\dot{\varepsilon}_{sj} \tag{A.56}$$

where $\dot{\varepsilon}_{ij}$ are from Eq. (A.44). Physical components of the strain rate tensor are defined as

$$\dot{\varepsilon}_{ij}^* = \dot{\varepsilon}_{ij}\sqrt{\hat{g}^{ii}\hat{g}^{jj}} \quad \text{(no summation)} \tag{A.57}$$

The procedure to reduce the strain rate tensor to its principal axes is similar to that described in Sect. A.2.3.

Now the rate of volume change of a parallelepiped constructed in the principal co-ordinates will be calculated. The rates of changes in its edges will be equal to $\dot{\varepsilon}_1\, d\eta^1$, $\dot{\varepsilon}_2\, d\eta^2$ and $\dot{\varepsilon}_3\, d\eta^3$ and so the rate of volume change will be

$$\dot{\varepsilon}_1 d\eta^1 d\eta^2 d\eta^3 + \dot{\varepsilon}_2 d\eta^2 d\eta^3 d\eta^1 + \dot{\varepsilon}_3 d\eta^3 d\eta^1 d\eta^2$$

The rate of relative volume change is

$$\dot{\Theta} = \dot{\varepsilon}_1 + \dot{\varepsilon}_2 + \dot{\varepsilon}_3 \tag{A.58}$$

Thus, the first strain rate invariant in Eq. (A.55) characterises the rate of volumetric strain.

A.3.4 Volumetric and Deviatoric Parts of the Strain Rate Tensor

Strain rate tensor T_ξ can be decomposed as

$$T_\xi = \dot{\Theta}\frac{U}{3} + D_\xi \tag{A.59}$$

where U is a unit matrix. Here the volumetric part $\dot{\Theta}\,U$ characterises the change in volume while the deviatoric part, D_ξ characterises the change in shape. For incompressible materials $\Theta = 0$ and the strain rate tensor is identical to the strain rate deviator tensor.

The second invariant of the strain rate deviator $\dot{I}_{2D}$ is a very important entity in the theory of plastic deformation. The concepts of equivalent strain rate ξ_e and equivalent shear strain rate $\dot{\gamma}_e$ are introduced as

$$\xi_e = \frac{2}{\sqrt{3}}\sqrt{\left|I_2\left(D_\xi\right)\right|},\ \dot{\gamma}_e = 2\sqrt{\left|I_2\left(D_\xi\right)\right|} \tag{A.60}$$

The factor 2/√3 is introduced to equate ξ_e to the tensile strain rate for the case of uniaxial loading. The value of ξ_e can be found using the following relations:

$$\xi_e = \frac{2}{\sqrt{3}}\sqrt{\xi_1^1\xi_2^2 + \xi_2^2\xi_3^3 + \xi_3^3\xi_1^1 - \xi_2^1\xi_1^2 - \xi_3^2\xi_2^3 - \xi_1^3\xi_3^1} = \sqrt{\frac{2}{3}\xi_j^i\xi_i^j} \tag{A.61}$$

where $\xi_i^j = \dot{\varepsilon}_i^j - (1/3)\dot{\Theta}\delta_{ij}$ are the components of the strain rate deviator.

A.3.5 On Some Scalar Characteristics of a Deformed State

The concept of nominal (engineering) strain which is ordinarily used for analysing tensile (compression) test data can be extended to 3D strain state by introducing the concept of strain intensity (effective strain), which can be calculated for finite strains in accordance with the following formula.

$$\varepsilon_e = \frac{2}{\sqrt{3}}\sqrt{\left|I_2\left(D_\varepsilon\right)\right|} = \sqrt{\frac{2}{\sqrt{3}}\left[\varepsilon_j^i\varepsilon_i^j - \frac{1}{3}\left(\varepsilon_i^i\right)^2\right]} \tag{A.62}$$

The numerical factor 2/√3 is introduced to set the value of ε_e as equal for the one dimensional case to $(L-L_0)/L_0$.

The concept of true strain which is usually used for understanding tensile (compression) test data can be extended to 3D strain state by using the Odqvist parameter.

$$s = \int_0^t \xi_e dt' = \int_0^t \sqrt{\frac{2}{3}\xi_j^i\xi_i^j}\,dt' \tag{A.63}$$

where ξ_e (effective strain rate) is the strain rate intensity; ξ^i_j are mixed components of the strain rate tensor. Strictly speaking, the elastic strain rate should be included in Eq. (A.63). However, for superplastic flow plastic strains are very large compared with which the elastic strains can be ignored without introducing serious errors.

Appendix B: Kinematics of Some Simple Deformation Modes

Some simple deformation modes are considered below: tension, fixed-end torsion, simple and pure shear, tension combined with torsion, free bulging of a sphere. The results are given in matrix form; the invariants of different tensors are calculated in order to illustrate clearly the differences among them. All calculations are based on the general theoretical consideration presented in Appendix A.

All tensors are written in component form, using both Eulerian and Lagrangian representations. The corresponding reference basis is always indicated to avoid confusion. For example, the covariant components of two different strain tensors can be the same, e.g., (B.4), in other cases the components can be different, see (B.14). The details of calculations are omitted; the results given below can be used straightway for example, while assigning practical exercises to students.

Assumptions made were as follows: material is incompressible; strain state is uniform (no flow localisation is present). Finite strain behaviour is considered.

B.1 Tension/Compression of a Cylindrical Rod

Let L_0, L respectively be the initial and current lengths of a rod. $e = (L{-}L_0)/L_0$. For a tension test $e > 0$, for a compression test $e < 0$. In the latter case barrelling is assumed to be absent.

Let $\overset{o}{\vec{e}}_1, \overset{o}{\vec{e}}_2, \overset{o}{\vec{e}}_3$ be the basis of the convective cartesian system of co-ordinates at some point O at the initial moment of time $t = 0$. Therefore $\overset{o}{\vec{e}}_1 = i,\ \overset{o}{\vec{e}}_2 = j,\ \overset{o}{\vec{e}}_3 = k$, where $\vec{i}, \vec{j}, \vec{k}$ are the unit basis vectors. At some moment of time $t > 0$ the point O moves to point O', the bases of the convective system of co-ordinates at point O' being $\hat{\vec{e}}_1, \hat{\vec{e}}_2, \hat{\vec{e}}_3$ (Fig. B.1). It is evident that the basis $\hat{\vec{e}}_1, \hat{\vec{e}}_2, \hat{\vec{e}}_3$ are orthogonal but not normalised, since the lengths of the unit vectors are not equal to 1.

$$\hat{\vec{e}}_1 = \left(\frac{L}{L_0}\right)\overset{o}{\vec{e}}_1,\quad \hat{\vec{e}}_2 = \sqrt{\frac{L_0}{L}}\,\overset{o}{\vec{e}}_2,\quad \hat{\vec{e}}_3 = \sqrt{\frac{L_0}{L}}\,\overset{o}{\vec{e}}_3 \tag{B.1}$$

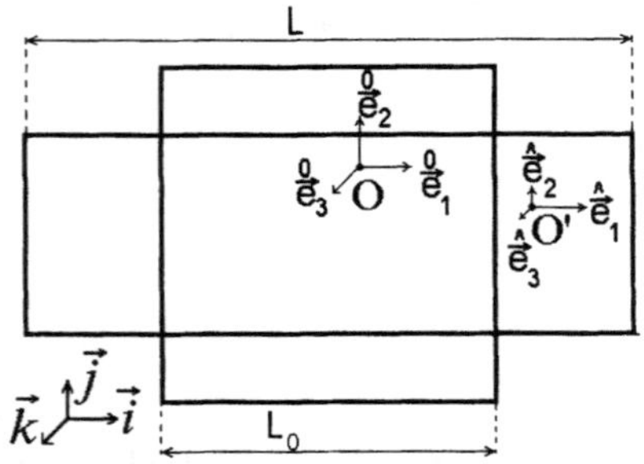

Fig. B.1. Uniaxial tension of a cylindrical rod (schematic)

The matrix of metric tensor will be given as follows.

$$\hat{g}_{ij} = \begin{bmatrix} (1+e)^2 & 0 & 0 \\ 0 & \dfrac{1}{(1+e)} & 0 \\ 0 & 0 & \dfrac{1}{(1+e)} \end{bmatrix} \tag{B.2}$$

where $e = (L - L_0)/L_0$. Taking into account that $\overset{o}{g}_{ij} = \delta_{ij}$, where δ_{ij} is Kronecker delta, one can write the incompressibility condition as

$$\det \hat{g}_{ij} = \det \overset{o}{g}_{ij} = \det \delta_{ij} = 1 \tag{B.3}$$

Strain tensors: Covariant components of strain tensors are the same for both Lagrangian and Eulerian representations. So they are not indicated here by 'o' or '^':

$$\varepsilon_{ij} = \frac{1}{2}\left(\hat{g}_{ij} - \overset{o}{g}_{ij} \right) = \begin{bmatrix} \dfrac{e+e^2}{2} & 0 & 0 \\ 0 & -\dfrac{e}{[2(1+e)]} & 0 \\ 0 & 0 & -\dfrac{e}{[2(1+e)]} \end{bmatrix} \tag{B.4}$$

Strain tensor in Eulerian representation:

$$\overset{o}{T}_{\varepsilon} = \varepsilon_{ij} \overset{o}{\vec{e}}{}^{i} \overset{o}{\vec{e}}{}^{j} \tag{B.5}$$

Strain tensor in Lagrangian representation

$$\hat{T}_{\varepsilon} = \varepsilon_{ij}\hat{\vec{e}}^{i}\hat{\vec{e}}^{j} \tag{B.6}$$

Eulerian representation: In this case, there is no difference between mixed, contravariant and covariant components. Therefore, the invariants of the strain tensor are

$$\overset{0}{I}_1 = \varepsilon_1^1 + \varepsilon_2^2 + \varepsilon_3^3 = e^2 \frac{(3+e)}{[2(1+e)]}$$

$$\overset{0}{I}_2 = \varepsilon_1^1\varepsilon_2^2 + \varepsilon_2^2\varepsilon_3^3 + \varepsilon_3^3\varepsilon_1^1 - \varepsilon_2^1\varepsilon_1^2 - \varepsilon_3^2\varepsilon_2^3 - \varepsilon_1^3\varepsilon_3^1 = -e^2 \frac{(3+6e+2e^2)}{[4(1+e)^2]} \tag{B.7}$$

$$\overset{0}{I}_3 = \det \varepsilon_j^i = e^3 \frac{\left(\frac{1+e}{2}\right)}{[4(1+e)^2]}$$

Substituting (B.7) in (A.38) one can see that the condition $\Theta = 0$ is satisfied. The principal strains are as follows:

$$\overset{0}{\varepsilon}_1 = e + \frac{e^2}{2} \quad \overset{0}{\varepsilon}_2 = \overset{0}{\varepsilon}_3 = -\frac{e}{[2(1+e)]} \tag{B8}$$

They are the roots of the following equation.

$$\left(\overset{0}{\varepsilon}_k\right)^3 - \overset{0}{I}_1\left(\overset{0}{\varepsilon}_k\right)^2 + \overset{0}{I}_2\overset{0}{\varepsilon}_k - \overset{0}{I}_3 = 0 \tag{B.9}$$

Let φ_o be the angle between some fibre and $\overset{o}{\vec{e}}_I$ at $t = 0$. Then the elongation of this fibre is given by

$$\varepsilon_{\varphi} = \frac{\sqrt{[(3+3e+e^2)\cos^2\varphi_o - 1]}}{(1+e)+1} - 1 \tag{B.10}$$

The fibre which does not change its length is

$$\varphi_o^* = arc\ \cos\left[\frac{1}{\sqrt{3+3e+e^2}}\right] \tag{B.11}$$

Lagrangian representation : The mixed components of the strain tensor can be determined in accordance with the following equations.

$$\hat{\varepsilon}^i_j = \hat{g}^{is}\varepsilon_{sj} \tag{B.12}$$

where $\hat{g}^{is}$ are the components of the matrix inverse to (B.2).

$$\hat{g}^{ij} = \begin{bmatrix} \dfrac{1}{(1+e)^2} & 0 & 0 \\ 0 & 1+e & 0 \\ 0 & 0 & 1+e \end{bmatrix} \tag{B.13}$$

Strain tensor in Lagrangian representation will then be given by

$$\hat{\varepsilon}^*_{ij} = \hat{\varepsilon}^i_j = \begin{bmatrix} \dfrac{\dfrac{e+e^2}{2}}{(1+e)^2} & 0 & 0 \\ 0 & -\dfrac{e}{2} & 0 \\ 0 & 0 & -\dfrac{e}{2} \end{bmatrix} \tag{B.14}$$

The strain tensor invariants are

$$\hat{I}_1 = -\frac{e^2\left(\dfrac{3}{2}+e\right)}{(1+e)^2}, \quad \hat{I}_2 = -\frac{e^2(3-e^2)}{[4(1+e)^2]}, \quad \hat{I}_3 = \frac{e^3\left(1+\dfrac{e}{2}\right)}{[4(1+e)^2]} \tag{B.15}$$

Substituting (B15) in (A38) one can see that the incompressibility condition $\Theta = 0$ is satisfied. Principal strains are

$$\hat{\varepsilon}_1 = \left(\frac{e+e^2}{2}\right)(1+e)^2, \quad \hat{\varepsilon}_2 = \hat{\varepsilon}_3 = -\frac{e}{2} \tag{B.16}$$

They are the roots of the following characteristic equation.

$$(\hat{\varepsilon}_k)^3 - \hat{I}_1(\hat{\varepsilon}_k)^2 + \hat{I}_2\hat{\varepsilon}_k - \hat{I}_3 = 0 \tag{B.17}$$

The validity of relationships (A.37) can be checked by direct substitutions.

Strain deviators : *Eulerian representation*: In this case one has

$$\tilde{\varepsilon}_{ij} = \varepsilon_{ij} - \left(\frac{1}{3}\right)\overset{0}{I}_1 \delta_{ij} \tag{B.18}$$

Since $\overset{o}{g}_{ij} = \delta_{ij}$, then the covariant and the mixed components are the same for the basis $\overset{o}{\vec{e}}_1, \overset{o}{\vec{e}}_2, \overset{o}{\vec{e}}_3$.

$$\tilde{\varepsilon}_{ij} = \begin{bmatrix} \dfrac{\dfrac{e+e^2+e^3}{3}}{1+e} & 0 & 0 \\ 0 & -\dfrac{\dfrac{e+e^2+e^3}{3}}{2(1+e)} & 0 \\ 0 & 0 & -\dfrac{\dfrac{e+e^2+e^3}{3}}{2(1+e)} \end{bmatrix} \tag{B.19}$$

Invariants of the deviator are

$$\overset{0}{I}_{2D} = -3\frac{\left(\dfrac{e+e^2+e^3}{3}\right)^2}{[4(1+e)^2]}; \qquad \overset{0}{I}_{3D} = \frac{\left(\dfrac{e+e^2+e^3}{3}\right)^3}{[4(1+e)^3]} \tag{B.20}$$

Strain intensity, according to (A.62), is obtained as

$$\overset{0}{\varepsilon}_i = \frac{\left(\dfrac{e+e^2+e^3}{3}\right)}{(1+e)} \tag{B.21}$$

Lagrangian representation : The strain deviator is

$$\hat{\tilde{\varepsilon}}_{ij} = \begin{bmatrix} \dfrac{e+e^2+e^3/3}{(1+e)^2} & 0 & 0 \\ 0 & -\dfrac{e+e^2+e^3/3}{2(1+e)^2} & 0 \\ 0 & 0 & -\dfrac{e+e^2+e^3/3}{2(1+e)^2} \end{bmatrix} \tag{B.22}$$

Invariants of the deviator are

$$\hat{I}_{2D} = -3\frac{\left(\dfrac{e+e^2+e^3}{3}\right)^2}{[4(1+e)^4]}; \qquad \hat{I}_{3D} = \frac{\left(\dfrac{e+e^2+e^3}{3}\right)^3}{[4(1+e)^6]} \tag{B.23}$$

Strain intensity, according to (A.62), is

$$\hat{\varepsilon}_i = \frac{\left(\dfrac{e+e^2+e^3}{3}\right)}{(1+e)^2} \tag{B.24}$$

Strain rate tensor: Strain rate tensor components ξ_{ij} are

$$\xi_{ij} = \frac{1}{2}\frac{d\hat{g}_{ij}}{dt} \tag{B.25}$$

Substituting (B.2) in (B.25)

$$\xi_{ij} = \begin{bmatrix} \dot{e}(1+e) & 0 & 0 \\ 0 & -\dfrac{\dot{e}}{[2(1+e)^2]} & 0 \\ 0 & 0 & -\dfrac{\dot{e}}{[2(1+e)^2]} \end{bmatrix} \tag{B.26}$$

where $\dot{e} = de/dt = (1/L_0)dL/dt$; the process of deformation is determined by the function $e(t)$. In particular, for a constant cross-head velocity test $e(t) = vt/L_0$, $\dot{e} = v/L_0$. For a tension test $v > 0$.

Mixed components of the strain rate tensor are

$$\xi^i_j = \begin{bmatrix} \dfrac{\dot{e}}{(1+e)} & 0 & 0 \\ 0 & -\dfrac{\dot{e}}{[2(1+e)]} & 0 \\ 0 & 0 & -\dfrac{\dot{e}}{[2(1+e)]} \end{bmatrix} \tag{B.27}$$

Due to incompressibility $\xi^k_k = 0$. (It should be noted that $\xi_{kk} \neq 0$).
Strain rate tensor invariants are

$$\hat{I}_1 = \xi^1_1 + \xi^2_2 + \xi^3_3 = 0 \quad \text{(incompressibility)} \tag{B.28}$$

$$\hat{I}_2 = \xi^1_1\xi^2_2 + \xi^2_2\xi^3_3 + \xi^3_3\xi^1_1 - \xi^1_2\xi^2_1 - \xi^2_3\xi^3_2 - \xi^3_1\xi^1_3 = -\frac{3\dot{e}^2}{[4(1+e)^2]} \tag{B.29}$$

$$\hat{I}_3 = \det \xi^i_j = \frac{\dot{e}^3}{[4(1+e)^3]} \tag{B.30}$$

Strain rate intensity is

$$\xi_e = \left(\frac{2}{\sqrt{3}}\right)\sqrt{|\hat{I}_2|} = \frac{|\dot{e}|}{(1+e)} \tag{B.31}$$

The value of accumulated strain is

$$\varepsilon_e = \int_0^t \xi_e(\tau)d\tau = \int_0^t \frac{|\dot{e}(\tau)|}{1+e(\tau)}d\tau \tag{B.32}$$

Equation (B.32) is valid for any arbitrary process $e(\tau)$, $0 \le \tau \le t$. Consider some simple cases of application of Eq. (B.32).

1. *Constant cross-head velocity test* : *v= constant > 0*. Current length $L = L_0 + vt$; v = const > 0. Then $e(t) = vt/L_0$, $\dot{e} = v/L_0$, Strain rate $\xi_e = v/L = v/(L_0+vt)$, Accumulated strain $\varepsilon_e = \ln(L/L_0)$.

2. *Constant cross-head velocity compression test* : v = constant < 0; Let $u = -v$. Then, current height $H = H_0 - ut$; where u = const > 0. Then $e(t) = vt/L_0 = -ut/L_0 < 0$, $\dot{e} = v/L_0 = -u/L_0$. Strain rate $\xi_e = v/H = -u/H = -u/(H_0 - ut)$. Accumulated strain $\varepsilon_e = \ln(H_0/H)$.
3. *Cyclic loading*: tension from L_0 to L_{max}; then compression from L_{max} to L_0. Accumulated strain $\varepsilon_e = 2N\ln(L_{max}/L_0)$, where N is the number of cycles.
4. *Cyclic loading*: compression from H_0 to H_{min}; then tension from H_{min} to H_0. Accumulated strain: $\varepsilon_e = 2N\ln(H_0/H_{min})$, where N is the number of cycles.

B.2
Simple shear

Let γ be the shear angle; $\delta = \tan\gamma$. Let the basis of the convective system of co-ordinates at point O(x_0, y_0, z_0) at time $t = 0$ be orthonormalised. (Fig. B.2):

$\overset{o}{\vec{e}}_1 = i$, $\overset{o}{\vec{e}}_2 = j$, $\overset{o}{\vec{e}}_3 = k$, where $\vec{i}$, $\vec{j}$, $\vec{k}$ are the unit vectors of Cartesian basis. At some moment of time $t > 0$ the point O transfers to point O'; the basis of the convective system of co-ordinates at point O' being $\hat{\vec{e}}_1, \hat{\vec{e}}_2, \hat{\vec{e}}_3$. One can obtain from Fig. B.2 the following equations for the transfer from the initial configuration to the current one

$$X = x_0 + y_0\delta \qquad Y = y_0 \quad Z = z_0 \; 1 \tag{B.33}$$

Then,

$$\hat{\vec{e}}_1 = \frac{\partial\vec{r}}{\partial x_0} = \overset{o}{\vec{e}}_1 \,, \hat{\vec{e}}_2 = \frac{\partial\vec{r}}{\partial y_0} = \overset{o}{\vec{e}}_2 + \delta\,\overset{o}{\vec{e}}_1 \,, \quad 2$$
$$\hat{\vec{e}}_3 = \frac{\partial\vec{r}}{\partial z_0} = \overset{o}{\vec{e}}_3 \tag{B.34}$$

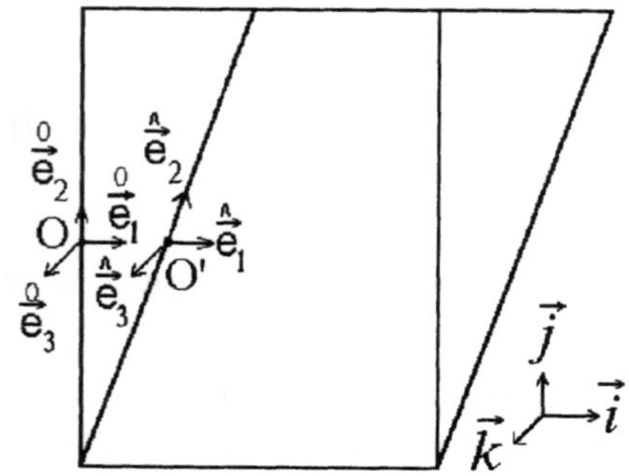

Fig. B.2. Schematic of simple shear

The displacement vector is $\hat{\vec{u}} = \hat{\vec{r}} - \hat{\vec{r}}_0 = \hat{\vec{u}}\ (y_0\delta; 0; 0)$ (initial basis). Then metric tensor is

$$\hat{g}_{ij} = \begin{bmatrix} 1 & \delta & 0 \\ \delta & 1+\delta^2 & 0 \\ 0 & 0 & 1 \end{bmatrix} \tag{B.35}$$

And so,

$$\hat{g}^{ij} = \begin{bmatrix} 1+\delta^2 & -\delta & 0 \\ -\delta & 1 & 0 \\ 0 & 0 & 1 \end{bmatrix} \tag{B.36}$$

Strain tensors: Covariant components for both the strain tensors are

$$\varepsilon_{ij} = \begin{bmatrix} 0 & \frac{\delta}{2} & 0 \\ \frac{\delta}{2} & \frac{\delta^2}{2} & 0 \\ 0 & 0 & 0 \end{bmatrix} \tag{B.37}$$

Eulerian representation:Invariants:

$$\overset{0}{I}_1 = \frac{\delta^2}{2}; \quad \overset{0}{I}_2 = -\frac{\delta^2}{4}; \quad \overset{0}{I}_3 = 0 \tag{B.38}$$

Principal strains

$$\overset{0}{\varepsilon}_1 = \frac{\delta^2}{2} + \sqrt{\frac{\delta^2}{4} + \frac{\delta^4}{16}}; \overset{0}{\varepsilon}_2 = 0; \overset{0}{\varepsilon}_3 = \frac{\delta^2}{2} - \sqrt{\frac{\delta^2}{4} + \frac{\delta^4}{16}} \tag{B.39}$$

Lagrangian representation: Strain tensor

$$\hat{\varepsilon}^i_j = \begin{bmatrix} -\frac{\delta^2}{2} & \frac{\delta}{2} & 0 \\ \frac{\delta}{2} & 0 & 0 \\ 0 & 0 & 0 \end{bmatrix} \tag{B.40}$$

Invariants:

$$\hat{I}_1 = -\frac{\delta^2}{2}; \quad \hat{I}_2 = -\frac{\delta^2}{4}; \quad \hat{I}_3 = 0 \tag{B.41}$$

Principal strains:

$$\begin{aligned} &\hat{\varepsilon}_1 = -\frac{\delta^2}{2} + \sqrt{\frac{\delta^2}{4} + \frac{\delta^4}{16}}; \\ &\hat{\varepsilon}_2 = 0, \quad \hat{\varepsilon}_3 = -\frac{\delta^2}{2} - \sqrt{\frac{\delta^2}{4} + \frac{\delta^4}{16}} \end{aligned} \tag{B.42}$$

Physical components:

$$\hat{\varepsilon}^*_{ij} = \begin{bmatrix} 0 & \left(\frac{\delta}{2}\right)\sqrt{1+\delta^2} & 0 \\ \left(\frac{\delta}{2}\right)\sqrt{1+\delta^2} & \frac{\delta^2}{2} & 0 \\ 0 & 0 & 0 \end{bmatrix} \tag{B.43}$$

Strain deviators
Eulerian representation:

$$\tilde{\varepsilon}_{ij} = \begin{bmatrix} -\frac{\delta^2}{6} & \frac{\delta}{2} & 0 \\ \frac{\delta}{2} & \frac{\delta^2}{3} & 0 \\ 0 & 0 & -\frac{\delta^2}{6} \end{bmatrix} \tag{B.44}$$

Invariants:

$$\overset{0}{I}_{2D} = -\frac{\delta^2}{4} - \frac{\delta^4}{12}, \quad \overset{0}{I}_{3D} = \left(\frac{\delta^4}{24}\right)\left[\frac{1+2\delta^2}{9}\right] \tag{B.45}$$

Strain intensity, according to Eq. (A.62), is

$$\overset{0}{\varepsilon}_i = \left(\frac{\delta}{\sqrt{3}}\right)\sqrt{\frac{1+\delta^2}{3}} \tag{B.46}$$

Lagrangian representation:

$$\tilde{\varepsilon}_{ij} = \begin{bmatrix} -\frac{\delta^2}{3} & \frac{\delta}{2} & 0 \\ \frac{\delta}{2} & \frac{\delta^2}{6} & 0 \\ 0 & 0 & \frac{\delta^2}{6} \end{bmatrix} \tag{B.47}$$

Invariants of the strain tensor:

$$\hat{I}_{2D} = -\frac{\delta^2}{4} - \frac{\delta^4}{12}, \quad \hat{I}_{3D} = -\left(\frac{\delta^4}{24}\right)\left[\frac{1+2\delta^2}{9}\right] \tag{B.48}$$

Strain intensity:

$$\hat{\varepsilon}_i = \left(\frac{\delta}{\sqrt{3}}\right)\sqrt{\frac{1+\delta^2}{3}} \tag{B.49}$$

One can see from (B.49) and (B.46) that $\overset{0}{\varepsilon}_i = \hat{\varepsilon}_i$.

Strain rate tensor

$$\xi_{ij} = \begin{bmatrix} 0 & \frac{\dot{\delta}}{2} & 0 \\ \frac{\dot{\delta}}{2} & \delta\dot{\delta} & 0 \\ 0 & 0 & 0 \end{bmatrix} \tag{B.50}$$

Physical components of the strain rate tensor:

$$\xi_{ij}^* = \begin{bmatrix} 0 & \left(\frac{\dot{\delta}}{2}\right)\sqrt{1+\delta^2} & 0 \\ \left(\frac{\dot{\delta}}{2}\right)\sqrt{1+\delta^2} & \delta\dot{\delta} & 0 \\ 0 & 0 & 0 \end{bmatrix} \tag{B.51}$$

Mixed components:

$$\xi^i_j = \begin{bmatrix} -\frac{\delta\dot{\delta}}{2} & \frac{\dot{\delta}(1-\delta^2)}{2} & 0 \\ \frac{\dot{\delta}}{2} & \frac{\delta\dot{\delta}}{2} & 0 \\ 0 & 0 & 0 \end{bmatrix} \tag{B.52}$$

Invariants of the strain rate tensor:

$$\hat{I}_1 = 0;\ \hat{I}_2 = -\frac{\dot{\delta}^2}{4};\ \hat{I}_3 = 0 \tag{B.53}$$

Principal strain rates:

$$\hat{\varepsilon}_1 = \frac{\dot{\delta}}{2};\ \hat{\varepsilon}_2 = 0;\ \hat{\varepsilon}_3 = -\frac{\dot{\delta}}{2} \tag{B.54}$$

Shear strain rate intensity:

$$H = 2\sqrt{\left|\hat{I}_2\right|} = \dot{\delta} \tag{B.55}$$

Accumulated shear strain:

$$\Gamma = \int H dt = \delta \tag{B.56}$$

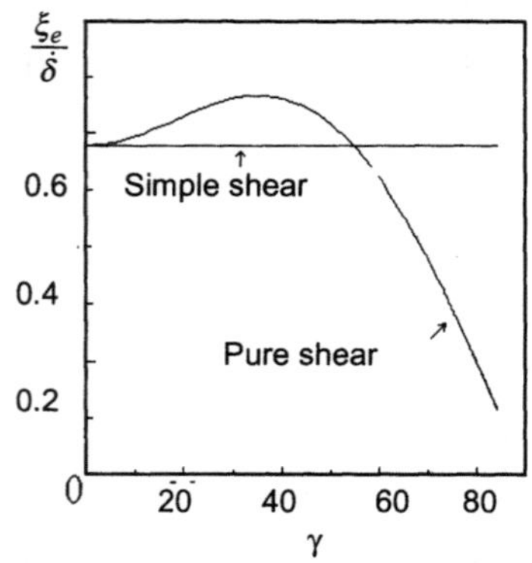

Fig. B.3. Dependence of the relative strain rate $\xi_e/\dot{\delta}$ on the shear angle γ. For simple and pure shear.

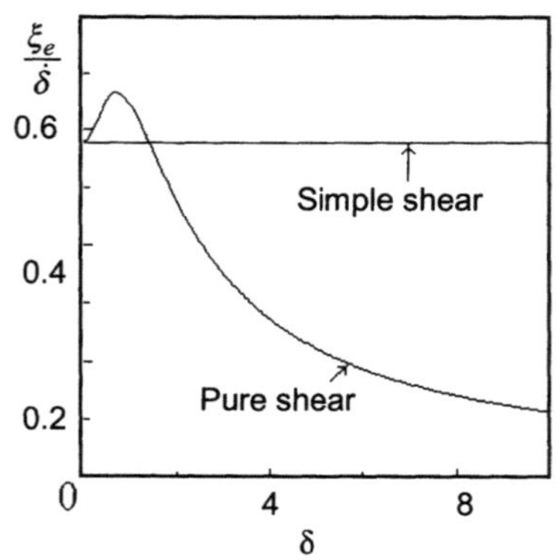

Fig.B.4. Dependence of the relative strain rate $\xi_e / \dot{\delta}$ on δ for simple and pure shear

Strain rate intensity:

$$\xi_e = \frac{H}{\sqrt{3}} = \frac{\dot{\delta}}{\sqrt{3}} \tag{B.57}$$

Accumulated strain:

$$\varepsilon_e = \frac{\Gamma}{\sqrt{3}} = \frac{\delta}{\sqrt{3}} \tag{B.58}$$

The dependences of the strain and strain rate, calculated in accordance with the above equations are presented in Figs. B3 and B4.

B.3 Pure Shear

Let γ be the shear angle; $\delta = \tan\gamma$. Let the basis of the convective system of co-ordinates at the point $M(x_0, y_0, z_0)$ at time $t = 0$ be orthonormalised (Fig. B.5): $\overset{o}{\vec{e}}_1 = i,\ \overset{o}{\vec{e}}_2 = j,\ \overset{o}{\vec{e}}_3 = k$, where $\vec{i}, \vec{j}, \vec{k}$ are the unit vectors of Cartesian basis. At some moment of time $t > 0$ the point O is shifted to point O'; the basis of the convective system of co-ordinates at point O' is $\hat{\vec{e}}_1,\ \hat{\vec{e}}_2,\ \hat{\vec{e}}_3$ One can obtain from Fig. B.5 the following equations for the transfer from the initial configuration to the current one:

$$X = x_0 + y_0 \sin\gamma, \qquad Y = y_0 \cos\gamma,\ Z = \frac{z_0}{\cos\gamma} \tag{B.59}$$

Then,

$$\hat{\vec{e}}_1 = \frac{\partial \vec{\gamma}}{\partial x_0} = \overset{o}{\vec{e}}_1, \quad e_2 = \frac{\partial \vec{\gamma}}{\partial y_0} = \overset{0}{\vec{e}}_1 \sin\gamma + \overset{o}{\vec{e}}_2 \cos\gamma,\ \hat{\vec{e}}_3 = \frac{\partial \vec{r}}{\partial z_0} = \frac{\hat{\vec{e}}_3}{\cos\gamma}, \tag{B.60}$$

The metric tensor is

$$\hat{g}_{ij} = \begin{bmatrix} 1 & \sin\gamma & 0 \\ \sin\gamma & 1 & 0 \\ 0 & 0 & \dfrac{1}{\cos^2\gamma} \end{bmatrix} \tag{B.61}$$

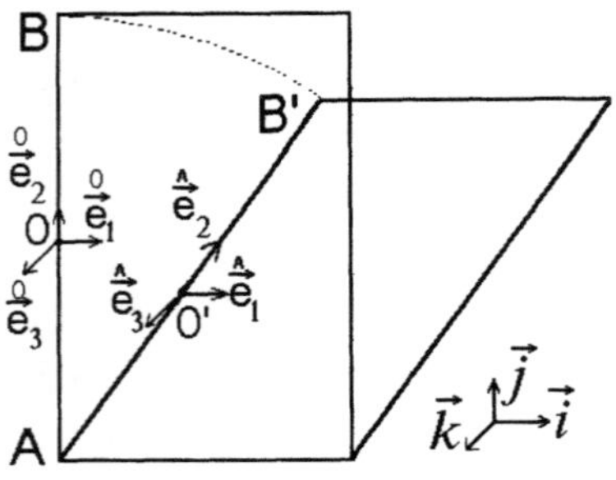

Fig. B.5. Schematic of pure shear

Then

$$\hat{g}^{ij} = \begin{bmatrix} \frac{1}{\cos^2\gamma} & -\frac{\sin\gamma}{\cos^2\gamma} & 0 \\ -\frac{\sin\gamma}{\cos^2\gamma} & \frac{1}{\cos^2\gamma} & 0 \\ 0 & 0 & \cos^2\gamma \end{bmatrix} \quad \text{(B.62)}$$

Strain tensors: Covariant components of strain tensors:

$$\varepsilon_{ij} = \begin{bmatrix} 0 & \left(\frac{1}{2}\right)\sin\gamma & 0 \\ \left(\frac{1}{2}\right)\sin\gamma & 0 & 0 \\ 0 & 0 & \left(\frac{1}{2}\right)\tan^2\gamma \end{bmatrix} \quad \text{(B.63)}$$

Eulerian representation: Invariants of strain tensor:

$$\overset{0}{I}_1 = \left(\frac{1}{2}\right)\tan^2\gamma, \overset{0}{I}_2 = -\left(\frac{1}{4}\right)\sin^2\gamma; \\ \overset{0}{I}_3 = -\left(\frac{1}{8}\right)\sin^2\gamma\tan^2\gamma \quad \text{(B.64)}$$

Principal strains:

$$\varepsilon_1 = \left(\frac{1}{2}\right)\sin\gamma,\ \varepsilon_2 = \left(\frac{1}{2}\right)\tan^2\gamma,\ \varepsilon_3 = -\left(\frac{1}{2}\right)\sin\gamma \quad \text{(B.65)}$$

Strain tensor:*Lagrangian representation*

$$\hat{\varepsilon}^i_j = \begin{bmatrix} -\left(\frac{1}{2}\right)\tan^2\gamma & \left(\frac{1}{2}\right)\left(\frac{\sin\gamma}{\cos^2\gamma}\right) & 0 \\ \left(\frac{1}{2}\right)\left(\frac{\sin\gamma}{\cos^2\gamma}\right) & -\left(\frac{1}{2}\right)\tan^2\gamma & 0 \\ 0 & 0 & \left(\frac{1}{2}\right)\sin^2\gamma \end{bmatrix} \tag{B.66}$$

Invariants:

$$\hat{I}_1 = \left(\frac{1}{2}\right)\sin^2\gamma - \tan^2\gamma, \qquad \hat{I}_2 = -\left(\frac{1}{4}\right)\tan^2\gamma\left[1 + 2\sin^2\gamma\right];$$
$$\hat{I}_3 = -\left(\frac{1}{8}\right)\sin^2\gamma\tan^2\gamma \tag{B.67}$$

Strain deviator
Eulerian representation:

$$\tilde{\varepsilon}_{ij} = \begin{bmatrix} -\left(\frac{1}{6}\right)\tan^2\gamma & \left(\frac{1}{2}\right)\sin\gamma & 0 \\ \left(\frac{1}{2}\right)\sin\gamma & -\left(\frac{1}{6}\right)\tan^2\gamma & 0 \\ 0 & 0 & \left(\frac{1}{3}\right)\tan^2\gamma \end{bmatrix} \tag{B.68}$$

Invariants of deviator:

$$\overset{0}{I}_{1,D} = 0, \overset{0}{I}_{2,D} = -\left(\frac{1}{4}\right)\sin^2\gamma - \left(\frac{1}{12}\right)\tan^4\gamma;$$
$$\overset{0}{I}_{3,D} = \left(\frac{1}{12}\right)\tan^2\gamma\left[\left(\frac{1}{9}\right)\tan^4\gamma - \sin^2\gamma\right] \tag{B.69}$$

Lagrangian representation:

$$\hat{\tilde{\varepsilon}}_{ij}=\begin{bmatrix} -\left(\frac{1}{6}\right)\left(\tan^2\gamma+\sin^2\gamma\right) & \frac{\sin\gamma}{2\cos^2\gamma} & 0 \\ \frac{\sin\gamma}{2\cos^2\gamma} & -\left(\tan^2\gamma+\sin^2\gamma\right)/6 & 0 \\ 0 & 0 & \left(\tan^2\gamma+\sin^2\gamma\right)/3 \end{bmatrix} \quad \text{(B.70)}$$

Invariants of deviator

$$\hat{I}_{1D}=0;\ \hat{I}_{2D}=-\left(\frac{1}{4}\right)\frac{\tan^4\gamma}{\sin^2\gamma}-\left(\frac{1}{12}\right)\left(\tan^2\gamma+\sin^2\gamma\right)^2;$$
$$\hat{I}_{3D}=\left(\frac{1}{108}\right)\left(\tan^2\gamma+\sin^2\gamma\right)^3-\left(\frac{1}{12}\right)\left(\tan^2\gamma+\sin^2\gamma\right)^2\frac{\sin^2\gamma}{\cos^4\gamma}; \quad \text{(B.71)}$$

Strain intensity:

$$\hat{\varepsilon}_i=\left(\frac{\delta}{\sqrt{3}}\right)\sqrt{1+\left(\frac{\delta^2}{3}\right)\left[4+\frac{2}{\left(1+\delta^2\right)}+\frac{1}{\left(1+\delta^2\right)^2}\right]} \quad \text{(B.72)}$$

Strain rate tensor

$$\xi_{ij}=\begin{bmatrix} 0 & \left(\frac{1}{2}\right)\dot{\gamma}\cos\gamma & 0 \\ \left(\frac{1}{2}\right)\dot{\gamma}\cos\gamma & 0 & 0 \\ 0 & 0 & \frac{\dot{\gamma}\sin\gamma}{\cos^3\gamma} \end{bmatrix} \quad \text{(B.73)}$$

Mixed components:

$$\hat{\xi}^i_j=\begin{bmatrix} -\left(\frac{1}{2}\right)\dot{\gamma}\tan\gamma & \frac{\dot{\gamma}}{(2\cos\gamma)} & 0 \\ \frac{\dot{\gamma}}{(2\cos\gamma)} & -\left(\frac{1}{2}\right)\dot{\gamma}\tan\gamma & 0 \\ 0 & 0 & \dot{\gamma}\tan\gamma \end{bmatrix} \quad \text{(B.74)}$$

Invariants:

$$\hat{I}_1 = 0, \quad \hat{I}_2 = -\frac{\left[\dfrac{\dot{\gamma}^2}{(4\cos^2\gamma)}\right]}{[1+3\sin^2\gamma]}; \qquad \text{(B.75)}$$
$$\hat{I}_3 = -\left(\frac{1}{4}\right)\dot{\gamma}^3 \tan\gamma$$

Characteristic equation:

$$\dot{\delta}_R^3 + \hat{I}_2\dot{\delta}_R - \hat{I}_3 = 0 \qquad \text{(B.76)}$$

Discriminant of cubic equation (B.76) is as follows:

$$D = \frac{1}{27}\hat{I}_2^3 + \frac{1}{4}\hat{I}_3^2 = \frac{\dot{\delta}^6}{64\left(1+\delta^2\right)^6}\left[\delta^2 - \frac{1}{27}\left(1+4\delta^2\right)^3\right] \qquad \text{(B.77)}$$

As is well known, if the discriminant of a cubic equation is negative ($D < 0$), it has three real roots. It is easy to show that $D \le 0$ for any δ; the dependence $D(\delta)$ is not monotonic: $D(\delta)$ decreases when $0 < \delta < \delta^*$ but increases when $\delta > \delta^*$. For $\delta = \delta^* \cong 0.354$, $D(\delta^*) = 0$.

Accumulated shear strain
Shear strain intensity is:

$$H = 2\sqrt{\left|\hat{I}_{2D}\right|} = \left(\frac{\dot{\gamma}}{\cos\gamma}\right)\sqrt{1+3\sin^2\gamma} = \left[\frac{\dot{\delta}}{(1+\delta^2)}\right]\sqrt{1+4\delta^2} \qquad \text{(B.78)}$$

Then the accumulated shear strain can be found by integrating Eq. (B.78):

$$\Lambda = \int_0^t H dt = \int_0^{\text{Ä}} \left\{\frac{\dot{\delta}}{(1+\delta^2)}\sqrt{1+4\delta^2}\right\} dt \qquad \text{(B.79)}$$

where $\Delta = \tan\Gamma$, $\Gamma = \max \gamma(t) = \gamma(T)$ $(0 \le t \le T)$, i.e., the process of deformation is assumed to be monotonic. Calculating the integral by means of standard procedure one can find after some transforms (it is necessary to apply the second Eulerian substitution)[1]

[1] The calculation of this integral is a good exercise for students

$$\Lambda = 2\ln\left|\frac{2+\zeta}{2-\zeta}\right| + \frac{\sqrt{3}}{2}\ln\left|\frac{\zeta^2 - 2\sqrt{3}\zeta + 4}{\zeta^2 + 2\sqrt{3}\zeta + 4}\right|, \tag{B.80}$$

where $\zeta = \left[\sqrt{1+\Delta^2} - 1\right] \Big/ \Delta$.

Then the strain rate intensity will be as follows:

$$\xi_e = \frac{\dot{\gamma}_e}{\sqrt{3}} = \frac{2}{\sqrt{3}}\sqrt{\left|\hat{I}_{2D}\right|} = \frac{\dot{\gamma}}{\sqrt{3}\cos\gamma}\sqrt{1+3\sin^2\gamma} = \frac{\dot{\delta}}{\sqrt{3}\left(1+\delta^2\right)}\sqrt{1+4\delta^2} \tag{B.81}$$

At last the accumulated strain is found as

$$\varepsilon_e = \frac{2}{\sqrt{3}}\ln\left|\frac{2+\zeta}{2-\zeta}\right| + \frac{1}{2}\ln\left|\frac{\zeta^2 - 2\sqrt{3}\zeta + 4}{\zeta^2 + 2\sqrt{3}\zeta + 4}\right| \tag{B.82}$$

where $\zeta = \left[\sqrt{1+4\Delta^2} - 1\right] / \Delta$

B.4 Bulging of a Sphere

Let a sphere of initial radius R_0 and uniform initial thickness h_0 ($h_0 \ll R_0$) be subjected to internal gas pressure. Then, the metric tensor of a convective system of co-ordinates will be given by

$$\hat{g}_{ij} = \begin{bmatrix} \left(\frac{\dot{R}}{R_0}\right)^2 & 0 & 0 \\ 0 & \left(\frac{\dot{R}}{R_0}\right)^2 & 0 \\ 0 & 0 & \left(\frac{\dot{h}}{h_0}\right)^2 \end{bmatrix} \tag{B.83}$$

where R, h are the current values of the radius and thickness respectively. Incompressibility condition can be written using Lagrangian variables as

$$\det \hat{g}_{ij} = \det \overset{o}{g}_{ij} = \det \delta_{ij} = 1, \tag{B.84}$$

where δ_{ij} is Kronecker delta. One can obtain from (B.83) and (B.84) that

$$h = h_0 \left(\frac{R_0}{R} \right)^2 \tag{B.85}$$

Equation (B.85) can also be derived from the condition of volume constancy $4\pi(R_0)^2 h_0 = 4\pi R^2 h$.

Covariant components of the strain rate tensor are

$$\xi_{ij} = \frac{1}{2}\frac{d\hat{g}_{ij}}{dt} = \begin{bmatrix} \frac{R\dot{R}}{R_0^2} & 0 & 0 \\ 0 & \frac{R\dot{R}}{R_0^2} & 0 \\ 0 & 0 & \frac{h\dot{h}}{h_0^2} \end{bmatrix} \tag{B.86}$$

Mixed components are

$$\xi_j^i = \xi_{sj}\sqrt{\hat{g}^{si}\hat{g}^{sj}} = \begin{bmatrix} \frac{\dot{R}}{R} & 0 & 0 \\ 0 & \frac{\dot{R}}{R} & 0 \\ 0 & 0 & \frac{\dot{h}}{h} \end{bmatrix} \tag{B.87}$$

From the incompressibility condition

$$\frac{\dot{h}}{h} = -2\frac{\dot{R}}{R} \tag{B.88}$$

Then, the equivalent strain rate is

$$\xi_e = \sqrt{\frac{2}{3}\xi_{ij}\xi_{ij}} = \frac{2\dot{R}}{R} = -\frac{\dot{h}}{h} \tag{B.89}$$

It is to be mentioned that if one considers similar equations for the process of die-less drawing of a wire one would obtain the same equations (assuming the hypothesis of plane sections remaining planar). However, the meaning of the equations would be different.

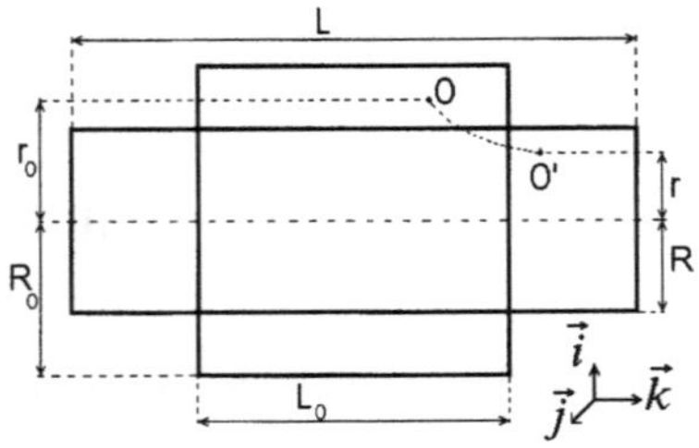

Fig. B.6. The initial and the current states of the cylindrical specimen subjected to combined loading by axial force and torque

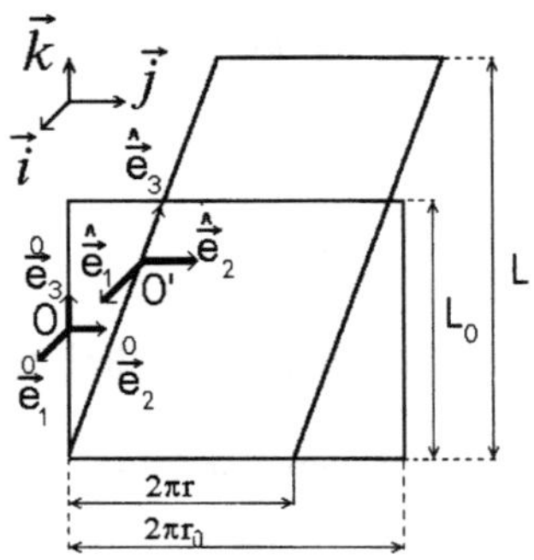

Fig. B.7. The evolution of the cylindrical surface containing the point O at the initial and the current moments of time

B.5 Finite Strain Kinematics under Combined Loading of a Cylindrical Rod by Axial Force and Torque

Let R_0 and L_0 be the radius and length of the specimen at the initial moment of time. The current state of the deformed material is characterised by specimen length L, angle φ, relative angular velocity ω and axial velocity v of the specimen ends (Fig. B.6). The two functions $v(t)$ and $\omega(t)$ determine the loading conditions.

It is assumed further that the material is incompressible and the deformation is uniform in the axial direction (along the specimen length). The position of some particle O is determined by its distance from the axis of symmetry r_0 at time $t = t_0$. At some moment of time $t > t_0$ point O moves to point O' which is at distance r from the axis of symmetry (Fig. B.6). Now the convective system of co-ordinates $\zeta^1, \zeta^2, \zeta^3$ are introduced which is 'frozen' into the material. Let the basis of the convective system of co-ordinates at the moment of $t = t_0$ at the point O $\overset{o}{\vec{e}}_1, \overset{o}{\vec{e}}_2, \overset{o}{\vec{e}}_3$ be Cartesian. At some moment of time $t > t_0$ the bases of the accompanying system of co-ordinates at point O, $\hat{\vec{e}}_1$, $\hat{\vec{e}}_2$, $\hat{\vec{e}}_3$ are not, in general, already orthonormalized. In order to find the components of the metric tensor g_{ij} one plots the evolution of the cylindrical surface including point O for moments of time $t = t_0$, $t > t_0$ (Fig. B.7). Taking into account that $|\overset{o}{\vec{e}}_1| = |\overset{o}{\vec{e}}_2| = |\overset{o}{\vec{e}}_3| = 1$, one can obtain

$$\left|\hat{\vec{e}}_1\right| = \left|\hat{\vec{e}}_2\right| = \frac{r}{r_0}, \quad \left|\hat{\vec{e}}_3\right| = \left(\frac{L}{L_0}\right)\sqrt{L^2 + r^2\varphi^2} \tag{B.90}$$

Taking into consideration the incompressibility condition $(r_0)^2 L_0 = r^2 L$ one can find the matrix of the metric tensor of the accompanying system of co-ordinates at point M' and the moment of time $t \geq t_0$ as follows:

$$\{g_{ij}\} = \begin{bmatrix} \frac{L_0}{L} & 0 & 0 \\ 0 & \frac{L_0}{L} & \frac{r_0 \varphi}{L} \\ 0 & \frac{r_0 \varphi}{L} & \frac{r_0^2 \varphi^2}{L L_0} \end{bmatrix} \tag{B.91}$$

At time $t = 0$ the matrix of the metric tensor is a unitary matrix, that is $\overset{0}{g}_{ij} = \delta_{ij}$, where δ_{ij} is Kronecker delta. Matrix inverse of (B.91) may be found by standard procedures:

$$\{\hat{g}^{ij}\} = \begin{bmatrix} \frac{L}{L_0} & 0 & 0 \\ 0 & \frac{L}{L_0} + \frac{r_0^2 \varphi^2}{L^2} & -\frac{r_0 \varphi L_0}{L^2} \\ 0 & -\frac{r_0 \varphi L_0}{L^2} & \left(\frac{L_0}{L}\right)^{20} \end{bmatrix} \tag{B.92}$$

Covariant components of the strain rate tensor may be found in accordance with the formulae $\xi = \left(\frac{1}{2}\right) d\overline{g}_{ij} / dt$

$$\{\xi_{ij}\} = \begin{bmatrix} -\frac{L_0 v}{2L^2} & 0 & 0 \\ 0 & -\frac{L_0 v}{2L^2} & \frac{r_0 \omega}{2L} - \frac{r_0 \varphi v}{2L^2} \\ 0 & \frac{r_0 \omega}{2L} - \frac{r_0 \varphi v}{2L^2} & \frac{L v}{L_0^2} + \frac{r_0^2 \varphi \omega}{L L_0} - \frac{r_0^2 \varphi^2 v}{2 L_0 L^2} \end{bmatrix}$$

where $v = dL/dt$ and $\omega = d\varphi/dt$. The mixed components of the strain rate tensor can be determined from $\xi^i{}_j = \xi_{sj} \hat{g}^{is}$.

$$\left\{\xi^i_j\right\}=\begin{bmatrix} -\dfrac{v}{2L} & 0 & 0 \\ 0 & -\dfrac{v}{2L}-\dfrac{r_0^2\varphi\omega L_0}{2L^3} & \dfrac{r_0\omega}{2L_0}-\dfrac{r_0^3\varphi^2\omega}{2L^3}-\dfrac{3r_0\varphi v}{2LL_0} \\ 0 & \dfrac{r_0L_0^2\omega}{2L^3} & \dfrac{v}{L}+\dfrac{r_0^2\varphi\omega L_0}{2L^3} \end{bmatrix} \tag{B.93}$$

It should be mentioned that due to the condition of incompressibility the trace $\xi^i_i = 0$. Consequently, the strain rate tensor coincides with its deviator. The invariants of the strain rate tensor are to be found by using its mixed components.

$$\begin{gathered} I_1 = 0,\ I_2 = -\frac{3v^2}{4L^2}-\frac{(r_0\omega)^2 L_0}{4L^3}, \\ I_3 = \frac{v^3}{4L^3}+\frac{(r_0\omega)^2 L_0 v}{8L^4} \end{gathered} \tag{B.94}$$

If one reduces the strain rate tensor to its principal axes one could find that the principal strain rates are as follows.

$$\xi_1 = -\frac{v}{2L};\xi_{2,3} = \frac{v}{4L}\pm\frac{1}{2}\sqrt{\frac{r_0^2\omega^2 L_0}{L^3}+\frac{9v^2}{4L^2}} \tag{B.95}$$

The equivalent shear strain rate $\dot{\gamma}$ is defined as twice the square root of the absolute value of the second invariant of the strain rate deviator.

$$\dot{\gamma} = 2\sqrt{|I_2|} = \left(\frac{1}{L}\right)\sqrt{r^2\omega^2+3v^2} \tag{B.96}$$

The equivalent strain rate ξ_e is as follows.

$$\xi_e = \sqrt{\frac{2}{3}\xi^i_j\xi^j_i} = \frac{1}{L\sqrt{3}}\sqrt{3v^2+r^2\omega^2} \tag{B.97}$$

Characteristics of the deformed state:

It is known that for infinitesimal strains there is only one strain tensor. The components of this tensor have obvious meaning: three diagonal components describe the linear deformations while the non-diagonal components describe the shear strains. Besides, for most of the metal working processes, the changes in the metal volume are negligible so that the sum of the diagonal components is equal to

0 (incompressibility condition). And so the strain tensor coincides with its deviator. For finite strain one can introduce a number of strain tensors. Every strain tensor allows an unambiouous description of the strain state, but the components of all tensors do not have a self-evident meaning as in the case of infinitesimal strains. The strain tensors in Eulerian and Lagrangian representations are further considered.

Strain tensors: Covariant components of the strain tensors in Eulerian and Lagrangian representations are the same:

$$\{\varepsilon_{ij}\}=\begin{bmatrix} -\frac{e}{[2(1+e)]} & 0 & 0 \\ 0 & -\frac{e}{[2(1+e)]} & \frac{\delta}{[2(1+e)]} \\ 0 & \frac{\delta}{[2(1+e)]} & \frac{e+e^2}{\frac{2+\delta^2}{[2(1+e)]}} \end{bmatrix} \tag{B.98}$$

where $\delta = r_0\varphi/L_0$ (δ is the accumulated shear strain for the case of simple torsion, $\delta = \varepsilon_{\text{tors}}\surd 3$); $e = (L - L_0)/L_0$ (e is engineering strain, $e{=}e_t$).

Eulerian representation: Here, the contravariant, mixed and covariant components are the same, since it has been assumed that in the initial state the basis of the accompanying system of co-ordinates are orthonormal. Then, one can find the invariants of the deformed state as

$$\overset{0}{I}_1 = \frac{\left(\frac{e^2}{2}\right)(3+e)}{(1+e)} + \frac{\delta^2}{[2(1+e)]}, \qquad \overset{0}{I}_2 = -\frac{e^2\left(3+6e+2e^2\right)}{[4(1+e)^2]} - \frac{\delta^2(1+2e)}{[4(1+e)^2]} \tag{B.99}$$

Lagrangian representation: Mixed components of the strain tensor are as follows. $\varepsilon_1^i = \varepsilon_{si} g^{is}$

$$\{\hat{\varepsilon}_j^i\} = \begin{bmatrix} -\frac{e}{2} & 0 & 0 \\ 0 & -\frac{e}{2}-\frac{\delta^2}{[2(1+e)]} & \frac{\delta}{[2(1+e)^2]} \\ 0 & \frac{\delta}{[2(1+e)]} & \frac{(e+e^2/2)}{(1+e)^2} \end{bmatrix} \tag{B.100}$$

Invariants of the strain tensor are

$$\hat{I}_1 = -\frac{e^2\left(e+\frac{3}{2}\right)}{(1+e)^2} - \frac{\delta^2(1-e)}{[4(1+e)^2]};$$
$$\hat{I}_2 = -\frac{e^3(3-e^2)}{[4(1+e)^2]} - \frac{\delta^2}{[2(1+e)^2]}; \quad \text{(B.101)}$$
$$\hat{I}_3 = \frac{e^3\left(1+\frac{e}{2}\right)}{[4(1+e)^2]} + \frac{e\delta^2}{[8(1+e)^2]}$$

Strain deviator

Strain tensor as well as any other tensor can be decomposed into spherical (hydrostatic) and deviatoric parts. The strain deviator is of special interest.

Eulerian representation

$$\tilde{\varepsilon}_{ij} = \begin{bmatrix} -a-b & 0 & 0 \\ 0 & -a-b & c \\ 0 & c & 2a+2b \end{bmatrix} \quad \text{(B.102)}$$

where $a = (e + e^2 + e^3/3)/[2(1 + e)]$; $b = \delta^2/\ [6(1+e)]$; $c = \delta/[2(1+e)]$. Invariants of the strain deviator in Eulerian representation:

$$\overset{0}{I}_{1D} = 0;\ \overset{0}{I}_{2D} = -\frac{3\left(\frac{e+e^2+e^3}{3}+\frac{\delta^2}{3}\right)^2}{[4(1+e)^2]} - \frac{\delta^2}{[4(1+e)^2]}$$
$$\overset{0}{I}_{3D} = \frac{\left(\frac{e+e^2+e^3}{3}+\frac{\delta^2}{3}\right)\left[2\left(\frac{e+e^2+e^3}{3}+\frac{\delta^2}{3}\right)^2 - \delta^2\right]}{[8(1+e)^4]} \quad \text{(B.103)}$$

Lagrangian representation

$$\left\{\hat{\tilde{\varepsilon}}_{ij}\right\} = \begin{bmatrix} -a'+b' & 0 & 0 \\ 0 & -a'-2b' & c' \\ 0 & c' & 2a'+b' \end{bmatrix} \quad \text{(B.104)}$$

where $a' = (e+e^2+e^3/3)/[2(1+e)^2]$; $b' = \delta^2/[6(1+e)^2]$; $c' = \delta/[2(1+e)^2]$. Invariants of the strain deviator in Lagrangian representation are

$$\hat{I}_{1D} = 0; \;\; \hat{I}_{3D} = \det \hat{\tilde{\varepsilon}}_j^i$$

$$\hat{I}_{2D} = -3\left\{\begin{bmatrix} \dfrac{\left(\dfrac{e+e^2+e^3}{3}\right)}{[2(1+e)^2]} \\ \\ \end{bmatrix}^2 - 3\delta^2 \frac{\left(\dfrac{e+e^2+e^3}{3}\right)}{[12(1+e)^4]} - \frac{3\delta^4}{[36(1+e)^4]} - \frac{\delta^2}{[4(1+e)^4]}\right. \tag{B.105}$$

The strain intensity in Eulerian representation is

$$\overset{0}{\varepsilon}_i = \sqrt{\left(\frac{e+e^2+e^3}{3} + \frac{\delta^2}{3}\right)^2 + \frac{\dfrac{\delta^2}{3}}{(1+e)}} \tag{B.106}$$

The strain intensity in Lagrangian representation

$$\hat{\varepsilon}_i = \sqrt{\left(\frac{e+e^2+e^3}{3}\right)^2 + \left(\frac{1+e+e^2+e^3}{3}\right)\frac{\delta^2}{3} + \frac{\dfrac{\delta^4}{9}}{(1+e)^2}} \tag{B.107}$$

Consider some particular cases. For the case of simple tension one can obtain the following.

$$\left.\overset{0}{\varepsilon}_i\right|_{\delta=0} = e\frac{\dfrac{1+e+e^2}{3}}{1+e}; \;\; \left.\hat{\varepsilon}_i\right|_{\delta=0} = e\frac{\dfrac{1+e+e^2}{3}}{(1+e)^2} \tag{B.108}$$

In Fig. B.8 the results of the calculations in accordance with Eqs. (B.108) are presented. One can see that the following inequalities are present in the interval $0 \le e \le 3$: $\overset{0}{\varepsilon}_i \le \varepsilon_t \le \hat{\varepsilon}_i$. However, with increasing e the lower curve in Fig. B.8 approaches the dashed curve, intersects it at $e \approx 3.7$ and further as $e \to \infty$ lies on the asymptote $\hat{\varepsilon}_i \to \varepsilon/3$. This is in agreement with Eq.(B.108).

For another particular case (simple torsion) one can obtain

$$\left.\overset{0}{\varepsilon}_i\right|_{\varepsilon=0} = \left.\hat{\varepsilon}_i\right|_{\varepsilon=0} = \frac{\delta}{\sqrt{3}}\sqrt{1+\frac{\delta^2}{3}} \tag{B.109}$$

In Fig. B.9 the results of the calculations are presented. For comparison the curve calculated in accordance with equation given in Chap.5 for a torsion test is also included. One can see that the following relations are valid. $\varepsilon_{11} \leq \varepsilon_{\text{tors}} \leq \overset{0}{\varepsilon}_i = \hat{\varepsilon}_i$ If one analyses Fig. B.8 and Fig. B.9, one will conclude that there is a problem in selecting a measure for the deformed state. Actually, the *same* deformed state may be characterised by *different* invariants. Thus, the strain intensity can be estimated on the basis of different strain tensors. If one uses some other strain tensor (other than what was considered above), e.g., the logarithmic strain tensor one will obtain another expression for the strain intensity. It should be mentioned that the non-uniqueness of the tensor description of the strain state gives rise to the problem of selection of the measures of a strain. This problem can not be solved within the framework of continuum mechanics, because all descriptions of the strain state will have equal legitimacy. So, the selection will have to depend on the judgement of the researcher (see Chap.5 also).

Accumulated strain: The value of the accumulated strain for combined loading by an axial force and a torque may be found by substituting (B.96) in (A.63).

$$\varepsilon_e = \int_0^t \xi_e dt = \int_0^t \frac{1}{L\sqrt{3}} \sqrt{3[v(t)]^2 + r^2 [\omega(t)]^2} dt \tag{B.110}$$

where the integral should be taken along the deformation trajectory. It is evident that the value of P depends on the specific functional forms of $v(t)$ and $\omega(t)$. In practice the most frequently used regime of loading is v = constant, ω = constant. In this case one can find that

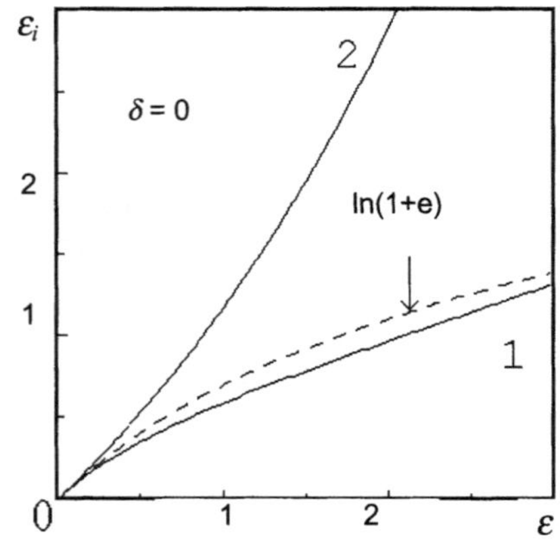

Fig. B.8. The dependences of the strain intensity in Lagrangian (1) and Eulerian (2) approaches on the relative strain. Dashed line – true strain.

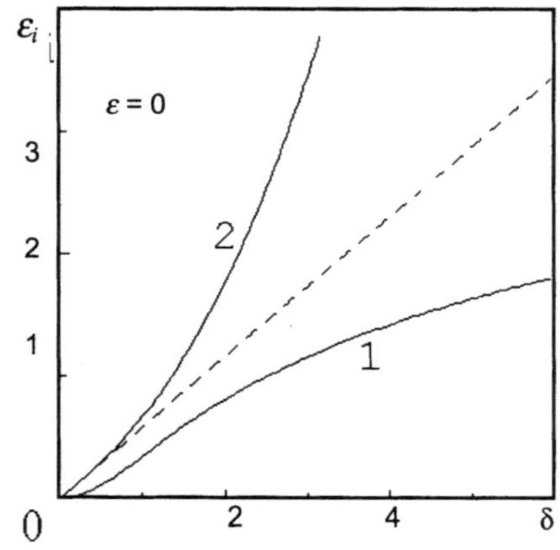

Fig. B.9. The dependences of $\varepsilon_{[11]}$ (curve 1) and the strain intensity (curve 2) on the value of δ. The dashed line shows the corresponding dependence for $\varepsilon_{\text{tors}}$

$$\varepsilon_e = 2\sqrt{1+\frac{\theta^2}{3}} - 2\sqrt{1+\frac{\theta^2}{3(1+e)}} + \ln(1+e) + \ln\frac{1+\dfrac{\theta^2}{6(1+e)}+\sqrt{1+\dfrac{\theta^2}{3(1+e)}}}{1+\dfrac{\theta^2}{6}+\sqrt{1+\dfrac{\theta^2}{3}}} \tag{B.111}$$

where $e = v_0 t/L_0$ and $\theta = r_0\omega_0/v_0$. It can be seen that Eq.(B.111) reduces as follows: as $\omega \to 0$, $\varepsilon_e \to \ln(1+e)$; as $e \to 0$, $\varepsilon \to r_0\varphi/L\sqrt{3}$. In other words, the expression includes simple tension and simple torsion as particular cases.

Equation (B.111) was derived above for finite strain behaviour. It has been used in a number of publications (see Chaps. 3 and 5).

Appendix C: On Dimensional Analysis

Dimensional analysis is not a branch of mechanics of solids. The methods of dimensional analysis are used in many branches of natural sciences. One will find below some elementary notes on this subject as it is used in mechanics of solids. Details can be found, for example, in [614, 615].

C.1 Basic Concepts

Study of any mechanical phenomenon is directly related to measurements of physical quantities. Dimensional analysis is the method of establishing relationships among physical variables based on their dimensions. Variables whose magnitude depends on the unit of measurement are considered to be quantities with dimensions (e.g., axial force, torque, etc.). Otherwise one deals with non-dimensional values (e.g., Poisson's ratio).

The units of measurement for different variables are, in general, interrelated. For example, the unit of velocity, $v = dl/dt$, is related to those of length, l, and time, t. It is essential to select the primary, or the principal units of measurement. For instance, the units of length L, time T and mass M are often introduced as the primary units (e.g., the well known CGS system). All other units then become secondary or derived as they can be expressed through the primary units. For example, one can write for axial force

$$P = \frac{ML}{T^2} = MLT^{-2} \tag{C.1}$$

Similarly, all other secondary units can be expressed as power relations of the kind

$$N = M^m L^l T^t \tag{C.2}$$

One may use a different set of primary units, e.g., the MKS system. In general, one can use an arbitrary set of primary units. However, in this case the above mentioned expression for velocity, for example, should be written as $v = k^* \, ds/dt$, where k^* is some dimensionless coefficient.

If one were to select the three primary units L, T and M, then one can express the dimension of any other mechanical variable as

$$[y] = L^p M^q T^r \tag{C.3}$$

where p, q, r are constants.

Consider the following simple example. It is known from experiments that a freely falling body obeys the relationship $l = gt^2/2$, where $g = 9.81$ m/s^2. One can obtain this dependence using dimensional analysis. Assume that the value of l depends on g and t

$$l = f(g,t) \tag{C.4}$$

One can find the functional form f in Eq.(C.4) empirically. But that method will require many experiments. Alternatively, one can apply the well known Buckingham π-theorem to solve this problem.

The formulation of the π-theorem is as follows. If some dimensional variable a is a function of a number of independent dimensional variables a_1, a_2, ...,a_n, and $k \leq n$ among them have primary dimensions, then this functional relationship can be represented as $\pi = f(\pi_1,\pi_2, \ldots\pi_{n-k})$, where π, π_1, π_2,..., π_{n-k} are non-dimensional combinations composed of n+1 dimensional variables a, a_1, a_2,..., a_n.

Therefore $[l] = L$, $[g] = LT^{-2}$, $[t] = T$, $n = 2$, $k = 2$. In accordance with π-theorem one can conclude that function (C.4) is $\pi = c$, where π is a non-dimensional variable composed of l, g, t, while c is a constant. One can obtain π as l/gt^2. Hence, Eq.(C.4) can be presented as $l/gt^2 = c$ or

$$l = cgt^2 \tag{C.5}$$

Thus, it is enough to carry out a single experiment in order to determine the value of $l/gt^2 = c$ and to obtain the approximate value of $c \approx 0.5$. Another example can be considered. Let T' be the period of oscillation of a pendulum, the length of which is L. Proceeding from dimensional analysis one can conclude that

$$T' = \varphi(\alpha)\sqrt{\frac{L}{g}} \tag{C.6}$$

where α is the angle, $g = 9.81$ m/s^2. The functional form of $\varphi(\alpha)$ on the right hand side of Eq. (C.6) can not be found from dimensional analysis. For small oscillations one can decompose $\varphi(\alpha)$ into a Taylor series and retain only the first term. Then, the following relationship can be obtained:

$$T' = C\sqrt{\frac{L}{g}} \tag{C.7}$$

where C is a constant that can be found from experiments.
The differential equation for a pendulum is

$$\frac{d^2\alpha}{dt^2} + \frac{g}{L}\sin\alpha = 0 \tag{C.8}$$

with initial conditions $\alpha(0) = \alpha_0$, $\dot{\alpha}(0) = \dot{\alpha}_0$.

For small oscillations ($|\alpha| << 1$) one can rewrite Eq.(C.8) as

$$\frac{d^2\alpha}{dt^2} + \frac{g}{L}\alpha = 0 \tag{C.9}$$

the solution of which is

$$\alpha(t) = \alpha_0 \cos\omega t + \frac{\alpha_0}{\omega}\sin\omega t, \text{ where } \omega = \sqrt{\frac{g}{L}} \tag{C.10}$$

One can see then that the period of oscillations is equal to

$$T' = \frac{2\pi}{\omega} = 2\pi\sqrt{\frac{L}{g}} \tag{C.11}$$

Comparing Eq.(C.7) and Eq.(C.11) one obtains $C = 2\pi$.

In the literature on superplasticity the standard power relation $\sigma = K\xi^m$ which describes the properties of a non-linear viscous fluid is often used. Therefore, we consider below some examples of viscous flow (linear and non-linear).

C.2 Viscous Flow

Dimensional analysis provides a fundamental theoretical basis for modelling mechanical states and processes. It is widely used in modelling different natural processes. One can model a material or a process. The following kinds of modelling can be considered. Geometrical modelling means that the material to be used remains the same, but the geometrical dimensions are different, e.g., model of a dam for generating hydro-electric power. Physical modelling assumes the use of another material for modelling, e.g., the rheological behaviour of steels at high temperatures can be modelled by standard uniaxial tests on lead at room temperature. In practice, various combinations of physical and geometrical modelling are often used. While doing this certain conditions of modelling should be observed. This will ensure similarity between the model and the material/process being modelled.

Consider the conditions of similarity for the isothermal flow of two newtonian viscous incompressible fluids of different densities and viscosities. Differential equations of motion for this flow are known as Navier–Stokes equations.

$$\frac{\partial v_i}{\partial t}+v_i\frac{\partial v_i}{\partial x_k}=F_i-\frac{1}{\rho}\cdot\frac{\partial p_H}{\partial x_i}+v\Delta v_i,(i,\ k=1,\ 2,\ 3) \tag{C.12a}$$

The compressibility condition is

$$\frac{\partial v_i}{\partial x_i}=0 \tag{C.12b}$$

where the summation convention of indices is assumed. Here $v=\mu/\rho$ is the coefficient of kinematic viscosity, μ the coefficient of dynamic viscosity, ρ the density, F_i the components of the vector of mass forces, v_j components of the velocity vector in Cartesian system of co-ordinates x^1, x^2, x^3 and Δ is Laplace's delta.

Reducing the Navier-Stokes equations to a normalised form using typical values of time, T, velocity, V, length, L, pressure, P and mass forces, F and denoting further by a prime symbol the normalised values of time, co-ordinates, velocities, pressures and forces,

$$\begin{aligned}&t=T\,t',V_i=Vv_i',v=V_0v',\ F_k=FF_k'\\&p_H=pp_H'\end{aligned} \tag{C.13}$$

Substituting Eq.(C13) in the Navier-Stokes equations one obtains

$$st\frac{\partial v_i'}{\partial t'}+v_j'\frac{\partial v_i'}{\partial x_j'}=\frac{1}{Fr}F_i'-Eu\frac{\partial p_H'}{\partial x_i'}+\frac{1}{\mathrm{Re}}\Delta v_i' \tag{C.14}$$

where the following classical dimensionless combinations of newtonian hydrodynamics are present.

$st=\dfrac{L}{vT}$ – Stroudel number , $Eu=\dfrac{P}{\rho v^2}$ – Euler number

$Fr=\dfrac{v^2}{FL}$ – Froude number , $Re=\dfrac{vL\rho}{\mu}=\dfrac{vL}{v}$ – Reynolds number

It is to be noted that for cases where two or more parameters have the same dimension, one may introduce their ratios, e.g., if L is the length of a tube and D its diameter, one can consider their ratio L/D.

Two physical phenomena are considered to be similar if the values characterising one can be calculated from those of the other by simply multiplying them with what are termed as coefficients of similarity. From this it follows that the following conditions should be satisfied

$$st_1=st_2,\quad Eu_1=Eu_2,\quad Fr_1=Fr_2,\quad Re_1=Re_2 \tag{C.15}$$

C.3
Non-Newtonian Flow

In case of non-newtonian flow, the natural time τ is usually added to the dimensional parameters (T, ν, L, F, ρ, ν). Such an approach can take into account the fact that stress is not a single valued function of strain rate; in general, its magnitude is determined by the history of loading. Evidently, full history of loading can not always be known. Therefore, the so-called principle of fading memory is introduced, i.e., the influence of the history of straining on the value of stress is weaker for a more distant history than for a recent one.

Consider, as an example, the classical Maxwell body (series connection of an elastic and a viscous element). The time dependence of flow stress at constant strain rate loading is given by

$$\sigma = \lambda \xi_0 \left[1 - exp\left(-\frac{Et}{\lambda} \right) \right] \tag{C.16}$$

where λ is the viscosity, E is Young's modulus. One can see that the characteristic time is given by $\tau_M = \lambda/E$. Evidently the above analysis is valid for instant loading. When the characteristic time τ_L increases with strain rate for $\tau_L << \tau_M$, the loading can be considered to be instantaneous and Eq.(C.16) is valid. If $\tau_L >> \tau_M$, then loading is quasi-stationary and the Maxwell body behaves like a viscous fluid.

The main dimensionless parameter of non-newtonian hydrodynamics is the Weissenberg number.

$$We = \tau \dot{\gamma} \tag{C.17}$$

where $\dot{\gamma}$ is the characteristic shear strain rate. $\dot{\gamma} = v/L$ and so

$$We = \frac{\tau v}{L} \tag{C.18}$$

In this case the Weissenberg number determines the ratio of the inertia forces to the shear forces. Sometimes, the Deborah number is considered.

$$De = \frac{\tau}{T_1} \tag{C.19}$$

where T_1 is the characteristic time scale of flow. Often, $T_1 = L/v$ and so

$$De = \frac{v\tau}{L} \tag{C.20}$$

C.4 Superplastic Flow

The following relation is often considered in the literature on superplasticity

$$\sigma = K\xi^m \tag{C.21}$$

instead of

$$\frac{\sigma}{\sigma_q} = \left(\frac{\xi}{\xi_q}\right)^m \tag{C.22}$$

Evidently, even an infinitesimally small variation in the value of m will automatically lead to a change in the dimension of K. Therefore, the mathematical form of Eq. (C.21) is not correct. It should be considered as a corrupted form of Eq.(C.22) where σ_q, ξ_q are the values of the reference flow stress and the corresponding (reference) strain rate. It is pertinent to note that the universal superplasticity curve $M/M_{max} - \log\xi/\xi_{opt}$ can be obtained directly using dimensional analysis. It is repeatedly noted in literature that for inducing superplastic flow in a polycrystalline material the average grain size should obey the relationship

$$d < d^* \tag{C.23}$$

where d^*=10–15 μm. The problem is that Eq.(C.23) is expressed in dimensional form. Recently, Astanin et. al. [140] have shown that superplasticity is lost when the thickness of a specimen is less than about 10 (average) grain diameters. This gives rise to a problem of introducing some dimensionless parameters connected with characteristic linear sizes.

C.5 Dimensionless Parameters for the Boundary Value Problem of Superplasticity

As shown above, dimensionless parameters usually termed 'similarity numbers' appear during the normalisation of Navier–Stokes equations. As a result, the numbers of Stroudel, Froude, Euler and Reynolds are obtained. Similarly, some dimensionless parameters and similarity numbers can be obtained during the normalisation of the system of equations corresponding to the boundary value problem (BP) of a technological, process. If one uses the standard power relation $\sigma = K\xi^m$ as the constitutive equation (CE) to set up the boundary value problem, one can obtain the Weissenberg number as a result of the normalisation of the system of equations. In general, one may obtain new similarity numbers. One will find below a scheme that allows non-dimensional criteria for one variant of the constitutive equation and boundary conditions to be obtained. This scheme seems to be useful in analysing superplastic flow.

Let the body to be deformed occupy volume *V*, the boundary Γ of which is assumed to be part-wise smooth. Let the following formulation of the BP be considered. Incompressibility condition

$$div\,\vec{v} = 0 \tag{C.24}$$

Equations of motion:

$$\rho\left(\frac{dv_i}{dt} - F_i\right) = \frac{\partial \sigma_{ij}}{\partial x_j} \tag{C.25}$$

CE: Vector properties

$$S_{ij} = \left(\frac{2\sigma_e}{3\xi_e}\right)\xi_{ij} \tag{C.26}$$

scalar properties:

$$\sigma_e = \Phi(s, \xi_e) \tag{C.27}$$

Boundary conditions (Γ is the total surface of the deforming body, $\Gamma = \Gamma_k \cup \Gamma_c \cup \Gamma_s$) Kinematic Boundary conditions are given on

$$\Gamma_k \subset \Gamma:\ \ \vec{v} = \vec{f}(x_b, t) \tag{C.28}$$

Mixed BC are given on $\Gamma_c \in \Gamma$:

$$v_i \frac{\partial F_c}{\partial x_i} = 0 \ \ or \ \ \left(\frac{\partial F_c}{\partial t}\right) + v_i\left(\frac{\partial F_c}{\partial x_i}\right) = 0; \tag{C.29}$$

$$\frac{\vec{\tau}_m}{\tau_m} = -\frac{\Delta\vec{w}}{|\Delta\vec{w}|},\ \tau_m \equiv |\vec{\tau}_m| = \varphi(\sigma_N, |\Delta\vec{w}|, \nu, ...) \tag{C.30}$$

Dynamic boundary conditions are given on Γ_N:

$$\Gamma_s \subset \Gamma:\ \ \sigma_{ij} n_j = q_i \tag{C.31}$$

where ν is the friction coefficient. Dynamic BC are given on N. See details in Section 4.1.2. It is assumed further that the initial conditions are the simplest (zero BC).

Let some typical values of the following parameters be known: length l_0, velocity v_0, stress σ_0, acceleration g_0. The characteristic time, τ, can then be introduced as

$$\tau = \left(\frac{v_0 t}{l_o}\right) \tag{C.32}$$

Let the normalised variables be denoted by the same symbols. Then Eq. (C.25) will remain the same, while Eq. (C.26) transforms as

$$\frac{dv_i}{d\tau} = \frac{\sigma_0}{\rho v_0^2}\frac{\partial \sigma_{ij}}{\partial x_j} + \frac{g_0 l_0}{v_0^2}F_i \tag{C.25a}$$

where there are two normalised parameters.

$$\kappa_1 = \frac{\sigma_0}{\rho (v_0)^2}, \kappa_2 = \frac{g_0 l_0}{(v_0)^2} \tag{C.33}$$

It is easy to show that Eqs. (C.26), (C.28) and the left hand side of Eqs. (C.30) do not lead to new dimensionless parameters during normalising.

Let ξ_s be the characteristic value of the strain rate. Then one has from (C.27)

$$\sigma_e = \Phi\left(s, \left[\frac{v_0}{\xi_s l_0}\right]\xi_e\right) \tag{C.27a}$$

and a new dimensionless parameter

$$\kappa_3 = \frac{\xi_s l_0}{v_0} \tag{C.34}$$

The following parameter can be derived from Eq.(C.30)

$$\kappa_4 = v \tag{C.35}$$

Finally, one can derive from Eq.(C.31)

$$\sigma_{ij} n_j = q_i \left(\frac{q_0}{\sigma_0}\right) \tag{C.36}$$

where q_0 is some typical value of the stress and so

$$\kappa_5 = \frac{q_0}{\sigma_0} \tag{C.37}$$

Now consider two processes: i) a natural process characterised by parameters κ_i^H; ii) a model process characterised by parameters κ_i^M. If $\kappa_i^M = \kappa_i^H$, then both the processes are described by the same equations, and so they are similar.

Consider the situation when these two processes are geometrically similar through some coefficient k'. It is assumed that the scalar properties Eq.(C.27) are the same and so $\kappa_3^M = \kappa_3^H$, or $(\xi_s l_0/v_0)^M = (\xi_s l_0/v_0)^H$. If $(\xi_s)^M = (\xi_s)^H$, then, taking into account $k = (l_0)^H/(l_0)^M$, one obtains $(v_0)^M = (v_0)^H/k$. Other equalities $\kappa_i^M = \kappa_i^H$ (i = 1, 2, 4, 5) lead to

$$(\sigma_0)^M = \frac{(\sigma_0)^H \rho^M}{\rho^H k^2}; \tag{C.38}$$

$$(g_0)^M = (g_0)^H \tag{C.39}$$

$$(q_0)^M = \frac{(q_0)^H \rho^M}{\rho^H k^2} \tag{C.40}$$

$$v^M = v^H \tag{C.41}$$

The equality $\tau^M = \tau^H$ leads to the equality of physical times $t^M = t^H$. The following difficulties in modelling can be mentioned now.

Equation (C.38) gives $\kappa_1 = \rho^M/\rho^H k^2$ and even when k is not very small one has to seek a very 'soft' material for the model process since the densities of materials do not differ considerably. Condition Eq.(C.39) can not be satisfied; fortunately, this condition is not important for most technological processes of metal working. (Mass forces are usually small compared with the deformation forces.) Equation (C.40) establishes the relation between surface loads.

The problem of modelling can be solved for the following two cases. If the stress does not depend on strain rate, the value of κ_3 becomes unimportant and the condition $\kappa_i^M = \kappa_i^H$ can be satisfied if one chooses $(v_0)^M = (v_0)^H$. The physical times are related as $t^M = t^H/k$ if the model material and that to be modelled are the same. In the second case of negligible influence of the inertia forces modelling is possible by the use of the same material; in this case the physical times will coincide $t^M = t^H$.

The following theorem proved by Kirpichev in 1874 may be of interest in studying the mechanics of superplastic flow: all isothermal slow flows are always similar if the mass forces can be neglected and the diagrams $\sigma_e = \Phi$ (s) are similar and the friction coefficients ν are identical.

Now consider non-isothermal plastic flow. In this case flow stress σ and contact friction shear stress τ_m will depend on temperature T:

$$\sigma_u = \Phi(s, \xi_e, T) \tag{C.42}$$

$$\tau_m = \varphi_c(v, \sigma_N, \xi_e, T, \ldots), \; on \quad \Gamma_c \tag{C.43}$$

Temperature distribution in the material to be deformed and the instrument are described by the following equations.

$$\frac{dT}{dt} = a^2 \Delta T + k_0 \sigma_u \xi_e \tag{C.44}$$

$$\frac{dT_2}{dt} = (a_2)^2 \Delta T_2 \tag{C.45}$$

As an example, the temperature BC of the following type is considered.

$$\left(\frac{\partial T}{\partial n}\right) = 0, \; \left(\frac{\partial T_2}{\partial n}\right) = 0 \; on \; \Gamma_s \; and \; \Gamma_c \tag{C.46}$$

$$T = T_2, \quad \lambda\left(\frac{\partial T}{\partial n}\right) - \lambda_2\left(\frac{\partial T_2}{\partial n}\right) = \frac{\tau_m |\Delta \vec{w}|}{J} \; on \; \Gamma_k \tag{C.47}$$

Initial conditions:

$$t = 0, \quad T = T_0 \Psi(x), \; T_2 = T_{20} \Psi(x) \tag{C.48}$$

Let T_0 and T_{20} be some typical values of temperature. Then one can obtain normalising Eqs. (C43) and (C44)

$$\begin{aligned} \frac{dT}{d\tau} &= \left(\frac{a^2}{v_0 l_0}\right) \Delta T + \frac{(\xi_s \sigma_0 l_0)\; \sigma_u \xi_e k_0}{\rho_0 c_0 v_0 T_0 J} \\ \frac{dT_2}{d\tau} &= \left[\frac{(a_2)^2}{v_0 l_0}\right] \Delta T_2 \end{aligned} \tag{C.49}$$

Correspondingly, the following dimensionless parameters can be derived:

$$\kappa_6 = \frac{a^2}{v_0 l_0}, \kappa_7 = \frac{(a_2)^2}{v_0 l_0}$$
$$\kappa_8 = \frac{\sigma_0}{\rho_0 c_0 T_0 J}, \ \kappa_9 = \kappa_3 \kappa_8 \tag{C.50}$$

Eq. (C.46) does not result in new dimensionless parameters, while Eqs. (C.49) give

$$T = \frac{T_2 T_{20}}{T_0}, \quad \frac{\partial T}{\partial n} - \frac{\lambda_2 T_{20}}{\lambda T_0}\frac{\partial T}{\partial n} = \frac{\sigma_0 l_0 v_0}{\lambda T_0 J}\tau_m |\Delta \vec{w}| \tag{C.51}$$

Correspondingly,

$$\kappa_{10} = \frac{T_{20}}{T_0}, \ \kappa_{11} = \frac{\lambda_2}{\lambda};$$
$$\kappa_{12} = \kappa_{10}\kappa_{11}, \ \kappa_{13} = \frac{\sigma_0 l_0 v_0}{\lambda T_0 J} \tag{C.52}$$

Thus, the number of parameters become significantly greater and their structure becomes more complicated. Then, full similarity of processes is not possible to achieve in practice. Let the system of parameters κ_i (i = 1, 2, ... ,13) be analysed assuming the following conditions

$$(\xi_s)^M = (\xi_s)^H, (T_0)^M = (T_0)^H, (T_{20})^M = (T_{20})^H \tag{C.53}$$

Equations (C.53) provide the similarity of Eq.(C.42) and the first of relationships Eq.(C.43) and somewhat simplify the analysis.

One can obtain from the conditions $\kappa_i^M = \kappa_i^H$ (i = 3, 6, 7)

$$(v_0)^M = \frac{(v_0)^H}{k}, \ a^M = \frac{a^H}{k}, (a_2)^M = \frac{(a_2)^H}{k} \tag{C.54}$$

The conditions $\kappa_i^M = \kappa_i^H$ (i = 8, 13) lead to

$$(\sigma_0)^M = \frac{k^2 (\sigma_0)^H \lambda^M}{\lambda^H} \tag{C.55}$$

Condition $\kappa_i^M = \kappa_i^H$ for i = 10 is satisfied automatically. For i = 11 one can derive that $(\lambda_2)^M = (\lambda_2)^H \lambda^M/\lambda^H$. Taking into account that $a^M = a^H/k$, one can derive that $(\lambda_2)^M = (\lambda_2)^H \rho^M/(c_0)^M / \rho^H/(c_0)^H k^2$. From $\kappa_i^M = \kappa_i^H$ (i = 1) it follows that $(\sigma_0)^M = (\sigma_0)^H \rho^M/\rho^H k^2$. Consequently one obtains

$$\lambda^{M} = \frac{\lambda^{H} \rho^{M}}{\rho^{H} k^{4}} \tag{C.56}$$

One can conclude from Eqs. (C.54)–(C.56) that a general choice of material for modelling is not possible in practice. However, the situation becomes favourable for the case of steady state (stationary) slow flow at high temperatures when the mass forces are negligible. In this case, one can exclude the parameters κ_1, κ_2, κ_3 as they are not important.

Assume that the material of the model is the same as that of the process to be modelled. Then, $\kappa_{\alpha}^{M} = \kappa_{\alpha}^{H}$, $\alpha =$ 4, 5, 8, 10, 11, 12 and the requirements for the temperatures will be the same at corresponding moments of time. This requirement can be satisfied by providing $\kappa_{\beta}^{M} = \kappa_{\beta}^{H}$, $\beta = 6, 9, 13$. Actually, for $\beta = 6$ one has

$$(v_0)^{M} = k(v_0)^{H} \tag{C.57}$$

while for $\beta = 13$ one has an identity. For $\beta = 9$ $(\xi_s)^M = k^2(\xi_s)^H$. Physical times are as follows:

$$t^{M} = \frac{t^{H}}{k^{2}} \tag{C.58}$$

Thus, the following rules should be observed while modelling high-temperature plastic flow: temperature fields should be similar while the process of plastic flow should be carried out faster in accordance with condition (C.58). The stresses and power characteristics are to be recalculated in accordance with the following formulae

$$P^{H} = P^{M}, \quad p^{H} = k^{2} p^{M}, \quad A^{H} = k^{3} A^{M}, \quad N^{H} = kN^{M} \tag{C.59}$$

where p is the specific load on the contact surface, P is the deformation force, A_l is the work of deformation, N is the power of deformation.

It can be shown that normal displacements of points on contact surfaces will be similar if the conditions of similarity are satisfied.

In conclusion, it is to be noted that the above derived conditions of similarity are rigorous. One has to select the most important characteristics to obtain unambiguous results. For example, if a lubricant is used in practice, then thermal contact may become non-ideal while the shear stress τ_m may depend on the relative sliding velocity $|\Delta\vec{w}|$, the viscosity μ and the thickness of lubricant δ_1. One can obtain then a new parameter $\zeta = \lambda R/l_0$ and an additional condition of similarity $R^M = R^H/k$ from the boundary condition $T - T_2 = \lambda R(\partial T/\partial n)$. Thermal resistance is difficult to control and so it will be difficult to provide a similarity condition. In contrast the value of

τ_m will be proportional to $\mu|\Delta\vec{w}|/\delta_l$, and from Eq.(C.59) it follows that $\Delta A^M = k\Delta A^H$. Thereafter one has to select the value of δ_1 from the condition $k(\mu/\delta_1)^M = (\mu/\delta_1)^H$, which is evidently difficult to ensure.

C.6 Physical Modelling of Superplastics

A study of the rheological behaviour of superplastics using a model alloy allows experiments at room temperature rather than at high temperatures. For example, the mechanical response of superplastic titanium alloys at 800–1000°C can be modelled using such materials as the lead–tin eutectic alloy at room temperature. One can also use transparent materials to examine visually the stress–strain distributions by using well known optical methods [614].

The authors of [615, 616] have considered the rolling of lead sheets for modelling high temperature rolling of steel sheets. The isothermal problem is considered assuming the constitutive equation

$$\sigma = A\varepsilon^n \xi^m \tag{C.60}$$

The parameters of the process are as follows: H_0, and h are the initial and the final sheet thicknesses, B_0 is the initial width of the sheet, R is the roll radius, f_c is the friction coefficient, v_B is the tangential velocity of the rolls.

In accordance with π-theorem one has

$$\theta = F\left(\varepsilon, \frac{R}{H_0}, \frac{B_0}{H_0}, n, m, f\right) \tag{C.61}$$

where θ is one of the following ratios:

$$\frac{\sigma}{A\left(\dfrac{v_B}{H_0}\right)^m}, \quad \frac{M}{B_0 H_0 R A\left(\dfrac{v_B}{H_0}\right)^m}, \frac{v_x}{v_B}, \frac{l}{B_0} \tag{C.62}$$

where σ is any component of the stress; l is some linear dimension (e.g., thickness, elongation.)

Two processes will be similar if the dimensionless quantities ε, R/H_0, B_0/H_0, n, m, f or ε, l/H_{cp}, B_0/H_0, n, m, f have the same values. It should be mentioned that for certain kinds of problems some other dimensionless parameters, e.g., E/σ_s, may have to be taken into account. Thus, the use of dimensional analysis for studying superplastic flow can be beneficial for the following reasons.

If one writes the constitutive equation using a normalised form, one obtains material constants having clear mechanical meaning. Firstly, such an approach simplifies the experimental determination of the material constants and reveals the domain of applicability of the constitutive equation under consideration. Secondly, while normalising the system of equations of the BP of mechanics of solids, one can obtain a full list of dimensionless parameters of the process (analogues of Reynold's number). This list can be used in the physical modelling of any process of metal working in the superplastic state. Finally, the numerical methods used to solve the normalised systems of equations are very effective because it is convenient to deal with dimensionless variables in the calculations. Also, the results obtained will be more general.

Appendix D: Group Properties of Thermoviscoplasticity

Many mathematical methods are used in mechanics of solids including such specialised ones like group analysis of the system of ordinary differential equations (which has been effectively used by Lee to solve differential equations through quadratures). Among specialists in mechanics interest was stimulated by the works of Ovsyannikov et al.[617, 618] who demonstrated that group methods can be effectively applied in seeking rigorous analytical solutions of differential equations as well as for their qualitative analysis. It was shown that a set of exact solutions for a given system of equations can be unambiguously determined using continuous groups of transforms. The application of the methods of group analysis in the theories of elasticity and plasticity is considered in [619] where a number of examples of exact solutions for the equations of the theory of plasticity are given.

D.1 About Single-Parameter Groups of Transforms

Consider the transformation T

$$z' = A(z) \tag{D.1}$$

which transfers the point $z = (z^1, \ldots, z^N)$ of Euclidean space $\boldsymbol{R}^N$ into a new position $z' = (z'^1, \ldots, z'^N)$ in the same space $\boldsymbol{R}^N$(co-ordinates z^i and z'^i of points z and z' are referred to the same system of co-ordinates). It is assumed that the transformation T is reversible, i.e., it has the inverse transform T^{-1} which transfers the point z' to z. The consecutive application of the transforms T and T^{-1} gives the identity transformation I which transfers the point z into itself ($T^{-1}T = TT^{-1} = I$).

Consider now other transformations which are included in the single-parameter set of transforms $\{T_a\}$:

$$z' = f(z, a) \tag{D.2}$$

where a is a real parameter. It will be assumed further that the identity transformation $T_0 = I$ corresponds to $a = 0$. One also assumes that $T_a \neq I$ for all $a \neq 0$. (If the identity transformation is obtained for some $a \neq 0$, then one can make a simple translation of the parameter a in order to satisfy these conditions.) Let T_a^{-1} be the transformation which is inverse to T_a. Correspondingly, let a^{-1} be accounted for by *a*, so that $T_a^{-1} = T_{a^{-1}}$.

For example, for the transformation of extension $z' = az$ the identity transmutation is obtained with $a = 1$. After the above mentioned translation, the following family of transformations is obtained.

$$z' = z + az \tag{D.3}$$

satisfying the condition $T_0 = \mathrm{I}$. Evidently, $a^{-1} = -\dfrac{a}{1+a}$, so that for an arbitrary value of *a* lying in the interval $-1 < a < \infty$ the inverse transformation $T_{a^{-1}}$ is defined. Consider now two values, *a* and *b*, belonging to the above mentioned interval and let the transformations be applied consecutively as follows.

$$z'' = z' + bz' = z + az + b(z + az) = z + (a + b + ab)\, z \tag{D.4}$$

Then, one can see that the result of application of two consecutive transformations belonging to the family Eq.(D.3) is identical to the result of the other transmutation with $c = a + b + ab$. Symbolically, this is indicated as $T_b T_a = T_{a+b+ab}$. It is said then that the transformation Eq.(D.3) forms a single-parameter group of transmutations.

In the general case, the transformations Eq.(D.2) form a single-parameter group of transmutations if they satisfy the condition:

$$T_b T_a = T_{\varphi(a,b)} \tag{D.5}$$

where the function $\varphi(a, b)$ is assumed to be sufficiently smooth. This means that a consecutive application of the transformations T_a and T_b is equivalent to the third transformation T_c with $c = \varphi(a, b)$. Taking into account that $T_0 T_a = T_a$ and that $T_b T_0 = T_b$ one concludes that

$$\varphi(a, 0) = a\,, \quad \varphi(0, b) = b\,.$$

For the translation group $z' = z + a$ one has $z'' = z' + b = z + a + b$ and

$$T_b T_a = T_{a+b} \tag{D.6}$$

It can be shown that one can always redesignate the parameter *a* in such a way to transform Eq.(D.5) to Eq.(D.6). Thus, the following main properties of the single-parameter group of transformations can be pointed out.

1) Existence of a unit element (identity transmutation) $T_0 = I$;
2) Existence of the inverse element $T_a^{-1} = T_{a^{-1}}$;
3) Associativity of multiplication: $T_c(T_b T_a) = (T_c T_b) T_a$.

The last property can be derived directly from the definition of multiplication as a consecutive fulfilment of transmutations. The three above listed properties consititute the basis for definition of group in the abstract theory of groups. Now consider the invariants of the group of transmutations. Let transformations Eq.(D.2) form a group and let Eq.(D.6) be satisfied. Let the function $f(z, a)$ be decomposed into a Taylor series with respect to the parameter in the vicinity of $a = 0$. Since $T_0 = I$, $f(z, 0) = z$, denoting

$$\zeta(z) = \left.\frac{\partial f(z,a)}{\partial a}\right|_{a=0} \tag{D.7}$$

One can rewrite Eq.(D.2) as

$$z' = z + \zeta(z)a + o(a) \tag{D.8}$$

Lee has proved a theorem that the function $f(z, a)$ is unambiguously defined by the first two terms of decomposition. It is said then that the group under consideration G is defined by its tangential vector field ζ (see Eq.(D.7)) and gives the tangential vector at the point z to the curve described by the point z' under transformation Eq.(D.2).

The function $F(z)$ is termed as an invariant of the group of transformations Eq.(D.8) if for all permissible z,a one has

$$F(f(z,a)) = F(z) \tag{D.9}$$

The following theorem is known. The function $F(z)$ is an invariant if and only if it satisfies to the equation

$$\zeta^i(z)\frac{\partial F(z)}{\partial z'} = 0 \tag{D.10}$$

where ζ^i are the components of the tangential vector $\zeta(z)$. The criterion of invariance Eq.(D.9) represents the homogeneous linear partial differential equation of the first order.

The characteristic equations for Eq.(D.10) are as follows.

$$\frac{dz'}{\zeta^1(z)} = \frac{dz^2}{\zeta^2(z)} = \ldots = \frac{dz^{N-1}}{\zeta^{N-1}(z)} \tag{D.11}$$

Therefore, any one parameter group of transmutations in $\boldsymbol{R}^N$ has N–1 functionally-independent invariants, all others being functions of these N–1 'primary' invariants. One can choose the left hand side parts of the first integrals of the characteristic system of equations.

Consider as an example the following group of extensions in $\boldsymbol{R}^3$

$$x' = xe^{a}, \qquad y' = ye^{2a}, \qquad z' = ze^{-2a}$$

In this case in accordance with Eq.(D.7) the tangential vector is $\xi = (x,2y,-2z)$ and Eq. (D.10) becomes

$$x\frac{\partial F}{\partial x} + 2y\frac{\partial F}{\partial y} - 2z\frac{\partial F}{\partial z} = 0$$

The characteristic system of equations is

$$\frac{dx}{x} = \frac{dy}{2y} = \frac{dz}{2z}$$

and the first integrals are $y/x^2 = C_1$, $x^2 z = C_2$. Consequently, the basis of invariants are formed by the functions $J_1 = y/x^2$ and $J_2 = x^2 z$ while the common invariant is

$$F = F(y/x^2, x^2 z)$$

If one introduces the differential operator

$$X = \zeta^{i}(z)\frac{\partial}{\partial z'} \tag{D.12}$$

then the criterion of invariance Eq.(D.10) can be rewritten as

$$XF = 0 \tag{D.12a}$$

This operator is termed an infinitesimal operator (or simply operator) of group G of transformations Eq.(D.8).Thus each single parameter group is fully defined by the first two terms of its Taylor decomposition or in other words, by its tangential vector field, which is referred to as an infinitesimal operator of the group.

D.2 Applications of Group Methods in Superplasticity

The main direction of application of the group methods in mechanics is in finding exact analytical solutions for systems of equations. Leonova [620, 621] has shown that one can also use them for an efficient analysis of experimental data. She has obtained a wide set of invariant combinations of parameters for the system of equations of thermo-visco-plasticity. This system includes standard equations (see Chap. 4 and Appendix C): equation of motion, incompressibility condition, the proportionality of stress and strain rate deviator, thermal conductivity equation containing temperature dependent coefficient of thermal conductivity and the term $K\sigma_e\xi_e$. (This is the source of the heat that arises due to plastic deformation.) The scalar CE is

$$\sigma_e = \Phi(\xi_e, T) \tag{D.13}$$

The main feature of her analysis is that concrete material functions need not be given. It is assumed that only the effective stress is a function of the effective strain rate and temperature. The most interesting aspect for superplasticity is that the invariant combinations of the parameters can be used for analysing the experimental data.

The following particular cases of the constitutive equation (D.13) can be mentioned.

1. $\Phi(\xi_e, T) \equiv 0$ – ideal incompressible flow.
2. $\Phi(\xi_e, T) = 2\mu\xi_e$ – newtonian flow.
3. $\Phi(\xi_e, T) = 2\mu(\xi_e)^n$ – non-newtonian flow.
4. $\Phi(\xi_e, T) = \Phi(v_J, T) = \sigma_s > 0$ for $v_J > 0$ with an additional condition of absence of deformation: $\xi_e \equiv 0$ for $\sigma_J < \sigma_s$ – a rigid plastic body.
5. $\Phi(\xi_e, T) = \sigma_s + 2\mu v_J$ for $v_J \geq 0$ – linear viscoplastic flow.

Leonova has completed a group analysis of the thermo-viscoplasticity equations and has given a set of practical recommendations concerning the choice of approximate expressions for the dependences of flow stress, σ, on temperature, T, and strain rate, ξ. These approximations are listed in Table D.1. For convenience two equivalent kinds of representation are used in Table D.1 which are related through the following relationships: $\Phi'(\xi,T) = a\Phi(a_1\xi+a_0, b_1T+b_0) + \Phi_0$; $f'(T) = f(b_1T+b_0)$. In Table D.1 α, β, γ are arbitrary constants; $m = -1/\beta$; $n = \gamma/\beta$; and Ψ is an arbitrary one parameter function.

Zener–Hollomon parameter $Z = \xi\, exp(Q/RT)$ is obtained from $\Phi_1 = \xi^n\Psi(T\xi^m)$ with $n = 1$, $m = 0$.

Considering $\sigma = \Phi_i(\xi, T)$ as a surface Π_i in Euclidean space T, ξ, σ with frame $\vec{e}_i (i = 1,2,3)$, $\vec{e}_i\vec{e}_j = \delta_{ij}$, one can see that the functions Φ_i give two parameter family of surfaces. The surface Π_i corresponding to Φ_i can be continuously transformed into any other surface Π_k ($\Pi_i \rightarrow \Pi_k$), the co-ordinate mesh on it being undisturbed. Table D.2 includes the transformations $w = w(\xi)$, $\theta = \theta(T)$, $s = s(\sigma)$, transferring all surfaces Π_i ($i = 1,..., 12$) into surface Π as follows:

$$\Phi_i^I : s = \gamma\theta + \varphi(w - \beta\theta); \Phi_i^{II} : s = nw + \varphi(\theta + mw) \tag{D.14}$$

where φ is an arbitrary one parameter function.

Table D.1 Equivalent representations of material functions [620, 621]

Representation 1		*Representation 2*	
$\Phi_1 = T^{\gamma}\Psi(\xi T^{-\beta})$,	$f = T^{\alpha}$;	$\Phi_1 = \xi^{n}\Psi(T\xi^{m})$,	$f = T^{\alpha}$;
$\Phi_2 = e^{\gamma T}\Psi(\xi e^{-\beta T})$,	$f = e^{\alpha T}$;	$\Phi_2 = \xi^{n}\Psi(T + m\ln\xi)$,	$f = e^{\alpha T}$;
$\Phi_3 = T^{\gamma}\Psi(\xi - \beta\ln T)$,	$f = T^{\alpha}$;	$\Phi_3 = e^{n\xi}\Psi(Te^{m\xi})$,	$f = T^{\alpha}$;
$\Phi_4 = e^{\gamma T}\Psi(\xi - \beta T)$,	$f = e^{\alpha T}$;	$\Phi_4 = e^{n\xi}\Psi(T + m\xi)$,	$f = e^{\alpha T}$;
$\Phi_5 = \gamma\ln T + \Psi(\xi T^{-\beta})$,	$f = T^{\alpha}$;	$\Phi_5 = n\ln\xi + \Psi(T\xi^{m})$,	$f = T^{\alpha}$;
$\Phi_6 = \gamma T + \Psi(\xi e^{-\beta T})$,	$f = e^{\alpha T}$;	$\Phi_6 = n\ln\xi + \Psi(T + m\ln\xi)$,	$f = e^{\alpha T}$;
$\Phi_7 = \gamma\ln T + \Psi(\xi - \beta\ln T)$,	$f = const$;	$\Phi_7 = n\xi + \Psi(Te^{m\xi})$,	$f = const$;
$\Phi_8 = \gamma T + \Psi(\xi - \beta T)$,	$f = const$;	$\Phi_8 = n\xi + \Psi(T + m\xi)$,	$f = const$;
$\Phi_9 = \Psi(T)\,\xi^{n}$;		$\Phi_9 = T^{\gamma}\Psi(\xi)$;	
$\Phi_{10} = \Psi(T)\,e^{n\xi}$;		$\Phi_{10} = e^{\gamma T}\Psi(\xi)$;	
$\Phi_{11} = \Psi(T) + m\ln\xi$;		$\Phi_{11} = \gamma\ln T + \Psi(\xi)$;	
$\Phi_{12} = \Psi(T) + m\xi$		$\Phi_{12} = \gamma T + \Psi(\xi)$.	

Table D.2 List of Transformations

	w	θ	s	$\varphi(z)$
Φ_1	$\ln\xi$	$\ln T$	$\ln\sigma$	$\ln\Psi(\exp z)$
Φ_2	$\ln\xi$	T	$\ln\sigma$	$\ln\Psi(\exp z)$
Φ_3	ξ	$\ln T$	$\ln\sigma$	$\ln\Psi(z)$
Φ_4	ξ	T	$\ln\sigma$	$\ln\Psi(z)$
Φ_5	$\ln\xi$	$\ln T$	σ	$\Psi(\exp z)$
Φ_6	$\ln\xi$	T	σ	$\Psi(\exp z)$
Φ_7	ξ	$\ln T$	σ	$\Psi(z)$
Φ_8	ξ	T	σ	$\Psi(z)$
Φ_9	$\ln\xi$	T	$\ln\sigma$	$\ln\Psi(z)$
Φ_{10}	ξ	T	$\ln\sigma$	$\ln\Psi(z)$
Φ_{11}	$\ln\xi$	T	σ	$\Psi(z)$
Φ_{12}	ξ	T	σ	$\Psi(z)$

Transformations listed in Table D.2 allows the experimental data to be treated in accordance with the same algorithm simultaneously for all functions Φ_i. Free parameters β and γ can be determined without analytical approximation of the function $\Psi(z)$, which remains arbitrary and which is determined fully by the experimental data alone. The sets of $\sigma = \sigma(\xi)$, T = constant; $\sigma = \sigma(T)$, ξ = constant are used as original experimental data. The details of these procedures are given in [620, 621]. Thus, one can investigate the closed system of equations of the formulation of the BP of mechanics of solids by group analysis methods. Such an approach allows the possibility of selecting the different classes of functions which can be successfully used in approximating experimental data obtained at elevated temperatures during superplastic flow or high temperature creep.

References

1. Grabski M (1975) Structural Superplasticity of Metals. Metallurgy Publishing House, Moscow (in Russian)
2. Smirnov OM (1979) Metal Working in a Superplastic State. Metallurgy Publishing House, Moscow (in Russian)
3. Padmanabhan KA, Davies GJ (1980) Superplasticity. Springer, Berlin Heidelberg
4. Pilling J, Ridley N (1989) Superplasticity in Crystalline Solids. The Institute of Metals, Camelot Press, UK
5. Kaibyshev OA (1992) Superplasticity in Alloys, Intermetallides and Ceramics. Springer, Berlin
6. Nieh TG, Wadsworth J, Sherby OD (1997) Superplasticity in Metals and Ceramics. Cambridge University Press, Cambridge, UK
7. Davies GJ, Edington JW, Cutler CP, Padmanabhan KA (1970) Superplasticity: A Review. J Mater Sci 5:1091–1102
8. Edington JW, Melton KN, Cutler CP (1976) Superplasticity. Prog Mater Sci 21:63–170
9. Langdon TG (1982) Mechanical properties of superplastic materials. Metall Trans 13A:689–701
10. Argyris J, Doltsinis J (1985) An apercu of superplastic forming. In: Sawczuk A, Bianchi G (eds.) Plasticity Today–Modelling, Methods and Applications. Elsevier, London, pp. 715–743
11. Hamilton CH, Ghosh AK, Wert JA (1985) Superplasticity in engineering – A review. Metals Forum 8:172–190
12. Paton NE (1985) Superplasticity in titanium alloys. In: Titanium Science and Technology. Proc V Int Conference, Munich, Vol. 2, Oberursel, Germany, pp. 649–672
13. Rowe GW, Developments in the use of superplastic alloys. In: Reid SR (ed.) Metal Forming and Impact Mechanics. Pergamon, Oxford, pp. 135–153
14. Sherby OD, Wadsworth J (1989) Superplasticity–Recent advances and future directions. Prog Mater Sci 33:169–221
15. Chokshi AH, Mukherjee AK, Langdon TG (1993) Superplasticity in advanced materials. Mater Sci Engng R10:237–274
16. Paton NE, Hamilton CH (eds.) (1982) Superplastic Forming of Structural Alloys. TMS–AIME, Warrendale, USA
17. Baudelet B, Suery M (eds.) (1985) Superplasticity. Editions du centre national de la recherche scientifique, Paris, France
18. Hamilton CH, Paton NE (eds.) (1988) Superplasticity and Superplastic Forming. TMS, Warrendale, USA
19. Hori S, Tokizane M, Furushiro N (eds.) (1991) Superplasticity in Advanced Materials. (ICSAM–91), JSRS, Osaka, Japan
20. Langdon TG (ed.) (1994) Superplasticity in Advanced Materials. (ICSAM–94), Mater Sci Forum, Vol. 170–172, Trans Tech Publications, Switzerland
21. Sakuma T, Aizawa T, Higashi K (eds.) (1997) Towards Innovation in Superplasticity I. Proc 1996 Int Symposium, Kyoto, Japan, Mater Sci Forum Vol. 233-234, Trans Tech Publications, Switzerland
22. Chokshi AH (ed.) (1997) Superplasticity in Advanced Materials. (ICSAM–97), Mater Sci Forum Vol. 243–245 Trans Tech Publications, Switzerland

23. Sakuma T, Aizawa T, Higashi K (eds.) (1999) Towards Innovation in Superplasticity II. Proc 1998 Int Symposium, Kobe, Japan, Mater Sci Forum Vol. 304–306, Trans Tech Publications, Switzerland
24. Mayo M, Kobayashi A, Wadsworth J (eds.) (1990) Superplasticity in Metals Ceramics and Intermetallics. MRS Symposium Proceedings Vol. 196, Pittsburg, USA
25. McNelley TR, Heikkenen HC (eds.) (1990) Superplasticity in Aerospace. TMS–AIME, Warrendale, USA
26. Ghosh AK, Bieler TR (eds.) (1995) Superplasticity and Superplastic Forming. TMS, Warrendale, USA
27. Aizawa T, Higashi K, Tokuda M (eds.) (1998) Microstructure, Micromechanics and Processing of Superplastic Materials. Proc Int Seminar (IMSP 97), Mie University Press, Tsu, Japan
28. Ghosh AK, Bieler TR (eds.) (1998) Superplasticity and Superplastic Forming. The Minerals, Metals & Materials Society, Warrendale, USA
29. Argyris J, Doltsinis J (1984) A primer on superplasticity in natural formulation. Comput Meth Appl Mech Engng 46:83–131
30. Chennot JL, Bellet M (1992) The viscoplastic approach for the finite element modelling of metal forming processes. In: Hartley I (ed.) Numerical Modeling of Material Deformation Processes Researches, Developments and Applications. Springer, Berlin, pp. 179–224
31. Wood RD, Bonet J (1993) Superplastic forming simulation using the finite element method. Metals, Materials and Processes 4:229–257
32. Gittus J, Zarka J, Nemat-Nasser S (eds.) (1986) Large Deformations of Solids Physical Basis and Mathematical Modelling. Elsevier Applied Science, London New York
33. Bengough GD (1912) A study of the properties of alloys at high temperatures. J Inst Metals 7:123–174
34. Pearson CE (1934) The viscous properties of extruded eutectic alloys of Lead–Tin and Bismuth–Tin. J Inst Metals 54:111–123
35. Backofen WA, Turner GR, Avery DH (1964) Superplasticity in an Al–Zn Alloy. Trans Am Soc Metals 57:980–990
36. Langdon TG (1991) Superplasticity a historical perspective. In: Hori S, Tokizane M, Furushiro N (eds.) Superplasticity in Advanced Materials (ICSAM–91). JSRS, Osaka, Japan, pp. 3–12
37. Lee D, Backofen WA (1967) Superplasticity in some titanium and zirconium alloys. Trans AIME, 239:1034–1040
38. Martin JA, Backofen WA (1967) Superplasticity in electroplated composites of lead and tin. ASM Trans Quart 60:352–359
39. Backofen WA, Azzarto FJ, Murty GS, Zehr SW (1968) Superplasticity Ductility. Papers presented at the seminar of the ASM, Metals Park, Ohio, USA, pp. 279–310
40. Ghosh A, Duncan JL (1970) Torsion test on supeplastic Tin–Lead alloy. Int J of Mech Sci 12: 499–511
41. Kuzmin SL, Lihatchev VA, Myshlyaev MM, Nikonov YuA, Senkov ON (1977) Superplasticity of mono- and polycrystalline aluminium under torsion. Physica Metallov i Metallovedenie 44:429–432
42. Arieli A, Mukherjee AK (1980) High temperature diffusion–controlled creep behavior of the Zn–22%Al eutectoid alloy tested in torsion. Acta Metall 28:1571–1581
43. Lihatchev VA, Myshlyaev MM, Senkov ON, Belyaev SP (1981) Creep of aluminum under superplasticity at torsion. Physica Metallov i Metallovedenie 52:407–416
44. Abubakirova RK, Lihatchev VA, Myshlyaev MM, Presnyakov AA, Chalaev DR (1981) Creep of superplastic alloy Zn-22%Al under torsion. Physica Metallov i Metallovedenie 51:201–210
45. Nabarro FRN (1987) Plastic deformation in tension and torsion. Phys Stat Solidi (a) 104:47–50

46. Mayo MJ, Nix WD (1989) Direct observation of superplastic flow mechanisms in torsion. Acta Metall 37:1121–1134
47. Khraisheh MK, Bayomi AE, Hamilton CH, Zbib HM, Zhang K (1995) Experimental observations of induced anisotropy during the torsion of superplastic Pb–Sn eutectic alloy. Scripta Metall 32:955–959
48. McQueen HJ, Jin N, Ryan ND (1995) Relationship of energy dissipation efficiency to microstructural evolution in hot working of AISI 304 steel. Mater Sci Eng A190:43–53
49. Zhou M, Clode MP (1997) Hot torsion tests to model the deformation behaviour of aluminum alloys at hot working temperatures. J Mater Process Technol 72:78–85
50. Murty GS, Singh UP, Iyengar NGR (1973) Indentation creep of the Sn-Pb eutectic alloy. J Inst Met 101: 337–340
51. Mayo MJ (1988) A study of superplasticity in indentation and torsion. PhD dissertation, Stanford University, USA
52. Mayo MJ, Nix WD (1988) A microindentation study of superplasticity in Pb, Sn, and Sn-28%Pb. Acta Metall 36:2183–2192
53. Chinh NQ, Junasz A, Tasnadi P, Kovacz I, Kobvacz-Csetenyi E (1996) The existence of the threshold stress in superplastic aluminum alloys. J Mat Sci Letters 15:406–408
54. Belk JA (1975) A quantitative model of the blow–forming of spherical surfaces in superplastic sheet metal. Int J Mech Sci 17:505–511
55. Ghosh AK, Hamilton CH (1980) On constant membrane stress test for superplastic metals. Metall Trans 11A:1915–1920
56. Hsu TC, Bidhendi IB (1982) A study of strain and strain rate dependent properties of a superplastic Zn–Al alloy under biaxial stresses. J Engng Mater and Technol 104:41–46
57. Enikeev FU, Kruglov AA (1995) An analysis of the superplastic forming of a thin circular diaphragm. Int J Mech Sci 37:473–483
58. Mahmudi R (1997) Stress strain dependence of work-hardening behaviour of aluminum sheet. J Mater Processing Technol 72:302–307
59. Padmanabhan KA, Davies GJ (1977) The superplastic behaviour of the Al-$CuAl_2$ eutectic during compressive deformation. Metal Sci 11:177–184
60. Zhang JM, Gao ZY, Zhang JY, Zhong ZY, Janschek P (1997) Strain rate hardening behaviour of superalloy IN718. J Mater Processing Technol 72:252–257
61. Huang C, Lai JKL, Lee CS (1998) Deformation characteristics of Ti-24Al-14Nb-3V-0.5Mo alloy during hot compression. J Mater Processing Technol 73:119–124
62. Garmong G, Paton NE, Chesnutt JC, Nevarez LF (1977) An evaluation of the ring test for strain-rate–sensitive materials. Metall Trans 8A:2026–2027
63. Wray PJ (1970) Tensile plastic instability at elevated temperatures and its dependence upon Strain Rate. J Appl Phys 41:3347–3352
64. Takayama Y, Furushiro N, Tozawa T, Kato H (1991) Effect of specimen shape on flow stress in superplastic materials. In: Hori S, Tokizane M, Furushiro N (eds.) Superplasticity in Advanced Materials (ICSAM–91) JSRS. Osaka, Japan, pp. 69–74
65. Johnson KI, Khaleel MA, Lavende CA, Pitman SG, Smith JT, Smith MT, Hamilton CH (1994) The effect of specimen geometry on the accuracy of constitutive relations in a superplastic 5083 aluminium alloy. Langdon TG (ed.) Superplasticity in Advanced Materials (ICSAM–94), Mater Sci Forum 170–172:627–632
66. Takayama Y, Furushiro N, Tozawa T, Kato H (1997) Usefulness of the R–type specimen method for evaluation of superplastic characteristics. Chokshi AH (ed.) Superplasticity in Advanced Materials (ICSAM–97). Mater Sci Forum 243–245:167–172
67. Hirohashi M, Nishizawa M (1999) Simple valuation of superplastic characteristic by tensile test using R–type specimen. In: Sakuma T, Aizawa T, Higashi K (eds.) Towards Innovation in Superplasticity II. Kobe, Japan, Mater Sci Forum 304–306:693–698
68. Hedworth J, Stowell MJ (1971) The measurement of strain rate sensitivity in superplastic alloys. J Mater Sci 6:1061–1069

69. Murty GS (1973) Stress relaxation in superplastic materials. J Mater Sci Letters 8:611–614
70. Arrowood R, Mukherjee AK (1987) Stress relaxation of a eutectic alloy in the superplastic condition. Mater Sci Eng 92:33–41
71. Akmoulin IA, Jonas JJ (1991) Production of fine-grained titanium alloys for supeprlastic forging by means of controlled recrystallization. In: Hori S, Tokizane M, Furushiro N (eds.) Superplasticity in Advanced Materials (ICSAM–91). JSRS, Osaka, Japan, pp. 471–477
72. Enikeev FU, Mazurski MI (1995) Determination of the strain rate sensitivity of a superplastic material during load relaxation test. Scripta Metall et Mater 32:1–6
73. Ha TK, Chang YW (1997) Load relaxation behaviour of superplastic Al alloys. Chokshi AH (ed.) Superplasticity in Advanced Materials (ICSAM–97). Mater Sci Forum 243–245:505–510
74. Rouxel T (1999) Stress relaxation and brittle–ductile transition in superplastic ceramics. In: Sakuma T, Aizawa T, Higashi K (eds.) Towards Innovation in Superplasticity II. Kobe, Japan, Mater Sci Forum 304–306:395–402
75. Seidensticker JR, Wiederhorn SM (1999) Investigation of superplasticity in 3Y–TZP by stress relaxation. Mater Sci Forum 304–306:403–410
76. Cadek J, Oikawa H Sustek V (1995) Threshold creep behaviour of discontinuous aluminum and aluminium alloy matrix composites: An overview. Mater Sci Engng A190:9–23
77. Viereck D, Lohe D, Vohringer O, Macherauch E (1992) Stress relaxation behaviour of the cobalt base superalloy CoNi23Cr22W14 In: Jano M, Inoue T (eds.) Mechanical behaviour of materials. Proc 6th Int Conference Kyoto, Japan, Vol. 3 Pergamon Press, pp. 607–612
78. Endo T, Hirano M, Yashida K (1991) Effect of thermo-mechanical history on the internal friction and superplasticity of Al-6Cu-0.5Zr alloy. In: Hori S, Tokizane M, Furushiro N (eds.) Superplasticity in Advanced Materials (ICSAM–91). JSRS, Osaka, Japan, pp. 157–162
79. Mulyukov RR, Mikhailov SB, Salimonenko DA, Valiakhmetov OR, Lutfullin PYa, Myshlyaev MM, Salishchev GA (1996) Damping properties of a titanium alloy with a submicrocrystalline structure. J Advanced Mater 3:73–76
80. Mulyukov R, Mikhailov S, Zaripova R, Salimonenko D (1996) Damping properties of 18Cr-10Ni stainless steel with submicrocrystalline structure. Mater Res Bulletin 31:639–645
81. Schneibel JH (1980) The influence of anelasticity on the transient behavior superplastic Sn–Pb eutectic after stress and strain rate changes. Acta Metall 28:1527–1535
82. Todd RI, Hazzledine PM (1988) The mechanism of superanelasticity and its implications. In: Paton NE, Hamilton CH (eds.) Superplasticity and Superplastic Forming. TMS, Warrendale, USA, pp. 33–37
83. Vale SH (1984) Anelasticity in fine–grained materials. Acta Metall 32:693–706
84. Schneibel JH, Hazzledine PM (1982) Superanelasticity in superplastic Sn–Pb alloys. Acta Metall 30:1223–1230
85. Higashi K, Ohnishi T, Nakatami Y (1985) Superplastic behavior of commercial aluminum bronze. Scripta Metall 19:821–823
86. Ma Y, Langdon TG (1994) Factors influencing the exceptional ductility of a superplastic Pb-62 pct Sn alloy. Metall Mater Trans 25A:2309–2311
87. Safiullin RV, Vasin RA, Enikeev FU (2000) Determination of the parameters of superplastic forming for long rectangular thin sheet titanium alloy Ti-6Al-4V. Acta Metall Sinica (English) 13:567–573.
88. Greshnov VM (1983) Infuence of the mechanical scheme of deformation on the mechanical properties and microstructure of superplastic alloys Zn-22%Al and VT9. Proc Russian Acad Sci Metals 6:158–162 (in Russian)
89. Tihonov AC (1978) Superplasticity in Metals and Alloys. Nauka, Moscow, p.142 (in Russian)
90. Vasin RA, Enikeev FU, Mazurski MI (1998) On the strain rate sensitivity values of superplastic materials. Plant Laboratory 9:50–55 (in Russian)

91. Vasin RA, Enikeev FU, Mazurski MI (1997) Determination of the strain rate sensitivity of a superplastic material at constant load test. Mater Sci Engng A 224:131–135
92. Vasin RA, Enikeev FU, Mazurski MI (1999) Method to determine the strain rate sensitivity of a superplastic material from the initial slopes of its stress–strain curves. J Mater Sci 33:169–171
93. Vasin RA, Enikeev FU, Safiullin RV (1999) Mathematical modeling of superplastic forming of a long rectangular box section. In: Sakuma T, Aizawa T, Higashi K (eds.) Towards Innovation in Superplasticity II. Kobe, Japan, Mater Sci Forum 304–306:765–770
94. Enikeev FU (1993) Plastic behavior of superplastic material moving between two coaxial rotating cylinders. Int J Mech Sci 35:81–88
95. Vasin RA, Enikeev FU (1998) Introduction to the mechanics of superplasticity. Part I. Gilem, Ufa, Russia (in Russian)
96. Enikeev FU (1997) Strain–rate sensitivity index m definition, determination, narrowness. In: Chokshi AH (ed.) Superplasticity in Advanced Materials (ICSAM–97), Mater Sci Forum 243–245:77–82
97. Vasin RA, Enikeev FU, Mazurski MI (1998) On the strain rate sensitivity values of superplastic materials. Mater Sci Eng 255:169–171
98. Sirenko AA, Murzinova MA, Enikeev FU (1995) On the universal relationship between specific characteristics of superplastic deformation. J Mater Sci Lett 14:773–774
99. Imayev RM, Imayev VM (1991) Mechanical behavior of TiAl submicrocrystalline intermetallic compund at elevated temperatures. Scripta Metall 25:2041–2046
100. Zaripov NG, Kaibyshev OA, Kolnogorov OM (1993) Structural superplasticity of the Bi_2O_3–based ceramics. Phys Solids 35:2114–2121 (in Russian)
101. Zaripov NG, Petrova LV, Kaibyshev OA, Efimov OYu, Pityulin AN (1993) The role of the metallic phase in high temperature deformation of titanium carbide–based ceramics. J Mater Sci Lett 12:502–504
102. Iwasaki H, Higashi K, Tahimura S, Komatubara T, Hayami S (1991) Superplastic deformation characteristics of 5083 aluminium alloy. In: Hori S, Tokizane M, Furushiro N (eds.) Superplasticity in Advanced Materials (ICSAM–91). JSRS, Osaka, Japan, pp. 447–452
103. Ghosh AK, Cheng CH (1991) Superplastic deformation in titanium aluminides and modeling of transient deformation. In: Hori S, Tokizane M, Furushiro N (eds.) Superplasticity in Advanced Materials (ICSAM–91). JSRS, Osaka, Japan, pp. 299–310
104. Stoner SL, Mukherjee AK (1991) Superplasticity in fine- grained nickel silicide. In: Hori S, Tokizane M, Furushiro N (eds.) Superplasticity in Advanced Materials (ICSAM–91). JSRS, Osaka, Japan, pp. 323–328
105. Enikeev FU, Padmanabhan KA, Bhattacharya SS (1999) Model for grain boundary sliding and its relevance to optimal structural superplasticity Part 5–A unique numerical solution. Mater Sci Technol 15:673–682
106. Imayev RM, Imayev VM, Salishchev GA (1992) Formation of submicrocrystalline structure in TiAl intermetalic compound. J Mater Sci 27:4465–4471
107. Hart EW (1967) Theory of the tensile test. Acta Metall 15:351–355
108. Boyer HE, Gall TL (1985) Metals Handbook Desk edition American Society for Metals Metal Park, Ohio, USA
109. Backofen W (1972) Deformation Processing. Addison–Wesley, Reading, Massachusetts p.326
110. Burke MA, Nix WD (1975) Plastic instabilities in tension creep. Acta Metall 23:793–798
111. Sagat S, Taplin DMR (1976): The stability of plastic flow in strain rate sensitive materials. Metal Sci 10:94–100
112. Ghosh AK (1977) Tensile instability and necking in materials with strain hardening and strain rate hardening. Acta Metall 25:1413–1424
113. Sherby OD, Klundt RH, Miller AK (1977) Flow stress, subgrain size and subgrain stability at elevated temperature. Metall Trans 8 A:843–850

114. Nichols FA (1980) Overview: Plastic instabilities and uniaxial tensile ductilities. Acta Metall 28:663–673
115. Kocks UF (1982) The initiation and development of deformation heterogeneities in superplastic materials. In: Paton NE, Hamilton CH (eds.) Superplastic Forming of Structural Alloys. TMS–AIME, Warrendale, USA, pp. 41–55
116. Senkov ON, Likhachev VA (1986) The Effect of grain growth on the plastic instabilities and uniaxial tensile ductilities in superplastic alloys. Phys Stat Solidi A 98(2):441–452
117. Baudlet B, Suery M (1988): Plastic stability and strain to fracture during superplastic deformation In: Hamilton CH, Paton NE (eds.) Superplasticity and Superplastic Forming. TMS, Warrendale, USA, pp. 135–148
118. Tayupov AR (1994) Superplastic deformation stability by complex hardening parameter. Scripta Metall 30:1387–1389
119. Korbel A, Ciura F (1997) The mechanical instability of the metal substructure and formation of pseudo-periodic substructure in thermodynamically stable and unstable phases. J Mater Processing Technol 64:231–238
120. Berces G, Chinh NQ, Juhasz A, Lendvai J (1998) Kinematic analysis of plastic instabilities occurring in microhardness tests. Acta Mater 46:2029–2037
121. Tresca H (1864) Memoire sur lecoulement des lorps solides sourmis a des forteszx pressions. Comptes rendus de l' Academie des Sciences Paris, 59:754
122. Bell JF (1973) Experimental Foundations of Solid Mechanics. Springer, Berlin Heidelberg New York
123. Iljushin AA (1946) On the theory of small elastic plastic deformations. Applied Mechanics and Mathematics (PMM) 10:347–356 (in Russian)
124. Kishkin ST (1950) Against the formalism in the theory of plasticity. Proc Acad Sci USSR. Otdelenie Tehnicheskih Nauk 2:266–278 (in Russian)
125. Ratner SI (1950) On the problems of the theory of plasticity. Proc Acad Sci USSR. Otdelenie Tehnicheskih Nauk 3:435–450 (in Russian)
126. Iljushin AA (1950) Notes on some papers concerning the critical comments on the theory of plasticity. Proc Acad Sci USSR. Otdelenie Tehnicheskih Nauk 6:940–951 (in Russian)
127. Bridgman PW (1952) Studies in Large Plastic Flow and Fracture with Special Emphasis on the Effects of Hydrostatic Pressure. McGraw–Hill, New York
128. Bridgman PW (1964) Properties of materials under superindustrial stresses. In: Collected Experimental Papers Vol VII. Harward University Press, Cambridge, Massachusetts, pp. 4090–4103
129. Zener C, Hollomon JH (1944) Effect of strain rate upon plastic flow of steels. J Appl Phys 15:22–32
130. Padmanabhan KA, Davies GJ (1973) On the activation energy for superplastic flow. Phys Stat Solidi 18A:295–302
131. Vasin RA (1997) Constitutive models in superplasticity: A review. Chokshi AH (ed.) Superplasticity in Advanced Materials (ICSAM–97), Mater Sci Forum 243–245: 173–178
132. Loveday MS, Dyson BF (1979) Creep deformation and cavitation damage in nimonic 80A under a triaxial tensile stress. In: Miller KJ, Smith KF (eds.) Proc III Int Conference Mechanical Behaviour of Materials, Vol 2. Cambridge University Press, Cambridge, pp. 213–222
133. Mott NF, Jones H (1936) The Theory of the Properties of Metals and Alloys. Oxford University Press, London: pp. 147–151
134. Boyle JT, Spence, J (1983) Stress Analysis for Creep. Butterworths, UK
135. Shesterikov SA (1997) Personal communication
136. Mazurski MI, Enikeev FU (1997) Superplasticity as universal structural–mechanical phenomenon. Chokshi AH (ed.) Superplasticity in Advanced Materials. (ICSAM–97), Mater Sci Forum 243–245:83–88

137. Mazurski MI, Enikeev FU (1998) A new theoretical concept for micrograin superplasticity providing the prediction of the optimum conditions for superplastic deformation. Phys Stat Solidi A206:519–534
138. Padmanabhan KA, Schlipf J (1996) Model for grain boundary sliding and its relevance to optimal structural superplasticity Part 1–Theory. Mater Sci Technol 12:391–399
139. Astanin VV, Faizova SN, Padmanabhan KA (1996) Model for grain boundary sliding and its relevance to optimal structural superplasticity Part 2–Evidence for co-operative grain/interphase boundary sliding and plane interface formation. Mater Sci Technol 12:489–494
140. Astanin VV, Padmanabhan KA, Bhattacharya SS (1996) Model for grain boundary sliding and its relevance to optimal structural superplasticity Part 3 – Effect of flow localisation and specimen thickness on superplasticity in alloy Supral 100. Mater Sci Technol 12:545–550
141. Venkatesh TA, Bhattacharya SS, Padmanabhan KA, Schlipf J (1996) Model for grain boundary sliding and its relevance to optimal structural superplasticity Part 4–Experimental verification. Mater Sci Technol 12:635–643
142. Karim A (1969) On the nature of superplastic deformation in the Mg–Al eutectic. Scripta Metall 3:887–992
143. Dunlop GL, Taplin DMR (1972) The tensile properties of superplastic aluminium bronze. J Mater Sci 7:84–92
144. Burton BA (1971) Creep yield stress for superplastic deformation. Scripta Metall 5:669–672
145. Geckinli AE, Barrett CR (1976) Superplastic deformation of the Pb–Sn eutectic. J Mater Sci 11:510–521
146. Booeshaghi F, Garmestani H (1998) On the existence of threshold stress. Scripta Mater 38:89–94
147. Vasin RA, Enikeev FU, Mazurski MI (1994) Applicability of Bingham-type constitutive models to superplastic materials at different loading conditions. In: Langdon TG (ed.) Superplasticity in Advanced Materials (ICSAM–94), Mater Sci Forum 170–172:675–680
148. Mohamed FA (1983) Interpretation of superplastic flow in terms of a threshold stress. J Mater Sci 18:582–592
149. Cadek J (1987) The back stress concept in power law creep of metals: A review. Mater Sci Engng, 94:79–92
150. Mohamed FA (1988) On the threshold stress for superplastic flow. J Mater Sci Letters 7:215–217
151. Bieler TR, Mishra RS, Mukherjee AK (1994) The role of threshold stresses and incipient melting in high strain rate superplasticity. In: Langdon TG (ed.) Superplasticity in Advanced Materials (ICSAM–94), Mater Sci Forum 170–172:65–70
152. Soliman MS (1994) Superplastic characteristics of the Pb–62%Sn eutectic alloy at room temperature. Scripta Metall 31:439–444
153. Murty GS, Banerjee S (1994) Evaluation of threshold stress from the stress–strain rate data of superplastic materials. Scripta Metall 31:707–712
154. Suh S, Dollar M (1994) On the threshold stress in mechanically alloyed NiAl. Scripta Metall 31:1663–1668
155. Yang ST, Mohamed FA (1995) On the characteristics of the threshold stress for superplastic flow in Zn–22 % Al. Metall Mater Trans 26A:493–499
156. Mishra RS, Bieler TR, Mukherjee AK (1997) Mechanism of high strain rate superplasticity in aluminum alloy composites. Acta Mater 45:561–568
157. Enikeev FU, Determination of the value of the threshold stress for superplastic flow. Mater Sci Engng (In press)
158. Bricknell RH, Bentley AP (1979) The activation energy for superplastic flow in Al–6Cu–0.4 Zr. J Mater Sci 14:2547–2554

159. Mohamed FA, Shei SA, Langdon TG (1975) The activation energies associated with superplastic flow. Acta Metall 23:1443–1450
160. Liauo CS, Huang JC (1998) Relationship between apparent and true activation energies. Scripta Mater 39:873–879
161. Bhattacharya SS, Padmanabhan KA (1991) On the aplicability of a boundary sliding controlled flow process to superplastic deformation. In: Hori S, Tokizane M, Furushiro N (eds.) Superplasticity in Advanced Materials (ICSAM–91). JSRS, Osaka, Japan, pp. 33–38
162. Padmanabhan KA (1977) A theory of structural superplasticity. Mater Sci Engng 29:1–18
163. Shtremel MA (1997) Strength of alloys Part II Deformation Moscow, MISIS p 526 (in Russian)
164. Drucker DC (1952) A More fundamental approach to plastic stress–strain relations. Proc First US Nat Cong Appl Mech, New York, pp. 487–491
165. Drucker DC (1959) A definition of stable inelastic material. J Appl Mech 26:101–106
166. Hencky H (1924) Zur theorie plastischer deformationen und der hierdurch im material hervorgerufenen nachspannungen. Zeitschrift fur Angewandte Mathematik und Mechanik, Berlin 4: 323–334
167. Zyczkowski M (1981) Combined Loading in the Theory of Plasticity. Polish Scientific Publishers, Warsaw, Poland
168. Iljushin AA (1961) On postulate of plasticity. Prikl Matematika i Mechanika 25:503–507(in Russian)
169. Iljushin AA (1963) Fundamentals of the General Mathematical Theory of Plasticity. Acad Sci USSR, Moscow (in Russian)
170. Korshunov AA, Enikeev FU, Mazurski MI, Salishchev GA, Dmitriev OV, Muravlev AV, Chistyakov PV (1996) Grain–Structure refinement in titanium alloy under different loading schedules. J Mater Sci 31:4635–4639
171. Bylia OI, Vasin RA, Ermachenko AG, Karavaeva MV, Muravlev AV, Chystyakov PV (1997) The influence of simple and complex loading on structure changes in two–phase titanium alloy. Scripta Metall 36:949–954
172. Valiev RZ, Muljukov HYa, Ahmadeev NA, Hafizov SB, Muljukov RR (1992) Microstructure and magnetic properties of ultrafine-grained nickel after severe plastic deformation in accordance with different loading schedules. In: Proc of 5th Conference 'Superplasticity in inorganic materials', Ufa, Russia: p.57. (in Russian)
173. Klepaczko J (1968) Generalised conditions for stability in tension tests. Int J Mech Sci 10:297–313
174. Campbell JD (1967) Plastic instability in rate-dependent materials. J Mech Phys Solids 15:359–370
175. Rossard C (1966) Formation de la striction dans la deformation achaud par traction. Revue de metallurgie 63:225–235
176. Estrin Y, Kubin LP (1988) Plastic instabilities:Classification and physical mechanisms. Res Mechanica 23:197–221
177. Zbib HM, Aifantis EC (1988) On the localisation and post-localisation behaviour of plastic deformation I on the initiation of shear bands. Res Mechanica 23:261–277
178. Okabe T, Hatayama T (1991) Effect of strain rate sensitivity on non-uniform deformation in superplastic materials. In:Hori S, Tokizane M, Furushiro N (eds.) Superplasticity in Advanced Materials (ICSAM–91). JSRS, Osaka, Japan, pp. 51–56
179. Ferron G, Idrissi HK, Zeghloul A (1992) A state variable modelling of plasticity and necking under uniaxial tension. Trans ASME, J Engng Mater Technol 114:378–383
180. Perzyna P (1984) Constitutive modelling of dissipative solids for post-critical behavior and fracture. Trans ASME, J Engng Mater Technol 106:410–419
181. Iljushin AA (1940) Deformation of viscoplastic solid Uchenye Zapiski MGU. Mehanika 39:3–81 (in Russian)

182. Ishlinsky AYu (1943) On the stability of viscoplastic flow of a rod and strip. Prikl Matematika i Mehanika 7:109–130 (in Rassian)
183. Erjanov JS, Egorov AK (1987) Stability of the non-uniform deformation of non-linear materials. Alma-Ata, Nauka: p. 278 (in Russian)
184. Malinin NN (1986) Creep in Metal Working. Mashinostroyenie Publishing House, Moscow
185. Georgievskij DV (1994) Sufficient integral estimates of stability of viscoplastic flow at shear Izv. RAN, Mehanika tverdogo 4:124–131 (in Russian)
186. Bell JF (1995) Laboratory expriments on thin walled tubes at large finite strain. Int J Plasticity 11:119–144
187. Truesdell C, Toupin RA (1960) The classical field theories In: Flugge S (ed.) Encyclopedia of Physics Vol III/1 Principles of Classical Mechanics and Field Theory. Springer-Verlag, Berlin Heidelberg, pp. 226–293
188. Iljushin AA (1990) Continuum Mechanics. Moscow State University Publishing House, Moscow (in Russian)
189. Astanin VV, Sirenko AA (1990) Superplasticity of Amg6 alloy foils. Procd USSR Acad Sci 4:132–136 (in Russian)
190. Iljushin AA, Lensky VS (1967) Strength of Materials. Translation from Russian, Pergamon Press, Oxford
191. Wagoner RH (1983) Novel Techniques in Metal Deformation Testing. Proc of Symp St Louis, Missuri, Metallurgical Society of AIME, Warrendale, USA
192. Timoshenko SP, Gudier JN (1951) Theory of Elasticity. 2nd edition, McGraw-Hill, New York
193. Kolosov GV (1909) On the application of the theory of complex variable functions for the plane stress problem in the theory of elasticity, PhD Thesis, Tartu, Russia
194. Inglis CE (1913) Trans Inst Nav Archit, pt 1, pp. 219–230
195. Rabotnov YuN (1969) Creep Problems in Structural Members. North–Holland, Amsterdam
196. Prager W (1961) Einfuhrung in die Kontinuumsmechanik. Birkhauser, Basel-Stuttgart. English translation: Introduction to Mechanics of Continua. Ginn & Co, New York
197. Drucker DC (1967) Introduction to Mechanics of Deformable Solids. McGraw-Hill Book Company, New York
198. Keunings R (1990) Progress and challenges in computational rheology. Rheologica Acta 29:556–570
199. Babamuratov KSh, Iljushin AA, Kabulov VK (1987) Method SN-EVM and its Applications to the Problems of the Theory of Plasticity. Fan Publications, Tashkent (in Russian)
200. Landau LD, Lifshitz EM (1988) Theoretical Physics Vol. VI Hydrodynamics. Nauka, Moscow (in Russian)
201. Reiner M (1958) Rheology. In: Flugge S (ed.) Encyclopedia of Physics. Vol. VI Elasticity and Plasticity. Springer-Verlag Berlin Heidelberg, pp. 434–550
202. Prandtl L (1928) Ein Gedenkenmodell zur kinetischen theorie der festen korper. Zeitschrift fur Angewandte Mathematik und Mechanik 8:85–106
203. Gohfeld DA, Sadakov OC (1984) Plasticity and Creep of Structural Members under Repeated Loading. Mashinistrojenije, Moscow (in Russian)
204. Saint-Venant B de (1871): Memoire sur L'establissement des equations differentieless. Journal des Mathematiques Pures et Appliquees 16:373–383
205. Saint-Venant B de (1872): Sur L' intensite des forces capables de deformer, avec continuite, des blocs ductiles cylindriques pleins ou evides, et places dans divereses circonstances, Comptes Rendus Hebdomadaires des Seanses de L'Academie des Sciences 74:1009–1015
206. Levy M (1870) Memoire sur les equations generales Comptes Rendus Hebdomadaires des Seanses de L'Academie des Sciences 70:1323–1325

207. Levy M (1871) On integrating partial differential equations describing the internal movement in solids. Comptes Rendus Hebdomadaires des Seanses de L'Academie des Sciences 73: 1098–1103
208. Huber MT (1904) Die spezifische formanderungsarbeit als mab der amstiengung eines materials – Lembrg , Czasopimo Techniczne 22:81
209 Mises R (1913) Mechanik der festen korper im plastish deformablen zustand. gottingen nachrichten, Mathematisch-Physicalische Klasse, Gottingen 1:582–592
210 Dieter GE (1988) Mechanical Metallurgy. SI Metric Edition. McGraw-Hill, Singapore
211 Nadai A (1923) Der beginn des fliessvorganges. In:Einem Tordierten stab. zeitschrift fur angewandte mathematik und mechanik 3:442–444
212 Iljushin AA (1943) Some problems of plastic deformations. Applied mechanics and mathematics (PMM) 7:245–272 (in Russian)
213 Iljushin AA (1947): Theory of plasticity at simple loading of the bodies exhibiting plastic hardening. Applied Mechanics and Mathematics (PMM) 11:291–296 (in Russian)
214 Sawczuk A (1989): Mechanics and Plasticity of Structures. Polish Scientific Publishers, Warsaw, Poland
215 Iljushin AA (1945) Relation between the Saint-Venant–Levy–Mises theory and the theory of small elastic-plastic deformations. Applied Mechanics and Mathematics (PMM) 9:207–218 (in Russian)
216 Prandtl L (1924) Spunnungsverteilung in platischen korpern. In: Proceedings of I International Congress of Applied Mechanics, Delft, pp. 43–54
217 Reuss A (1930) Berucksichtigung der elastichen formanderungen in der plastizitatstheorie. Zeitschrift fur Angewandte Mathematik und Mechanik, Berlin 10:266–274
218 Drucker DC (1964) On the postulate of stability of material in the mechanics of continua. J Metals 3:235–249
219 Vasin RA, Enikeev FU, Mazurski MI (1995) On materials having descending stress–strain diagram. Proc of Russian Acad Sci. Mechanics of solids 2:181–182
220 Vasin RA (1990) Constitutive equations in the theory of plasticity. In: Itogi nauki I tehniki VINITI. Mechanics of solids 21:3–75 (in Russian)
221 Taylor GI, Elam CF (1923) Distortion of an aluminum crystal during a tensile test. Proc Roy Soc. London A102:643
222 Taylor GI (1938) Plastic strains in metals. J Inst Metals 62:306–324
223 Schmidt E (1924) Yield point of crystals–critical shear stress law. I Int Cong on Applied Mechanics. Delft, p 342
224 Taylor GI (1938) Analysis of Plastic Strain in a Cubic Crystal. Timoshenko Anniversary Volume. Macmillan, p.218
225 Batdorf SB, Budiansky BA (1949) Mathematical theory of plasticity based on the concept of slip. NACA Techn Note. No 1871
226 Batdorf SB, Budiansky BA (1954) Polyaxial stress strain relations of a strain hardening metal. Trans ASME. J Appl Mechanics 21:323–326
227 Cicalla P (1950) Sulle deformazione plastiche. Atti Della Accademia Nazionalle dei Lincei. Rendiconti Classe di Scienze Fisiche. Matematiche e Naturali 8:583
228 Cicalla P (1950) Sobrela teoria de batdorf y budiansky de lka deformacion plastica. Rev Univ Cordoba (Argent) 12:2
229 Bishop JFW, Hill R (1951) A theory of the plastic distortion of a polycrystalline aggregate under combined stress. Phil Mag 42: 414–427
230 Bishop JFW, Hill R (1951) A theoretical derivation of the plastic properties of a polycrystalline face centered metal. Phil Mag 42: 1298
231 Lin TH (1954) A proposed theory of plasticity based on slip. Proc II US Nat Congr Appl Mech. New York
232 Lin TH (1958) On stress strain relations based on slip. Proc III US Nat Congr Appl Mech. New York

233 Lin TH (1971) Physical theory of plasticity. Adv Appl Mech. 11:255–311
234 Hill R, Rice GR (1972) Constitutive analysis of elastic–plastic crystals at arbitrary strain. J Mech and Phys Solids 20:401–413
235 Asaro RG (1983) Crystal plasticity. Trans ASME, J Appl Mech 54:921–934
236 Odqvist FKG (1974) Mathematical Theory of Creep and Creep Rupture. Clarendon Press, Oxford
237 Zubchaninov VG (1990): Fundamentals of the Theory of Elasticity and Plasticity. Higher School Publishing, Moscow, (in Russian)
238 Ludvik P (1909) Element der Technologischen Mechanik, Julius springer,Berlin
239 Malinin NN (1975) Applied Theory of Plasticity and Creep. Mashinostroenie, Moscow (in Russian)
240 Lensky VS (1958) Experimental verification of the laws of isotropy and retardation principle in the case of combined loading. Proc USSR Acad Sci. OTN (11):15–24 (in Russian)
241 Lensky VS (1962) A hypothesis of local determinacy in the theory of plasticity. Proc USSR Acad Sci. Mehanika I Mashinostroyenie 5:154–158 (in Russian)
242 Malyi VI (1966) On the evenness of the components of the stress vector–functional in the theory of plasticity. Vestnik Moscowskogo Universiteta, Moscow, Matematika i Mehanika 6:80–84 (in Russian)
243 Malyi VI (1967) Expansion of the stress functional with respect to a small parameter. Vestnik Moscowskogo Universiteta, Moscow, Matematika i Mehanika 2:73–80 (in Russian)
244 Vasin RA (1971) Some problems of stress–strain relations in the case of combined loading. Izd-vo Moscowskogo Universiteta, Moscow, Uprugost i neuprugost 1:59–126
245 Ogibalov PM (1971) The Iljushin laws and methods in the mechanics of deformable solids and media. Mechanics of Polymers 1:36–40 (in Russian)
246 Levitas VI (1987) Large Elastoplastic Deformations of Materials under High Pressure. Naukova Dumka, Kiev, p. 224
247 Pozdeev AA, Trusov PV, Nyashin YuI (1986) Large Elastoplastic Deformations: Theory, Algorithms, Applications. Nauka, Moscow :232
248 Vasin RA (1988) On the experimental investigation of the functional of placticity within the framework of the theory of elastoplastic proceses. In: Plasticity and Fracture of Solids. Nauka, Moscow, pp. 40–57 (in Russian)
249 Bhattacharya SS, Padmanabhan KA (1989) Similarities and differences in the approaches to structural supeprlasticity and high temperature creep. Trans Indian Inst Metals 42 (suppl), pp. S123–S137
250 ASM Metals Handbook (Supplement), Creep and creep rupture tests., Supplement. August 1955. Met Progr 68(2A):175–184
251 Green WW (1971) Creep and stress rupture testing. In: Bunshaw RF (ed.) Techniques of Metals Research V Part1 Interscience Publishers, New York
252 Andrade EN da C (1910) On the viscous flow in metals and allied phenomena. Proc Roy Soc A84:1–12
253 Bayley RW (1929) Creep of steel under simple and compound stresses and the use of high initial temperature in steam power plant. Trans Tokyo Sectional Meeting of the World Power Conference, Tokyo
254 Bayley RW (1935) The utilisation of creep test data in engineering design. Proc Inst Mech Engng, p.131
255 Leckie FA (1980) Advances in creep mechanics. In: Ponter ARS, Hayhurst DR (eds.) Creep in Structures (IUTAM III Symposium. Leicester, UK) Springer-Verlag, Berlin, pp. 13–47
256 Andrade EN da C (1962) The validity of $t^{1/3}$ law of flow of metals. Phil Mag 7:84
257 Andrade EN da C (1914): Creep and recovery. Proc Roy Soc. London 90A:329–342

258 Garofalo F (1960) Properties of Crystalline Solids. ASTM Special Technical Publications No.283,p. 82

259 Shesterikov SA, Melnikov GP, Arshakuni AL (1980) On the selection of the constitutive equaitons for creep. Strength of Materials 6:77–81 (in Russian)

260 Soderberg CR (1936) The interpretation of creep tests for machine design. Trans ASME 58: 8

261 Odqvist FKG (1953) Influence of primiary creep on stresses in structural parts. Trans Roy Inst Tech No 66. Stockholm

262 Hoff NJ (1956) Creep buckling. The Aeronautical Quarterly VII, pp. 1-20

263 Davenport CC (1938) Correlation of creep and relaxation properties of copper. J Appl Mech 5:A55–A60

264 Kachanov LM (1960) Creep Theory. Fizmatgiz, Moscow (in Russian)

265 Malinin NN (1951) Certain unidimensional problems of transient creep. Inzhenerny Sbornik 10. Izd-vo USSR Acad Sci 10:17–34

266 Nadai A (1931) Plasticity. McGraw-Hill, New York London

267 Kuhlmann–Wilsdorf D (1989) Theory of plastic deformation: properties of low energy dislocation structures. Mater Sci Engng A113:1–41

268 Landau LD, Lifshitz EM (1987) Theoritical Physics Vol. VII Theory of Elasticity. Nauka,Moscow (in Russian)

269 Wasserman AM, Danilkin VA, Korobov OC (1985) Methods to Control and to Investigate Light Alloys. Metallurgy Publishers, Moscow (in Russian)

270 Saltykov SA (1970) Stereometrical Metallography. Metallugy Publishing House, Moscow (in Russian)

271 Weertman J, Weertman JR (1968) Temperature dependent mechanical properties. In: Physical Metallurgy. Ed. Cahn R.W.Vol. 3 pp. 216–247

272 Rabotnov Yu N (1979): Mechanics of Deformable Solids. Nauka, Moscow (in Russian)

273 Sidoroff F (1988) Internal variables and phenomenological for metals plasticity. Rev Phys Appl 23:649–660

274 Brown SB, Kwon HK, Anand L (1989) An internal variable constitutive model for hot working of metals. Int J Plast 5:95–130

275 Kwon HT, Won CY (1998) An internal variable theory of structural superplasticity. Acta Mater 46:2741–2749

276 Smirnov OM (1991) Superplastic material rheology affected by structure evolution. In: Hori S, Tokizane M, Furushiro N, (eds.) Superplasticity in Advanced Materials (ICSAM–91). JSRS, Osaka, Japan, pp 813–818

277 Perevezentsev VN, Rybin VV, Chuvil'deev VN (1992) The theory of structural superplasticity. Acta Metall 40:887–894, also 895–905, 907–914 and 915–924

278 Gong C, Tokuda M, Xu B (1998) The constitutive equations of superplastic materials In: Aizawa T, Higashi K, Tokuda M, (eds.): Proc Int Seminar on Microstructure, Micromechanics and Processing of Superplastic Materials (IMSP 97), Mie University Press, Tsu, Japan, pp. 247–255

279 Hamilton CH, Kannan K, Khaleel MA, Smith MT, Vetrano JS (1998) Factors affecting superplastic ductility and elongation. In: Ghosh AK, Bieler TR, (eds.): Superplasticity and Superplastic Forming. The Minerals, Metals & Materials Society, Warrendale, USA, pp. 43–51

280 Wang C, Hou G (1994): A study on Post-SPF mechanical characteristics of Ti–6Al–4V alloy. Mater Sci Forum 170–172:213–218

281 Al–Naib TYM, Duncan JL (1970): Superplastic metal forming. Int J Mech Sci 12:463–477

282 Ghosh AK, Hamilton CH (1982) Influence of material parameters and microstructure on superplastic forming. Metall Trans 13A:733–743

283 Enikeev FU (1993) Mathematical modelling of the rheological behavior of materials during superplastic forming processes. PhD Thesis. Institute for Metals Superplasticity Problems, Ufa, Russia, (in Russian)
284 Drucker DC (1984) Material response and continuum relations or from microscales to macroscales. Trans of ASME. J Engng Mater Technol 106:286–289
285 Lee EH (1986) Interaction between physical mechanisms and the structure of continuum theories. In: Gittus J, Zarka J, Nemat-Nasser S (eds.), Large Deformations of Solids Physical Basis and Mathematical Modelling. Elsevier Applied Science, London New York, pp. 143–161
286 Sidoroff F, Teodosiu C (1986) Microstructure and phenomenological models for metals. In: Gittus, J, Zarka J, Nemat-Nasser S (eds.) Large Deformations of Solids Physical Basis and Mathematical Modelling. Elsevier Applied Science, London New York, pp. 163–186
287 Hart EW (1984) A micromechanical basis for constitutive equations with internal state variables. Trans of ASME. J Engng Mater Technol 106:322–325
288 Aifantis EC (1984): On the microstructural origin of certain inelastic models. Trans ASME. J Engng Mater Technol 106:326–330
289 Hoffman O, Sachs G (1953) Introduction to the Theory of Plasticity for Engineers. McGraw–Hill, New York
290 Johnson W (1997) Reflections and experience with materials processing. J Mater Processing Technol 63:1–17
291 Lin TH (1984) A physical theory of plasticity and creep. Trans ASME. J Eng Mater Technol 106: 290–294
292 Odqvist FKG (1981) Historical survey of the development of creep mechanics from its beginnings in the last century to 1970. In: Ponter ARS, Hayhurst DR (eds.) Creep in Structures (IUTAM 3rd Symposium. Leicester, UK) Springer-Verlag, Berlin Heidelberg New York, pp. 1–12
293 Wilshire B (1981): The approach to creep and fracture of metals and alloys. In: Ponter ARS, Hayhurst DR (eds.) Creep in Structures. (IUTAM III Symposium,Leicester, UK) Springer-Verlag, Berlin Heidelberg New York, pp. 72–87
294 Ohashi Y, Ohno N, Kawai M (1982): Evaluation of creep constitutive equations for Type 304 stainless steel under repeated multiaxial loading. Trans ASME. J Engng Mater Technol 104: 159–164
295 So H, Lin YF, Kzempl E (1997) Comparison of flow patterns between plasticine and aluminium alloys in hot precision forging. J Mater Processing Technol 72:39–48
296 Konovalov AV (1997) Constitutive eqautions for elastoplastic medium under large plastic deformations. Proc Russian Acad Sci. Mechanics of Solids 5:139–147
297 Klee KD, Paulin J (1980) On numerical treatment of large elastic–viscoplastic deformations. Arch Mech 32:333–345
298 Swiatkowski K (1997) Physical modelling of metalworking processes using wax–based model materials. J Mater Processing Technol 72:272–276
299 Merzer AM (1982) Steady and transient creep behaviour based on unified constitutive equations. Trans ASME, J Engng Mater Technol 104:18–25
300 Penny RK, Leckie FA (1968) The mechanics of tensile testing. Int J Mech Sci 10:265–273
301 Wu HC, Wang TP (1982) Analysis of test system misalignment in the creep. Trans ASME, J Eng Mater Technol 104:280–284
302 Gunasekera JS, Havranek J, Littlejohn MH (1982) The effect of specimen size on stress–strain behaviour in compression. Trans ASME, J Eng Mater Technol 104:274–279
303 Barnes HA, Walters K (1985) The yield stress myth? Rheologica Acta 24:323–326
304 Molinari A, Canova GR, Ahzi S (1987) A self-consistent approach of the large deformation Polycrystal Viscoplasticity. Acta Metall 35:2983–2994
305 Mosher DA, Dawson PR (1996) A state variable constitutive model for supeprlastic Ti-6Al-4V based on grain size. Trans ASME. J Engng Mater Technol 118:162–168

306 Ha TK, Chang YW (1998): An internal variable theory of structural supeprlasticity. Acta Mater 46:2741–2749
307 Guionnet C (1992) Modelling of ratchetting in biaxial experiments. Trans ASME. J Engng Mater Technol 114:56–62
308 Papo A (1988) Rheological models for gypsum plaster pastes. Rheologica Acta 27:320–325
309 Hyde TH, Sun W, Tang A (1998) Determination of material constants in creep continuum damage constitutive equations. Strain 34:83–90
310 Sharafutdinov GZ (1987) On the constitutive eqautions of viscoelasticity and viscoplasticity. Proc Russian Acad Sci Mechanics of Solids 3 :125–133
311 Narain A (1986) On K–BKZ and other viscoelastic models as continuum generalisation of the classical spring–dashpot models. Rheological Acta 25:1–14
312 Tokuda M (1990) Metal plasticity: A systematic formulation of inelastic constitutive equations for polycrystalline metals. In: Inoue T, Kitagawa H, Shima S (eds.) Computational Plasticity. (Current Japanese Materials Researches) Vol 7. Elsevier Applied Science, London New York, pp. 25–42
313 Kolmogorov VL (1986) Mechanics of Metal Working, Metallugy Publishing House, Moscow
314 Tretjyakov AV, Zyuzin VI (1973) Mechanical Properties of Metals and Alloys under Metal Working. Metallugy Publishing House, Moscow
315 Diulgerov N, Istatkov A, Mitev N, Spirov I (1988) Superplastic low Manganese Zinc–Manganese alloy. In: Hamilton CH, Paton NE (eds.) (1988): Superplasticity and Superplastic Forming. TMS, Warrendale, USA, pp 419–427
316 Swaminatham K, Padmanabhan KA (1990) Tensile flow and fracture behaviour of a superplastic Al-Ca-Zn Alloy. J Mater Sci 25:4579–4586
317 Padmanabhan KA, Davies GJ (1970) Numerical analysis of supeprlasticity data for use in metal forming applications. J Mech Phys Solids 18:261–275
318 Theodore ND, Padmanabhan KA (1990) Numerical optimisation of superplastic deformation. J Mater Sci 25:2133–2143
319 Zehr SW, Backofen WA (1968) Superplasticity in Lead–Tin alloys. Trans Am Soc Metals 61:300–313
320 Bhattacharya SS, Satishnarayana GV, Padmanabhan KA (1995) A generic analysis for high temperature power law deformation: the case of linear ln(Strain Rate)–ln(Stress) relationship. J Mater Sci 30:5850–5866
321 Panchenko EV (1977) Investigations in Plasticity and Metalworking. TPI Publ, Tula, USSR, pp. 121–132
322 Bhattacharya SS, Padmanabhan KA (1995) A mechanistic model for boundary sliding controlled optimal supeprlastic flow: II Experimental verification. J Mater Proc and Manuf Sci 4:117–161
323 Yadava RK, Padmanabhan KA (1979) On the unsteady and steady state regions of superplastic flow. Mater Sci Engng 37:127
324 Chung DW, Cahoon JR (1979) Superplasticity in aluminium-silicon eutectic. Metal Sci 13: 635–640
325 Holt DL, Backofen WA (1966) Superplasticity in the Al–33Cu eutectic alloy. ASM Trans Quart 59:755–768
326 Matsuki K, Minami K, Tokizawa M, Murakami Y (1979) Superplastic behaviour in nominally single phase and two–phase Al-Cu alloys. Metal Sci 13:619–626
327 Galimov AK (1998) Private Communication
328 Langdon TG (1994) An evaluation of the strain contributed by grain boundary sliding in superplasticity. Mater Sci Engng A174: 225–230
329 Valiev RZ, Kaibyshev OA (1983) On the quantitative evaluation of superplastic flow mechanisms. Acta Metall 31:2121–2128

330 Salishchev GA, Galejev RM, Imayev RM (1991) Dynamic recrystallization during superplastic flow of materials. In: Hori S, Tokizane M, Furushiro N, (eds.) Superplasticity in Advanced Materials (ICSAM–91). JSRS, Osaka, Japan, pp. 163–168
331 Levchenko VS, Portnoy VK, Novikov II (1991) Unusual low grain boundary sliding in aluminum alloy with classical features of micrograin superplasticity. In: Hori S, Tokizane M, Furushiro N, (eds.) Superplasticity in Advanced Materials (ICSAM–91). JSRS, Osaka, Japan, pp. 39
332 Blackwell PL, Bate PS (1993) The absense of relative grain translation during superplastic deformation of an Al–Li–Mg–Cu–Zr alloy. Metall Trans 24A:1085–1093
333 Novikov II, Nikiforov AO, Polkin VI, Levchenko VS (1996) Mechanisms of superplastic deformation of aluminum alloy Amg4. Izvestiya Vuzov, Tcvetnaya Metallyrgiya 1:43–48 (In Russian)
334 Blackwell PL, Bate PS (1996): Behavior of internal markers in Ti–6Al–4V deformed in superplastic tension. Metall and Mater Trans 27A:3747–3748
335 Blackwell PL, Bate PS (1999) Superplastic deformation without relative grain translation. In: Sakuma T, Aizawa T, Higashi K (eds.) Towards Innovation in Superplasticity II. Kobe, Japan, Mater Sci Forum 304–306:189–194
336 Vasin RA, Enikeev FU, Mazurski MI (1998) Mechanical modelling of the universal superplastic curve. In: Aizawa T, Higashi K, Tokuda M, (eds.) Proc of Int Seminar on Microstructure, micromechanics and processing of superplastic materials (IMSP 97). Mie University Press, Tsu, Japan, pp. 223–230
337 Ghosh AK (1982) Characterization of superplastic behaviour of metals. In: Paton NE, Hamilton CH (eds.) Superplastic Forming of Structural Alloys. TMS–AIME, Warrendale, USA, pp. 85–103
338 Ghosh AK, Hamilton CH (1979) Mechanical behavior and hardening characteristics of a superplastic Ti–6Al–4V alloy. Metall Trans 10A:699–706
339 Smirnov OM (1988) Rheological criteria for rational use of superplasticity in metal working by pressure In: Hamilton CH, Paton NE (eds.): Superplasticity and Superplastic Forming. TMS, Warrendale, USA, pp. 179–183
340 Enikeev FU, On the material constant determination for Smirnov's Model. Proc. Russian Acad Sci, Metally (submitted) (in Russian)
341 Enikeev FU, Application of Smirnov's model in describing the mechanical reponse of superplastics under step strain rate tests. Proc Russian Acad Sci, Metally (submited) (in Russian)
342 Tsepin MA, Yershov AN, Tayupov AR, Smirnov OM (1991): Application of the finite elements method for analysis of superplastic deformation. In: Hori S, Tokizane M, Furushiro N (eds.). Superplasticity in Advanced Materials. (ICSAM–91), JSRS, Osaka, Japan, pp. 743–747
343 Smirnov OM (1997) Superplasticity of metals: phenomenology based on rheological properties and structural dynamics In: Chokshi AH (ed.) Superplasticity in Advanced Materials (ICSAM–97). Mater Sci Forum 243–245:443–452
344 Smirnov OM (1999) Rheological behaviour of superplastic nanocrystalline and amorphous materials. In: Sakuma T, Aizawa T, Higashi K (eds.) Towards Innovation in Superplasticity II. Kobe, Japan, Mater Sci Forum 304–306:341–348
345 Haasen P (1978) Physical Metallurgy. Cambridge University Press, Cambridge, pp334–338
346 Vasin RA, Enikeev FU, Mazurski MI, Munirova OS, Mechanical modelling of the thermomechanical behaviour of superplastic materials. Proc Russian Acad Sci, Metally (submitted)(in Russian)
347 Ashby MF, Verrall RA (1973) Difusion accomodated flow and superplasticity. Acta Metall 21:149–163
348 Alden TH (1967) The origin of superplasticity in the Sn-5%Bi alloy. Acta Metall 15:469–480

349 Mukherjee AK (1971) The rate controlling mechanism in superplasticity. Mater Sci Engng 8:83–89
350 Gifkins RC (1973) Superplasticity, creep and grain boundary sliding. Scripta Metall 7:27–33
351 Shorshorov MH, Tikhonov AS, Bulat SI (eds.) (1983) Superplasticity of Metalic Materials. Nauka, Moscow (in Russian)
352 Hayden HW, Gibson RC, Merrick HF, Brophy JH (1967) Superplasticity in the Ni–Fe–Cr system. ASM Trans Quart 60:3–14
353 Coble RL (1963) A model for boundary diffusion controlled creep in polycrystalline materials. J Appl Phys 34:1679–1682
354 Nabarro FRN (1948) Report on the Conference on Strength of Solids London: Physical Society, pp.75–81
355 Herring G (1951) Diffusion viscosity of a polycrystalline solid. J Appl Phys 21:437–445
356 Avery DH, Backofen WA (1965) A structural basis for superplasticity. Trans ASM 58:551–562
357 Packer CM, Sherby OD (1967) An interpretation of the superplasticity phenomenon in two–phase alloys. Trans ASM 60:21–28
358 Cook G (1934): J Inst Met 54:134
359 Ghosh AK (1994) A new physical model for superplastic flow. Langdon TG (ed.) Superplasticity in Advanced Materials (ICSAM–94), Mater Sci Forum 170–172:39–46
360 Hamilton CH (1989) Simulation of static and deformation enhanced grain growth effects on superplastic ductility. Metall Trans 20A: 2783–2792
361 Wilkinson DS, Caceres CH (1984) On the mechanism of strain enhanced grain growth during superplastic deformation. Acta Metall 32:1335–1345
362 Wilkinson DS, Caceres CH (1984) An evaluation of available data for strain-enhanced grain growth during superplastic flow. J Mater Sci Lett 3:395–399
363 Kaibyshev OA, Pshenichniuk AI, Astanin VV (1998) Superplasticity resulting from cooperative grain boundary sliding. Acta Mater 46:4911–4916
364 Pshenichniuk AI, Astanin VV, Kaibyshev OA (1998) The Model of grain boundary sliding stimulated by intragranular slip. Philos Mag 77:1093–1106
365 Astanin VV, Kaibyshev OA, Pshenichniuk AI (1997) Cooperative processes during superplastic deformation. In: Chokshi AH (ed.) Superplasticity in Advanced Materials. (ICSAM–97), Mater Sci Forum 243–245:41–46
366 Astanin VV, Kaibyshev OA, Faizova SN (1991) Cooperative grain boundary sliding under superplastic flow. Scripta Metall Mater 25:2663–2668
367 Zelin MG, Dunlap MR, Rosen R, Mukherjee AK (1993) The direct observation of cooperative grain boundary sliding and migration during superplastic deformation of Lead-Tin eutectic in shear. J Appl Phys 74:1–11
368 Astanin VV, Kaibyshev OA (1994) Cooperative grain boundary sliding and superplastic flow nature. In: Langdon TG (ed.) Superplasticity in Advanced Materials. (ICSAM–94), Mater Sci Forum 170–172:23–28
369 Zelin MG, Mukherjee AK (1995) Cooperative phenomena at grain boundaries during superplastic flow. Acta Metall 43:2359–2372
370 Zelin MG (1996) Cooperative grain boundary sliding in materials with non–uniform microstructure. J Mater Sci Lett 15:2068–2070
371 Larin SA, Perevezentsev VN, Chuvil'deev VN (1990) Influence of the grain growth on the strain hardening of superplastic materials. Physica Metallov i Metallovedenie1:180–188
372 Bylia OI, Unpublished work
373 Padmanabhan KA, Nitsche R, Hahn H (1995) On the deformation of nanocrystalline metals and ceramics. Fourth European Conference on Advanced Materials and Processes (EUROMAT 95). Symposium G, Special and Functional Materials. Associazone Italiana Di Metallurgia, Milano, Italy, pp. 289–298

374 Hahn H, Padmanabhan KA (1997) A model for the deformation of nanocrystalline Materials. Philos Mag 76B:559–571
375 Hahn H, Mondal P, Padmanabhan KA (1997) Plastic deformation of nanocrystalline materials. Nanostructured Mater 9:603–606
376 Padmanabhan KA (1999) Deformation of structural superplastics, nanostructured materials and metallic glasses: A unified approach. Proc Int Symp Mechanically Alloyed, Metastable and Nanostructured Materials (ISMAAM–99). Dresden, Germany, (in press)
377 Belyaev NM (1943) Application of plastic deformation theory to calculation of components at high temperatures. Proc USSR Acad Sci. OTN, No 7
378 Vasin RA, Muravlev AV, Chistyakov PV (1993) On the rheological properties of the superplastic titanium alloy VT9. In: Israilov M. Sh., Shmakov AP, Lensky VS (eds.) Elasticity and anelasticity Part I. Moscow State University Publishing House, Moscow, pp 163–171 (In Russian)
379 Vasin RA, Iljushin AA (1983) On a representation of the laws of elasticity and plasticity for plane problems. Proc USSR Acad Sci. Mechanics of Solids 4:114–118 (In Russian)
380 Iljushin AA (1967) A theory of long-term strength. Proc USSR Acad Sci. Mechanics of Solids No 3, pp. 21–35
381 Iljushin AA (1961) On the foundations of the general mathematical theory of plasticity. In: Voprosy teorii plastichnosti USSR Acad Sci, pp. 3–29 (in Russian)
382 Ghosh AK, Hamilton CH (1978) Process modelling – Fundamentals and applications to metals, Proc ASM, Metals Park, Ohio, pp. 303–331
383 Chandra N, Chandy K (1991) Superplastic process modeling of plane strain components with complex shapes. J Mater Shaping Technol 9:27–37
384 Jovane F (1968) An approximate analysis of the superplastic forming of a thin circular diaphragm: Theory and Experiments. Int J Mech Sci 10:403–427
385 Inoue T, Kitagawa H, Shima S (1990) Computational Plasticity. (Current Japanese Materials Researches) Vol. 7 Elsevier Applied Science, London New York
386 Levanov AN, Kolmogorov VL, Burkin SP (1976): Contact Friction in Metalworking Processes. Metallugy Publishing House, Moscow (in Russian)
387 Levanov AN (1997) Improvement of metal forming processes by means of useful effects of plastic friction. J Mater Processing Technol 72:314–316
388 Efimov AB, Romanjuk AB, Chumachenko EN (1995) On the principalities of contact friction in metalworking processes. Proc Russian Acad Sci, Mechanics of Solids: 82–98 (in Russian)
389 Mirzandjanzade AH, Mirzoyan AA, Gevinyan GM, Seidrza MK (1966) Hydraulics of the Clay and Cement Solutions. Nedra, Moscow, (in Russian)
390 Loitcanskij LG (1987) Mechanics of Fluids and Gases. Nauka, Moscow (in Russian)
391 Oh H-K (1995) Determination of ductile fracture (ductility) at any stress state by means of the uniaxial tensile test. J Mater Processing Technol 53:582–587
392 Kolmogorov VL (1997) On the history of the determination of ductile fracture (ductility) of metals. J Mater Processing Technol 72:190–193
393 Gubkin SI (1960) Plastic Deformation of Metals. Metallurgizdat, Moscow
394 Smirnov-Alaaev GA, Rosenberg VM (1956): Theory of Plastic Deformations of Metals. Mashgiz, Moscow (in Russian)
395 Presnyakov AA (1958) Determination of Metals Plasticity. Izd-vo AN Kaz SSR Ala-Ata (in Russian)
396 Kachanov LM (1958) On the time of fracture under creep. Izv. an SSSR, OTN, 8:pp.26–31
397 Rabotnov Yu N (1959) Mechanism of prolonged fracture. In: Problems of Strength of Materials and Structures. Izd-Vo an SSSR, Vol.95 (In Russian)

398 Kolmogorov VL (1995) Model of metal fracture in cold deformation and ductility restoration by annealing. In: Ghosh SK, Predeleanu M (eds.). Materials Processing Defects. Elsivier, Amsterdam.
399 Burdukovsky VG, Kolmogorov VL, Migachev VA (1995) Prediction of resources of materials of machine and construction elements in the process of manufacture and exploitation. J Mater Process Technol 55:22–295
400 Kolmogorov VL (1996) Friction and wear model for a heavily loading sliding pair report 1 Metal fracture. J Wear 194:71–79
401 Sedov LI (1959) Similarity and Dimensional Methods in Mechanics. Translation from Russian. Infosearch, London
402 Sedov LI (1972) A Course in Continuum Mechanics. Translation from Russian, Netherlands
403 Bogatov AA, Mijiritckij OI, Smirnov SV (1984) Resource of Plasticity of Metals at Metalworking Processes. Metallurgiya, Moscow (in Russian)
404 Ashby MF, Tomkins B (1979) Micromechanisms of fracture and elevated temperature fracture mechanics. ICM 3, Cambridge, Vol.1, pp. 47–62
405 Kurakami S, Ohno N (1981) A continuum theory of creep and creep damage. In: Ponter ARS, Hayhurst DR (eds.) Creep in Structures. IUTAM III Symposium, Leicester, UK, Springer-Verlag, Berlin Heidelberg New York, pp. 422–444
406 Cocks ACF, Ashby MF (1981) Creep fracture by void growth. In: Ponter ARS, Hayhurst DR (eds.) Creep in Structures. (IUTAM III Symposium, Leicester, UK) Springer-Verlag, Berlin Heidelberg New York, pp. 368–387
407 Utyashev FZ, Enikeev FU, Latysh VV (1996) Comparison of deformation methods for ultrafine-grained structure formation. Annales De Chimie 21:379–389
408 Leonova EA, Kadymov MD (1987) On the analytical representation of the material properties for the problems of thermoviscoplasticity. In: Elasticity and Anelasticity. Moscow State University, Moscow, pp. 171–178 (in Russian)
409 Vasidzu K (1987) Variational Methods in the Theory of Elasticity and Plasticity. Mir, Moscow (in Russian)
410 Lions D, Duvo G-L (1986) Inequalities in the Mechanics and Physics. Nauka, Moscow (in Russian)
411 Glovinski R, Lions G-L, Tremoljer P (1979) Numerical Investigation of the Variational Inequalities. Mir, Moscow (in Russian)
412 Pozdeev AA, Nyashin Yui, Trusov PV (1982) Residual Stresses: Theory and Applications. Nauka, Moscow (in Russian)
413 Pobedrya BE (1995) Numerical Methods in the Theories of Elasticity and Plasticity. Moscow State University, Moscow (in Russian)
414 Wilkins ML (1967) Calculations of elastoplastic flow. In: Numerical Methods in Hydrodynamics. Mir, Moscow, pp. 212–263 (in Russian)
415 Biba NV, Lishnij AI, Sadykhov OB, Stiebounov SA (1994) Design and analysis of SPF technology with FORM–2D system. Mater Sci Forum 170–172:687–692
416 Kobayashi S, Oh SI, Altan T (1989) Metalforming and the Finite Element Method. Oxford University Press, Oxford
417 Lee CH, Kobayashi S (1973) New solutions to rigid-plastic deformation problems using matrix method. Trans ASME. J. Eng. Ind : pp. 865–873
418 Zienkiewicz OC, Godbole PO (1974) Flow of plastic and visco-plastic solids with special reference to extrusion and forming processes. Int J Num Methods in Eng 8:3–16
419 Gun G Ya (1980) Theory of Metalworking. Metallurgiya, Moscow (in Russian)
420 Gun G Ya (1984) Mathematical Modelling of the Metal Working Processes. Metallurgiya, Moscow (in Russian)

421 Duncan JL, Gordon G (1989) Utilisation of superplastic effects in plasticity and modern metal forming technology. Blazynsky (ed.) Elsevier Applied Science, London New York, pp 149–167
422 Chumachenko EN, Logashina IV, Chumachenko SE (1994) Automatization of calculations when developing the technological regimes of the isothermal deforming. Langdon TG (ed.) Superplasticity in Advanced Materials. (ICSAM–94), Mater Sci Forum 170-172:657–662
423 Chumachenko EN, Smirnov OM, Chumachenko SE (1997) Stress–strain state of a blank in a superplastic bulk forming under mutual influence of material rheology and contact friction. In: Chokshi AH (ed.) Superplasticity in Advanced Materials. (ICSAM–97), Mater Sci Forum 243–245:745–750
424 Romanjuk SN (1992) Mathematical modelling, control and optimization of metal working processes taking into account contact friction. PhD Thesis, Moscow
425 Doltsinis I St (1993) Numerical analysis and design of industrial superplastic forming. J de Physique IV. Colloque C7. Supplement Journal De Physique III 3:1187–1197
426 Rama SC, Chandra N (1991) Development of a pressure prediction method for superplastic forming processes. Int J Non-Linear Mechanics 26:711–725
427 Chandra N (1988) Analysis of superplastic metal forming by a finite element method. Int J Numer Meth Engng 26:1925–1944
428 Chandra N, Haisler WE, Goforth RE (1987) A finite element solution method for contact problems with friction. Int J Numer Meth Engng 27:447–495
429 Chandra N, Rama SC, Goforth RE (1991) Process modelling of superplastic forming processes using four different computational methods. In: Hori S, Tokizane M, Furushiro N (eds.). Superplasticity in Advanced Materials (ICSAM–91). JSRS, Osaka, Japan, pp. 837–844
430 Chandra N, Rama SC, Goforth RE (1991) Variable optimum strain rate for dynamically recrystallising Aluminium-Lithium alloy. In: Hori S, Tokizane M, Furushiro N (eds.). Superplasticity in Advanced Materials (ICSAM–91). JSRS, Osaka, Japan, pp. 765–770
431 Rama SC, Chandra N, Goforth RE (1992) Computational process modelling of dynamically recrystallizing superplastic materials. In: Chennot JL, Wood RD, Zienkiewicz OC. (eds.) Numiform'92, Balkema Publishers, Roterdam, Netherlands, pp. 867–872
432 Chandra N, Rama SC (1992) Application of finite element method to the design of superplastic forming processes. J Engng for Industry, Trans ASME 114:452–458
433 Rama SC, Chandra N, Goforth RE (1993) Numerical modelling of 3–D superplastic sheet forming processes. In: Chandra N, Garmestani H, Goforth RE (eds.): Advances in Superplasticity and Superplastic Forming. The Minerals & Materials Society, Warrendale, USA, pp. 133–154
434 Chandra N, Rama SC, Rama J (1994) Design and analysis of 3–D superplastic forming processes. In: Langdon TG (ed.) Superplasticity in Advanced Materials. (ICSAM–94), Mater Sci Forum 170–172:577–582
435 Xing HL, Wang ZR (1997) Finite element analysis and design of thin sheet superplastic forming. J Mater Process Technol 72:1–7
436 Lee KS, Huh H, Choy YJ (1997) Finite element simulation of superplastic punch forming with a thickness control ring. J Mater Process Technol 72: 684–689
437 Akkus N, Manabe K, Kawahara M, Nishimura H (1997) A finite element modelling for superplastic bulging of titanium alloy tube and pressure path optimization. In: Chokshi AH (ed.) Superplasticity in Advanced Materials. (ICSAM–97), Mater Sci Forum 243–245: 729–734
438 Bellet M, Chennot JL (1989) Numerical modelling of thin sheet superplastic forming. In: Thompson Brookfield VT (eds.) Numiform'89, Balkema Publishers, Rotterdam, Netherlands, pp. 401–406
439 Chennot JL, Bellet M (1990) A velocity approach to elasto-plastic and elasto-viscoplastic calculation by the finite element method. J Engng for Industry 112:150–154

440 Chennot JL (1992) Recent contributions to the finite element modeling of metal forming processes. J Mater Processing Technol 34:9–18
441 Bonet J, Wargadipura AHS, Wood RD (1989) A pressure cycle control algorithm for superplastic forming. Compu Appl Numerical Methods 5:121–128
442 Bonet J, Bhargava P, Wood RD (1994) SPF forming simulation of 3–dimensional DB components by the FEM method. In: Langdon TG (ed.) Superplasticity in Advanced Materials. (ICSAM–94), Mater Sci Forum 170–172:589–594
443 Bonet J, Wood RD (1994) Energy dissipation strategies for the prediction of pressure cycles in the numerical analysis of superplastic forming processes. In: Langdon TG (ed.) Superplasticity in Advanced Materials. (ICSAM–94), Mater Sci Forum 170–172:607–612
444 Bonet J, Wood RD, Zienkiewicz OC (1988) Finite element modelling of the superplastic forming of thin sheet. In: Hamilton CH, Paton NE (eds.). Superplasticity and Superplastic Forming. TMS, Warrendale, USA, pp. 291–295
445 Kannan K, Johnson CH, Hamilton CH (1997) The Role of flow properties and damage accumulation in superplastic ductility of Al-Mg-Mn alloy. In: Chokshi AH (ed.) Superplasticity in Advanced Materials. (ICSAM–97), Mater Sci Forum 243–245:125–130
446 Sadeghi RS, Pursell ZS (1994) Finite element modelling of superplastic forming. In: Langdon TG (ed.) Superplasticity in Advanced Materials. (ICSAM–94), Mater Sci Forum 170–172:571–576
447 Sadeghi RS, Pursell ZS (1997) Finite element modelling of superplastic forming using analytical contact surfaces. In: Chokshi AH (ed.) Superplasticity in Advanced Materials. (ICSAM–97), Mater Sci Forum 243–245:719–728
448 Wood RD, Bonet J (1996) J Mater Processing Technol 60:45–53
449 Chennot JL, Agassant JF, Montmitonnet P, Vergnes B, Billon N (1998) Proc first ESAFORM Conference on Material Forming. Sophia-Antipolis (France), Ecole Des Mines De Paris, CEMEF (CBNRS UMR 7635)
450 Zienkiewicz OC (1971) The Finite Element Method in Engineering Science. McGraw-Hill, London
451 Zienkiewicz OC (1984) Flow formulation for numerical solution of metal forming processes. In: Pittman JFT, Zienkiewicz OC, Wood RD, Alexander JM. (eds.) Numerical Analysis of Forming Processes, pp. 1–44
452 Kim YJ, Yang DY (1985) A rigid-plastic finite element formulation considering the effect of geometric change and its application hydrostatic bulging. Int J Mech Sci 27:453–463
453 Kruglov AA, Lutfullin RY, Kaibyshev OA (1992) Simulation of superplastic forming of spherical vessel using the finite element method. In: Chennot JL, Wood RD, Zienkiewicz OC. (eds.) Proc IV Int Conference on Numerical Methods in Industrial Forming Processes. Numiform 92. Balkena Publishers, Rotterdam, Netherlands, pp. 857–860
454 Parteder E, Bunter R (1998) Determination of flow curves by means of a compression test under sticking friction conditions using an iterative finite element procedure. J Mater Processing Technology 74:227–233
455 Gong C, Xu B, Tokuda M (1997) Dynamic explicit finite element analysis of superplastic forming processes. In: Aizawa T, Higashi K, Tokuda M. (eds.). Proc Int Seminar on Microstructure, micromechanics and processing of superplastic materials (MSP 97). Mie University Press, Tsu, Japan, pp. 279–288
456 Storakers B (1996) Finite plastic deformation of a circular diaphragm under hydrostatic pressure. Int J Mech Sci 38:619–628
457 Brandon JF, Lecent H, Oytana C (1979) A new formulation for the bulging of viscous sheet metals. Int J Mech Sci 21:379–386
458 Akkus N, Kawahara M, Nishimura H (1994) A technological analysis of superplastic deformation of titanium alloy pipes into spherical shape. In: Langdon TG (ed.) Superplasticity in Advanced Materials (ICSAM–94), Mater Sci Forum 170–172:633–638

459 Wang NM, Shammany MR (1969) On the plastic bulging of a circular diaphragm by hydrostatic pressure. J Mech Phys Solids 17:43–61

460 Yang HS, Mukherjee AK (1992) An analysis of the superplastic forming of a circular sheet diaphragm. Int J Mech Sci 34:283–297

461 Shaw MC (1984) A critical review of mechanical failure criteria. J Engng Mater Technol 106:219–226

462 Stroganov GB, Novikov II, Boitcov VV, Pshirkov VF (1979) Employment of Superplasticity in Metal Working. Mechanical Engineering Publishing, Moscow (in Russian)

463 Boyer RR (1996) An overview on the use of titanium in the aerospace industry. Mater Sci Eng A213:103–114

464 Padmanabhan KA (1971) PhD Thesis. University of Cambridge, UK

465 Padmanabhan KA (1973) Closed-die forging of superplastic Al-$CuAl_2$ eutectic alloy and its technological implications. Trans Indian Inst Metals: 41–48.

466 Altan T (1973) Forging equipment, materials and practices. Metals and Ceramic Information Center

467 Oyama T, Sherby OD, Ruano OA, Kum DW, Wadsworth J (1984) Superplasticity. In: S Agarwal (ed.) ASM Technical Publication 8401. American Society for Metals, Metals Park, Ohio

468 Moore JB, Tequesta J, Athey RL (1976) US Patent 3:519,503

469 Zhang D (1988) The application of superplastic forming for making plastic injection mould-cavity design and superplastic pressing. In: Hamilton CH, Paton NE. (eds.). Superplasticity and Superplastic Forming. TMS, Warrendale, USA, pp. 541–544

470 Karpin AP, Rygkov VG (1989) Proc IV Conference on Superplasticity in Metals. Ufa, Russia, p. 212. (in Russian)

471 Ermatchenko AG, Karavaeva MV, Zaripov AA (1995) Production of uniquely shaped precision forging with specified structure out of titanium alloys. In: Titanium 95–Science and Technology. (Proc VIII Int Conference London) pp. 848–855

472 Ermatchenko AG, Karavaeva MV (1996) High temperature precision forming of titanium blades. J Mater Engng Performance 5:589–592

473 Somani MC, Sundaresan R, Kaibyshev OA, Ermatchenko AG (1998) Deformation processing in superplasticity regime-production of aircraft engine compressor discs out of titanium alloys. Mater Sci Engng A243:134–139

474 Tang S (1973) Note on superplastic forging of circular disks. J Franklin Institute 296:207–212

475 Tang S (1973) Drawing and extrusion of superplastic metals through cone–shaped dies. J Franklin Institute 295:357–372

476 Tang S (1972) Steady extrusion of superplastic metallic alloys. Trans ASME, Series B, J Engng Industry 94:188–193

477 Ganago OA, Subich VN, Stepanov BA, Safonov AV (1980) A Study on the compression with torsion of a thin layer. Izv. VUZov, Mashinistroyenie 6:110–113

478 Subich VN (1984) Contact interaction of the axial forces and torque under forging with torsion. Izv. VU Zov, Mashinistroyenie 4:110–113 (in Russian)

479 Johnson W, Kudo H (1962) The Mechanics of Metal Extrusion. Manchester University Press, Manchester

480 Avitzur B (1968) Metal Forming: Processes and Analysis. McGraw-Hill, New York

481 Kachanov LM (1971) Foundations of the Theory of Plasticity. North Holland, Amsterdam

482 Tang S Mechanics of Superplasticity. R Krieger Publishing Co, New York

483 Tang S (1971) Steady extrusion of superplastic metallic alloys. Trans ASME, J Basic Engng. 97B:1131–1135

484 Jovane F, Shabaik AH, Thomsen EG (1969) Some extrusion studies of eutectic alloy of Pb and Sn. Trans ASME, J Engg Industry 91:680–686

485 Hirohashi M, Asanuma H (1988) Combined extrusion of superplastic Al-Zn alloys. In: Hamilton CH, Paton NE (eds.). Superplasticity and Superplastic Forming. TMS, Warrendale, USA, pp 545–549
486 Kellett B, Carry P, Mocellin A (1988) Extrusion of ZrO_2 at elevated temperatures. In: Hamilton CH, Paton NE (eds.). Superplasticity and Superplastic Forming. TMS, Warrendale, USA, pp. 625–630
487 Thomsen EG, Yang CT, Kobayashi S (1965) Mechanics of Plastic Deformation in Metal Processing. Macmillan Company, New York
488 Johnson RH (1969): Design Engineering, March p.33
489 Kodzi K, Sakiguti H (1975) Method of die-less wire drawing. Metal Engng 15:59–62
490 Kodzi K (1979) Prototype of continiuous die-less wire drawing machine and some experiments on determination of wire mechanical properties. J Jap Soc Technol Plast 20:814–819
491 Dikson R (1987) Company is working on the development of die-less wire drawing. Wire J Int 20:25–28
492 Dieless wire Drawing by Tension. USSR Patent No 1619540 of 19/06/1989
493 Method of Dieless Wire Drawing. USSR Patent No 1838984 of 13/10/1992
494 Shilian H, Baosheng W, Li C (1997) Engineering calculation of metal extrusion deformation force in the condition of superplasticity. In: Chokshi AH (ed.) Superplasticity in Advanced Materials. (ICSAM–97), Mater Sci Forum 243–245:751–756
495 Utyashev FZ, Enikeev FU, Petrov EN, Smotrov NV (1997) An approximate solution of the problem of roll forming of automotive wheels. Int Conference Mechanics and Technologies in Metal Working Processes. Orel, Russia, pp. 16–18.
496 Poluhin PI, Gun GYa, Galkin AM (1986) Strength of Materials under Plastic Deformation. Metallurgy Publishing House, Moscow (in Russian)
497 Ermatchenko AG. Personal communication
498 Trifonov VG. Personal communication
499 Laycock DB (1982) Superplastic forming of sheet metal. In: Paton NE, Hamilton CH. (eds.) Superplasticity and Superplastic Forming. TMS-AIME, Warrendale, USA, pp 257–271
500 Williamson JR (1982) Aerospace applications of SPF and SPF/DB. In: Paton NE, Hamilton CH. (eds.) Superplasticity and Superplastic Forming. TMS-AIME, Warrendale, USA, pp. 291–306
501 Jintao H, Jilin D, Sanshan C (1988) Development of superplastic forming technology in china. In: Hamilton CH, Paton NE (eds.) Superplasticity and Superplastic Forming. TMS, Warrendale, USA, pp. 571–582
502 Friedrich HE, Winkler PJ (1991) Fundamental questions concerning the application of superplastic forming and superplastic forming/diffusion bonding. In: Hori S, Tokizane M, Furushiro N, (eds.) Superplasticity in Advanced Materials (ICSAM–91). JSRS, Osaka, Japan, pp. 601–610
503 Tsuzuki T, Takahashi A, Sakamoto A (1991) Application of superplastic forming for aerospace components. In: Hori S, Tokizane M, Furushiro N (eds.) Superplasticity in Advanced Materials (ICSAM–91). JSRS, Osaka, Japan, pp. 611–620
504 Fernandez P (1991) A new superplastic aluminium alloy formall^{R} 570. In: Hori S, Tokizane M, Furushiro N. (eds.) Superplasticity in Advanced Materials (ICSAM–91). JSRS, Osaka, Japan, pp. 675–680
505 Hirt G, Winkler PJ (1991) Superplastic forming of Ti-6Al-4V at lower temperatures using hydrogen. In: Hori S, Tokizane M, Furushiro, N (eds.) Superplasticity in Advanced Materials (ICSAM–91). JSRS, Osaka, Japan, pp. 783–788
506 Barnes AJ (1994) Superplastic forming of aluminium alloys. Langdon TG (ed.) Superplasticity in Advanced Materials (ICSAM–94), Mater Sci Forum 170–172:701–714

507 Osada K, Yoshida H (1994) Recent applications of superplastic materials in Japan. Langdon TG (ed.) Superplasticity in Advanced Materials (ICSAM–94), Mater Sci Forum 170–172:715–724

508 Winkler PJ (1988) Diffusion bonding and superplastic forming, two complementary manufacturing techniques. In: Hamilton CH, Paton NE. (eds.) Superplasticity and Superplastic Forming. TMS, Warrendale, USA, pp. 491–506

509 Stephen D (1987) Superplasticity. NATO/AGARD Lecture series. Springfield VA. National Technical Information Services 154(P7): 1–37

510 Weisert ED, Stacher GW (1982) Concurrent superplastic forming diffusion bonding of titanium. In: Paton NE, Hamilton CH. (eds.) Superplasticity and Superplastic Forming. TMS–AIME, Warrendale, USA, pp. 273–289

511 Kennedy J (1988) Diffusion bonding and superplastic forming of 7475 aluminium alloy. In: Hamilton CH, Paton NE. (eds.) Superplasticity and Superplastic Forming. TMS, Warrendale, USA, pp. 523–527

512 Rolland B (1988): SPF–DB applications for military aircraft. In: Hamilton CH, Paton NE. (eds.) Superplasticity and Superplastic Forming. TMS, Warrendale, USA, pp. 601–602

513 Beck W (1988) DB/SPF Cooler outlet duct for aircraft application. In: Hamilton CH, Paton NE. (eds.) Superplasticity and Superplastic Forming. TMS, Warrendale, USA, pp. 685–697

514 McDarmaid DS, Partridge PG, Wisbey A (1991) A review of the mechanical properties after superplastic forming and diffusion bonding. In: Hori S, Tokizane M, Furushiro N (eds.) Superplasticity in Advanced Materials (ICSAM–91). JSRS, Osaka, Japan, pp. 621–632

515 Li Z, Guo H (1991) Superplastic forming/diffusion bonding combined technology of curved panel of titanium alloy. In: Hori S, Tokizane M, Furushiro N (eds.) Superplasticity in Advanced Materials (ICSAM–91). JSRS, Osaka, Japan, pp. 705–708

516 Kaibyshev OA, Lutfullin Rya, Safiullin RV, Fatkullin SN (1994) Problems and promises of integral technology based on the combination of superplastic forming and diffusion bonding (SPF/DB). In: Langdon TG (ed.) Superplasticity in Advanced Materials. (ICSAM–94), Mater Sci Forum 170–172:737–742

517 Safiullin RV, Rudenko OA, Enikeev FU, Lutfullin Rya (1997) Superplastic forming of sandwich cellular structures from titanium alloy. In: Chokshi AH (ed.) Superplasticity in Advanced Materials. (ICSAM–97), Mater Sci Forum 243–245:769–774

518 Acharyulu SLN, Kaibyshev OA, Sundaresan R, Bhatia ML, Safiullin RV, Lutfullin RYa (1997) Mechanical properties of the wing slat produced by SPF/DB technology. In: Chokshi AH (ed.) Superplasticity in Advanced Materials. (ICSAM–97), Mater Sci Forum 243–245:695–700

519 Higashi K (1994) Deformation mechanisms of positive exponent superplasticity in advanced aluminum alloys with nano or near nano scale grained structure. In: Langdon TG (ed.) Superplasticity in Advanced Materials. (ICSAM–94), Mater Sci Forum 170–172:131–140

520 Mabuchi M, Higashi K, Langdon TG (1994) An investigation of the role of a liquid phase in Al–Cu–Mg metal matrix composites exhibiting high strain rate superplasticity. Acta Metall Mater 42:1739–1745

521 Raj R (1996) A mechanistic basis for high strain rate superplasticity of aluminum based metal matrix composites. Mater Sci Engng A212:14–21

522 Blandin JJ, Baudelet B, Dendievel R (1997) Mechanical approach of high strain rate superplastic behaviour of composite materials. In: Sakuma T, Aizawa T, Higashi K (eds.) Towards Innovation in Superplasticity I. Kyoto, Japan, Mater Sci Forum 233–234:3–20

523 Higashi K, Mabuchi M (1997) Recent works in high strain rate superplasticity. In: Sakuma T, Aizawa T, Higashi K (eds.) Towards Innovation in Superplasticity I. Kyoto, Japan, Mater Sci Forum 233–234:155–162

524 Enikeev FU (1994) An analytical model for superplastic bulge forming of domes. In: Langdon TG (ed.) Superplasticity in Advanced Materials. (ICSAM–94), Mater Sci Forum 170–172:681–686
525 Cornfield GG, Johnson RH (1970) Forming of superplastic sheet materials. Int J Mech Sci 12:479–490
526 Chockalingam KS, Neelakantan M, Devaraj S, Padmanabhan KA (1985) On the pressure forming of two superplastic alloys. J Mater Sci 20:1310–1320
527 Viswanathan D, Venkataswamy S, Padmanabhan KA (1988) Theoretical and experimental studies on the pressure thermoforming of hemispheres of alloy Ti-6Al-4V. In: Hamilton CH, Paton NE. (eds.) Superplasticity and Superplastic Forming. TMS, Warrendale, USA, pp. 321–326
528 Guo ZX, Pilling J, Ridley N (1988) Bulge–forming of domes: A comparison of theoretical prediction and experiment. In: Hamilton CH, Paton NE (eds.) Superplasticity and Superplastic Forming. TMS, Warrendale, USA, pp. 303–308
529 Guo ZX, Ridley N (1989) Modelling of superplastic bulge forming of domes. Mater Sci Eng A114:97–104
530 Guo ZX, Ridley N (1990) Testing models for superplastic bulge forming of domes. Mater Sci Technol 6:510–515
531 Caceres CH, Wilkinson DS (1982) Grain growth at low strain rates in a superplastic Cu alloy. Scripta Metall 16:1363–1365
532 Caceres CH, Wilkinson DS (1984) Large strain behavior of a superplastic copper alloy. Acta Metall 32:415–422
533 Wilkinson DS (1985) Grain size effects in superplasticity. In: Baudelet B, Suery M (eds.) Superplasticity. Editions Centre National de la Recherche Scientifique. Paris, France, pp. 6.1–6.13
534 Wilkinson DS (1988) Microstructural instability during superplastic Flow. In: Hamilton CH, Paton NE (eds.) Superplasticity and Superplastic Forming. TMS, Warrendale, USA, pp. 81–89
535 Stowell MJ (1982) Cavitation in superplasticity. In: Paton NE, Hamilton CH (eds.) Superplastic Forming of Structural Alloys. TMS–AIME, Warrendale, USA, pp. 321–336
536 Ridley N, Pilling J (1985) Cavitation in superplastic alloys experimental. In: Baudelet B, Suery M (eds.) Superplasticity. Editions du Centre National de la Recherche Scientifique, Paris, France, pp. 8.1–8.17
537 Blandin JJ, Suery M (1988) Evolution of cavitation during superplastic deformation. In: Hamilton CH, Paton NE (eds.) Superplasticity and Superplastic Forming. TMS, Warrendale, USA, pp. 191–195
538 Pilling J, Ridley N (1988) Cavitation in superplastic alloys and the effect of hydrostatic pressure. Res Mechanica 23:31–63
539 Chokshi AH, Langdon TG (1989) Cavitation and fracture in the superplastic Al–33%Cu eutectic alloy. J Mater Sci 24:143–153
540 Bampton CC, Mahoney MW, Hamilton CH, Ghosh AK, Raj R (1983) Control of superplastic cavitation by hydrostatic pressure. Metall Trans 14A:1583–1591
541 Yang HS, Ahmed HK, Roberts WT (1989) Process control of superplastic forming under superimposed hydrostatic pressure. Mater Sci Engng A122:193–203
542 Holt DL (1970) An analysis of the bulging of a superplastic sheet by lateral pressure. Int J Mech Sci 12:491–497
543 Al–Hassani STS, Clemens GGW, Al–Naib, TYM (1977): The free bulge forming of Zn-Al superplastic sheet from a circular die. In: Alexander JM (ed.) Machine Tool Design and Research (Proc18th Int Conference) Macmillan, London, pp. 361–368
544 Ragab AR (1983) Thermoforming of superplastic sheet in shaped dies. Metal Technol 10:340–348

545 Song YQ, Zhao J (1986) A mechanical analysis of the superplastic free bulging of metal sheet. Mater Sci Engng 84:111–125
546 Chandra N, Kannan D (1992) Superplastic sheet metal forming of a generalised cup Part I–Uniform thinning, Part II–Non-uniform thinning. J Mater Engng Perfomance 1:801–822
547 Usugi T, Akkus N, Kawahara M, Nishimura H (1999) An analytical model of the superplastic bulge forming of sheet metal. Sakuma T, Aizawa T, Higashi K (eds.) Towards Innovation in Superplasticity II. Kobe, Japan, Mater Sci Forum 304–306:735–740
548 Akkus N, Suzuki K, Kawahara M, Nishimura H (1999): Influence of preforming on the final thickness distribution of the superplastically deformed domes. Sakuma T, Aizawa T, Higashi K (eds.) Towards Innovation in Superplasticity II. Kobe, Japan, Mater Sci Forum 304–306:759–764
549 Timoshenko SP (1959) The Theory of Plates and Shells. McGraw-Hill, New York
550 Swadling SJ (1979) Fabrication of titanium at high temperatures. AGARD Conference on Advanced Fabrication Processes. Proceedings No. 256, NATO
551 Kruglov AA, Lutfullin Rya, Tayupov AR (1994) Forming of spherical vessels out of superplastic preforms. Langdon TG (ed.) Superplasticity in Advanced Materials (ICSAM–94), Mater Sci Forum 170–172:769–774
552 Kaibyshev OA, Kruglov AA, Tayupov AR, Lutfullin Rya (1991) Superplastic forming of spherical vessels out of sheet alloys Blacksmith and Stamping Industry 8:19–20 (in Russian)
553 Feodosjev VI (1986) Strength of Materials. Nauka, Moscow (in Russian)
554 Chandra N, Roy B (1988) Membrane element analysis of axisymmetric and non–Axisymmetric superplastic metal forming processes. In: Hamilton CH, Paton NE (eds.) Superplasticity and Superplastic Forming. TMS, Warrendale, USA, pp. 283–289
555 Story JM (1988) Incorporation of sliding friction into a closed–form model of plane strain superplastic forming. In: Hamilton CH, Paton NE (eds.) Superplasticity and Superplastic Forming. TMS, Warrendale, USA, pp. 297–302
556 Ghosh AK, Hamilton CH (1979) Superplastic forming of a long rectangular box section – analysis and experiment. In: Altan T, Gegel H (eds.) Process Modelling. ASM, Metals Park, Ohio, USA, pp. 303–338
557 Malinin NN (1990) Creep theories in metal working. In: Sih GS (ed.) Plasticity and Failure Behavior of Solids. Kluwer Academic Publishers, Netherlands, pp. 31–59
558 Berdin VK, Enikeev FU, Fatkullin SN (1992) Method SPF/DB in producing hollowness blades from Ti-based alloys. In: Proc V Conference Superplasticity of inorganic materials.Ed.O.A.Kaibyshev. Institute for Metals Superplasticity Problems, Ufa, Russia (in Russian) p. 117.
559 Hamilton CH, Paton NE, John M (1980) Apparatus for superplastic forming. US Patents 4: 233,829 Nov 18, Int Cl B21B 9/00
560 Hamilton CH, Ghosh AK (1980) Determination of strain rate sensitivity index in superplastic Ti–6Al–4V. Metall Trans 11A:1494–1496
561 Murzinova MA, Mazurski MI, Salishchev GA, Afonichev DD (1997) Application of reversible hydrogen alloying for formation of submicrocrystalline structure in titanium alloys. Int J Hydrogen Energy 22:201–204
562 Zaripova RG (1991) Formation of ultrafine-grained structure and properties of stainless steels. Ph. D Thesis. Institute for Metals Superplasticity Problems, Ufa, Russia
563 Salishchev GA, Valiakhmetov OR, Valitov VA, Mukhtarov SK (1994) Submicrocrystalline and nanocrystalline structure formation in materials and search for outstanding superplastic properties. In: Langdon TG (ed.) Superplasticity in Advanced Materials. (ICSAM–94), Mater Sci Forum170–172:121–130
564 Salishchev G, Zaripova R, Galeev R, Valiakhmetov O (1995) Nanocrystalline structure formation during severe plastic deformation in metals and their deformation behaviour. Nanostructured Mater 6:913–916

565 Langford G, Cohen M (1969) Strain hardening of iron by severe plastic deformation. Trans ASM 82:623–638
566 Yutori T, Katsumata M, Kanetsuki Y (1989) Ultrathin wire with ultimate strength of about 500 kg/mm^2. Bull Jap Inst Metals 28:113–315
567 Languillaume J, Chmelik F, Kjapelski G (1993) Microstructures and hardness of ultrafine-grained Ni_3Al. Acta Metall Mater 41:2953–2962
568 Rybin VV (1986) Large Plastic Deformations and Failure of Metals. Metallugy Publishing House, Moscow (in Russian)
569 Panin VE, Lihachev VA, Graneev YuV (1985) Structure Levels of Deformation in Solids. Nauka, Russia (in Russian)
570 Valiev RZ, Krasilnikov NA, Tsenev NK (1991) Plastic deformation of alloys with Submicron–grained structure. Mater Sci Engng A137:35–40
1 Valiev RZ, Korznikov AV, Mulyukov RR (1993) Structure and properties of ultrafine –grained metals produced by severe plastic deformation. Mater Sci Engng A168:141–148
572 Korznikov AV, Ivanisenko YuV, Safarov IM (1994) Mechanical properties of the over-eutectoid steel having nanocrystalline structure. Nanostructured Mater 4:159–165
573 Gertsman VYu, Birringer R, Valiev RZ, Gleiter H (1994) On the structure and strength of ultrafine-grained copper produced by severe plastic deformation. Scripta Metall 30:229–234
574 Valiev RZ, Ivanisenko YuV, Rauch EF, Baudlet B (1996) Microstructure evolution in Armco-Iron due to severe plastic deformation. Acta Mater 44:4705–4712
575 Iwahashi Y, Horita Z, Nemoto M, Langdon TG (1997) An investigation of microstructural evolution during equal channel angular pressing. Acta Mater 45:4733–4741
576 Senkov ON, Froes FH, Stolyarov VV, Valiev RZ, Liu J (1998) Microstructure of aluminum-iron alloys subjected to severe plastic deformation. Scripta Mater 38:1511–1516
577 Nakashima K, Horita Z, Nemoto M, Langdon TG (1998) Influence of channel angle on the development of ultrafine grains in Equal Channel Angular pressing. Acta Mater 46:1589–1599
578 Furukawa M, Iwahashi Y, Horita Z, Nemoto M, Tsenev NK, Valiev RZ, Langdon TG (1997) Structural evolution and the Hall–Petch relationship in an Al-Mg-Li-Zr alloy with ultra-fine grain size. Acta Mater 45:4751–4757
579 Iwahashi Y, Horito Z, Nemoto M, Langdon TG (1998) The process of grain refinement in Equal Channel Angular pressing. Acta Mater 46:3317–3331
580 Kuznetsov RI, Bykov VI, Chernyshev VP (1985) Plastic deformation of solids under pressure I–Apparatus and techniques. Ural Branch of the Academy of Sciences of USSR, Sverdlovsk, Russia, (in Russian)
581 Segal VM, Reznikov VI, Kopylov VI (1994) Processes of Plastic Structure Formation in Metals. Science and Engineering Publishing House, Minsk, Belorussia (in Russian)
582 Goforth RE, Segal VM, Hartwig KT, Ferrase S (1995) Production of submicron-grained structure in aluminum 3003 by ECAE. In: Ghosh AK, Bieler TR. (eds.) Superplasticity and Superplastic Forming. TMS. Warrendale, USA, pp.25–32
583 Segal VM (1995) Materials processing by simple shear. Mater Sci Engng A197:157–164
584 Ahmadeev NH, Kopylov VI, Mulyukov RR, Valiev RZ (1992) Formation of submicrocrystalline structure in copper and nickel using severe plastic deformation. Izv Akad Nauk RAN, Metally, 5: 96–101 (in Russian)
585 Puch HLD (1970) Mechanical Behavior of Materials under Pressure. Elsevier, Amsterdam London New York
586 Alexandrov IV, Wang YD, Zhang Kluk, Valiev RZ (1996) X–Ray analysis of textured nanocrystalline materials. In: Proc Eleventh Int Conference on Textures in Materials. Int Academic Publishers, Beijing 2:929–940

587 Enikeev FU (1999) Finite strain kinematics of superplastic deformation of cylindrical specimens subjected to combined torsion and axial force. Proc Russian Acad Sci Series–Metals, No.2, pp.89–98 (in Russian)
588 Ghosh AK, Duncan JL (1970) Torsion test on superplastic tin–lead alloy. Int J Mech Sci 12:499–511
589 McQueen HJ, Jonas JJ (1984) Recent advances in hot working: fundamental dynamic softening mechanisms. American Society for Metals. J Appl Metalworking 3:233–241
590 Mazurski MI, Enikeev FU (1999) On some principles of ultrafine-grained structure formation by means of metal working techniques. Blacksmith and Stamping Industry 5 (in Russian) (in press)
591 Sakai T, Jonas JJ (1984) Dynamic recrystallization: mechanical and microstructural considerations. Acta Metall 32:189–209
592 USSR Patent No 1741960, 21 J 5/100. Method of the plastic structure formation and the device allowing one to carry it out
593 Greshnov VM, Golubev VM, Rtishchev AV (1997) New technological scheme of extrusion of metals. Blacksmith and stamping industry: 8–10 (in Russian)
594 Mazurski MI, Enikeev FU, Korshunov AA (1994) Method of torsion under pressure. Patent of Russia No 2021064
595 Alexandrov VK, Anoshkin NF, Bochvar GA (1979) Semi-Manufactured Articles made of Titanium Alloys. Metallurgy Publishing House, Moscow (in Russian)
596 Metallography of Titanium Alloys. Reference Book (1980) Metallurgy Publishing House, Moscow (in Russian)
597 Weiss I, Welsch GE, Froes FH, Eylon D (1984) Mechanisms of microstructure refinement in Ti–6Al–4V alloy. In: Proc Int Conference on Titanium 5. Munich, pp. 1503–1510
598 Percovski NZ, Semenova NM, Brun MYa, Mazolevskaya OA (1984) Electron microscopy study on the polygonization and recrystallization in two-phase titanium alloys. Phys Met Metallog 57:737–743 (in Russian)
599 Kaibyshev OA, Lutfullin RYa, Salishchev GA (1988) Influence of the superplastic deformation conditions on the transformation of lamellar microstructure in titanium alloy VT9. Phys Met Metallog 66:1163–1171 (in Russian)
600 Salishchev GA, Lutfullin RYa (1988) The transformation of Ti alloy laminar microstructure into a microduplex one. In: Hamilton CH, Paton NE (eds.) Superplasticity and Superplastic Forming. TMS. Warrendale, USA, pp. 103–107
601 Salishchev GA, Valiakhmetov OR, Galeyev RM (1993) Formation of submicrocrystalline structure in the titanium alloy VT8 and its influence on mechanical properties. J Mater Sci 28:2898–2902
602 Salishchev GA, Zaripova RG, Zakirova AA, McQueen HJ (1996) Submicro-crystalline materials structure formation and properties. In: McQueen HJ, Ryan ND, Konopleva EV (eds.) Hot Workability of Steels and Light Alloys-Composites,CIM, pp. 217–226
603 Salishchev GA, Galeev RM, Malisheva SP, Valiakhmetov OR (1997) Low temperature superplasticity of submicro-crystalline Titanium Alloys. In: Chokshi AH (ed.) Superplasticity in Advanced Materials (ICSAM–97). Mater Sci Forum 243–245:585–590
604 Salishchev GA, Mazurski MI, Levin IE (1990) Influence of the phase non-equilibrium on the globularization of lamellar microstructure in titanium alloy under hot deformation. Phys Metals Metallography 68:149-151 (in Russian)
605 Dmitriev OV (1992) On the possibility to obtain the superplastic properties in metals for the sake of multi-stage upsetting. Bulletin of Moscow State University. Series 1. Mathematics and Mechanics 3:66–69 (in Russian)
606 Bernstein ML (1974) Structure of Deformed Metals. Metallurgy Publishing House, Moscow (in Russian)
607 Polukhin P, Gorelik S, Vorontcov V (1983) Physical Principles of Plastic Deformation. Mir Publishers, Moscow

608 Sellars CM, McG. Tegart WJ (1972) Hot workability. Int Metall Rev 17:1–24
609 Jonas JJ, Sakai T (1984) Deformation processing and structure. G Krauss (ed.). ASM, Metals Park, Ohio, USA, p. 201
610 Vasin RA, Enikeev FU, Mazurski MI (1995) Constitutive equations for a polycrystalline material including its structural characteristics. In: Mathematical Modelling of Systems and Processes. Perm, Russia 3:19–23
611 Hasebe T, Sakane M, Ohnami M (1992): High temperature low cycle fatigue and cyclic constitutive relation of MAR-M247 directionally solidified superalloy. Trans ASME, J Engng Mater Technol. 114:162–167
612 O'Donnel RG, Parker BA (1986): Constitutive relations describing flow behaviour in aluminum alloys. Res Mechanica 19:17–33.
613 Woo CW, Asundi A, Deng W (1995): On rate-dependent plasticity–linear viscoelastic constitutive equations. Int. J. Solids and Structures 32:1793–1803.
614 Stempen V, Tsepin MA, Ershov AN, Semenko KM (1985): Rheological characteristics of the material for physical modelling of metalworking processes. VINITI No. 5985–85 (in Russian).
615 Shlomchak GG, Fen GA, Kutsai VG (1980): Similarity criteria for metal rolling. Izv. VUZov. Chernaya Metallurgiya 3C:79–81.
616 Shlomchak GG, Mamuzitch I, Vodopivec F (1994): Rheological similarity of metals and alloys. J Mater Processing Technol 40:315–325
617 Ovsyannikov LV (1978): Group Analysis of the Differential Equations. Nauka, Moscow (in Russian).
618 Ovsyannikov LV, Ibragimov NH (1975): Group Analysis of the Differential Equations of the Mechanics. In: Itogi nauji i tehniki. Obshaya mehanika, Nauka, Moscow C:5–52 (in Russian).
619 Annin BD, Bytev VO, Senashov SI (1985): Group Properties of the Equations of Elasticity and Plasticity. Nauka, Sibirskoe otdelenie, Novosibirsk (in Russian).
620 Leonova EA (1993): Invariant properties of the equations of thermo-viscoplasticity under the lack of information concerning the material properties. In: Iljushin AA (ed.) Uprugost i Neuprugost. Izd-vo MGU, Moscow, pp.55–87 (in Russian).
621 Leonova EA (1997): Representation of the experimental functions at the constitutive equations of thermo-visco-plasticity on the basis of the analysis of invariant properties. Mathematical modelling of systems and processes. Perm, No. 5:56–64 (in Russian).

Index

MIX
Papier aus verantwortungsvollen Quellen
Paper from responsible sources
FSC® C105338

If you have any concerns about our products,
you can contact us on
ProductSafety@springernature.com

In case Publisher is established outside the EU,
the EU authorized representative is:
Springer Nature Customer Service Center GmbH
Europaplatz 3, 69115 Heidelberg, Germany

Printed by Libri Plureos GmbH
in Hamburg, Germany